New Frontiers in Medical Device Technology

WILEY SERIES IN MICROWAVE AND OPTICAL ENGINEERING

KAI CHANG, Editor

Texas A&M University

FIBER-OPTIC COMMUNICATION SYSTEMS • *Govind P. Agrawal*

COHERENT OPTICAL COMMUNICATIONS SYSTEMS • *Silvello Betti, Giancarlo De Marchis, and Eugenio Iannone*

HIGH-FREQUENCY ELECTROMAGNETIC TECHNIQUES: RECENT ADVANCES AND APPLICATIONS • *Asoke K. Bhattacharyya*

COMPUTATIONAL METHODS FOR ELECTROMAGNETICS AND MICROWAVES • *Richard C. Booton, Jr.*

MICROWAVE SOLID-STATE CIRCUITS AND APPLICATIONS • *Kai Chang*

MULTICONDUCTOR TRANSMISSION LINE STRUCTURES • *J. A. Brandão Faria*

MICROSTRIP CIRCUITS • *Fred Gardiol*

HIGH-SPEED VLSI INTERCONNECTIONS: MODELING, ANALYSIS, AND SIMULATION • *A. K. Goel*

HIGH-FREQUENCY ANALOG INTEGRATED-CIRCUIT DESIGN • *Ravender Goyal, Editor*

OPTICAL COMPUTING: AN INTRODUCTION • *Mohammad A. Karim and Abdul Abad S. Awwal*

MICROWAVE DEVICES, CIRCUITS, AND THEIR INTERACTION • *Charles A. Lee and G. Conrad Dalman*

ANTENNAS FOR RADAR AND COMMUNICATIONS: A POLARIMETRIC APPROACH • *Harold Mott*

SOLAR CELLS AND THEIR APPLICATIONS • *Larry D. Partain*

ANALYSIS OF MULTICONDUCTOR TRANSMISSION LINES • *Clayton R. Paul*

INTRODUCTION TO ELECTROMAGNETIC COMPATIBILITY • *Clayton R. Paul*

NEW FRONTIERS IN MEDICAL DEVICE TECHNOLOGY • *Arye Rosen and Harel Rosen, Editors*

OPTICAL SIGNAL PROCESSING, COMPUTING, AND NEURAL NETWORKS • *Francis T. S. Yu and Suganda Jutamulia*

New Frontiers in Medical Device Technology

Edited by
ARYE ROSEN, Ph.D.
Drexel University

HAREL D. ROSEN, M.D.
UMDNJ—Robert Wood Johnson Medical School

A WILEY-INTERSCIENCE PUBLICATION
JOHN WILEY & SONS, INC.
NEW YORK/CHICHESTER/BRISBANE/TORONTO/SINGAPORE

This text is printed on acid-free paper.

Library of Congress Cataloging in Publication Data:

New frontiers in medical device technology/edited by Arye Rosen, Harel D. Rosen.
p cm.—(Wiley series in microwave and optical engineering)
Includes bibliographical references.
ISBN 0-471-59189-0
1. Medical technology. 2. Diagnostic imaging. 3. Lasers in medicine. 4. Microwave imaging in medicine. 5. Microwave heating.
I. Rosen, Arye. II. Rosen. Harel D. III. Series.
R855.3.N48 1995
610′. 28 – – dc20 94-26892

Printed in the United States of America

10 9 8 7 6 5 4 3 2 1

CONTRIBUTORS

Kenneth Carr, D. Eng., Microwave Medical Systems, Inc., Acton, MA

Richard L. Coren, Ph.D., Department of Electrical and Computer Engineering, Drexel University, Philadelphia, PA

Arnold J. Greenspon, M.D., Division of Cardiology, Department of Medicine, Jefferson Medical College, Philadelphia, PA

Joel M. Krauss, M.D., Mount Sinai Hospital, New York, NY

Reuben S. Mezrich, M.D., Ph.D., Department of Radiology, Robert Wood Johnson Medical School, New Brunswick, NJ

Arye Rosen, Ph.D., Department of Electrical and Computer Engineering, Drexel University, Philadelphia, PA

Harel D. Rosen, M.D., University of Medicine and Dentistry of New Jersey, Robert Wood Johnson Medical School, New Brunswick, NJ

William P. Santamore, Ph.D., Thoracic and Cardiovascular Surgery Division, University of Louisville, Louisville, KY

Fred Sterzer, Ph.D., MMTC, Inc., Princeton, NJ

Alexander A. Stratienko, M.D., Chattanooga Heart Institute, Chattanooga, TN

Leonard S. Taylor, Ph.D., Department of Electrical Engineering and Radiation Oncology, University of Maryland, College Park, MD

David Vilkomerson, Ph.D., EchoCath, Inc., Princeton NJ

Paul Walinsky, M.D., Division of Cardiology, Department of Medicine, Jefferson Medical College, Philadelphia, PA

Preface

My (A. Rosen's) decision to embark on the project of editing a book addressed to the engineering community as well as the medical community was influenced by the interest expressed by audiences at the medical sessions of the various engineering conferences that I have organized. An overwhelming number of participants—including most of the contributors to this book—in a medical session during the 1991 Electro Conference in New York were previously approached by representatives of John Wiley & Sons. However, the proposal for this book actually was drafted following the persistence of Professor Kai Chang from Texas A&M University. Having pursued education and careers in both engineering and physiology, I welcomed the opportunity to edit this book, which is intended, in a small way, to bridge the gap between the two disciplines. In my many presentations to both the engineering and the medical communities I advocated the need to train biomedical engineering graduate students in a medical school environment. It has always been my view that after receiving an undergraduate degree in engineering, the graduate-level studies in biomedical engineering should be based in a medical school rather than an engineering school. The biomedical scientist would then become an instant inventor in the consumer environment and would work in collaboration with the medical researcher, who nowadays is often also an engineer at heart.

The book is devided into three major sections:

- The utilization of radio frequency (RF) and microwaves in medicine, in particular the chapters on
  - RF microwaves in angioplasty (Chapter 1)
  - RF techniques in electrophysiology (Chapter 2)
  - RF and microwaves in urology for the treatment of benign prostatic hyperplasia (BPH), and possibly cancer (Chapter 3)
  - Microwave balloon technology for deep heating of tissue (Chapter 4)
  - Microwaves in surgery (Chapter 5)

- The utilization of lasers in
  - Ophthalmology (Chapter 6)
  - Cardiology—laser angioplasty (Chapters 2 and 7)
- Imaging and sensing
  - Magnetic resonance imaging (MRI) (Chapter 8)
  - Acoustics (Chapter 9)
  - Microwave radiometry in sensing of cancer tissue (Chapter 10)
  - The effects of low- and high-frequency radiation on tissue (Chapter 11) Chapter 11 was contributed at the request of the publishers.

We wish to thank all members of the editorial board for their advice and suggestions and Mr. George Telecki, our Wiley editor, for his encouragement. We also wish to express special appreciation to our wife and mother, Daniella Rosen, for her invaluable assistance in typing, illustrating, proofreading, and otherwise managing this project, and to our son and brother, Gilad D. Rosen, for his technical support.

ARYE ROSEN AND HAREL D. ROSEN

January 1995

Contents

New Frontiers in Medical Device Technology

CHAPTER ONE

Microwave Balloon Angioplasty

ARYE ROSEN, Ph.D., *Department of Electrical and Computer Engineering, Drexel University, Philadelphia, PA*
PAUL WALINSKY, M.D., *Division of Cardiology, Department of Medicine, Jefferson Medical College, Philadelphia, PA*

1.1 INTRODUCTION

The purpose of this chapter is to discuss the results of our experimental research (conducted on animals both in vitro and in vivo) on microwave balloon angioplasty (MBA), and to demonstrate the benefits of MBA as it relates to (1) reducing the rate of restenosis, (2) welding dissections, (3) treating intracoronary thrombi, and (4) creating biological stents that can potentially eliminate elastic recoil. Laser irradiation has been the most extensively studied form of thermal energy; however, other delivery systems provide alternative means of thermal energy delivery in conjunction with angioplasty, with potentially better results.

Microwave balloon angioplasty has the following characteristics that may make it effective for use in coronary heart disease:

1. Two modalities—pressure (balloon) and heat (RF; with volume heating) —are utilized.
2. The energy delivery cable can be constructed sufficiently small and flexible for delivery to the coronary tree.

New Frontiers in Medical Device Technology, Edited by Arye Rosen and Harel Rosen
ISBN 0-471-59189-0

3. Target temperatures can be varied according to specific requirements because of a flexible energy delivery rate.
4. The energy delivery system can be coupled to a temperature feedback system for on-line temperature monitoring and regulation.
5. The energy delivery system can generate varied fields of delivery if selective heating is beneficial.
6. The generator that is required is relatively small and inexpensive.

The subject of RF and microwaves in cardiology is relatively new, developed seriously only in the last 5 years (see also Chapter 2). Much of this recent progress is due mainly to the experience gained from utilizing microwaves in the treatment of cancer.

Although angioplasty has become an acceptable modality in cardiology, significant problems and unresolved issues still exist in conventional angioplasty[1,2]. The development of thermal angioplasty has become an area of great interest. Until recently, however, it was restricted to laser intervention. In as many as 30% of successful conventional angioplasty cases restenosis develops in the first 6 months postoperatively. In many other cases vessel recoil, dissection, or thrombus develops. A second or third balloon inflation may therefore be required.

These complications exaccerbate some of the instability of the vessel and can result in acute closure. As early as 1987[3] we suggested the use of microwave energy to perform thermal angioplasty. In addition, we had speculated, and demonstrated in animal experiments, that microwave energy penetrating the media would (1) reduce artery recoil and intimal proliferation (2) weld or anastomose dissections, and (3) treat intracoronary thrombi, all of which are major problems in both balloon and laser balloon angioplasty. The literature pertinent to microwave thermal balloon angioplasty (MBA) is reviewed in Section 1.2.

1.2 BACKGROUND AND HISTORY OF MICROWAVE CATHETER ANGIOPLASTY

This section presents a background history of balloon angioplasty and discusses the use of microwaves in medicine with reference to the microwave system used in MBA research.

Each year in the United States many people die suddenly from acute myocardial infarction, and many more suffer from chronic heart problems. A major contributing factor in both acute and chronic heart disease is a reduction in nutrient blood flow to the muscles of the heart resulting from diminished blood flow through the coronary blood vessels. The reduction in flow may be caused by deposits of atherosclerotic plaque on the walls of the blood vessels, causing a gradual narrowing of the lumen or channel of the affected blood vessel (Fig. 1.1).When the lumen is sufficiently narrowed, the rate of blood flow may be so diminished that a thrombus or clot forms spontaneously by a variety of

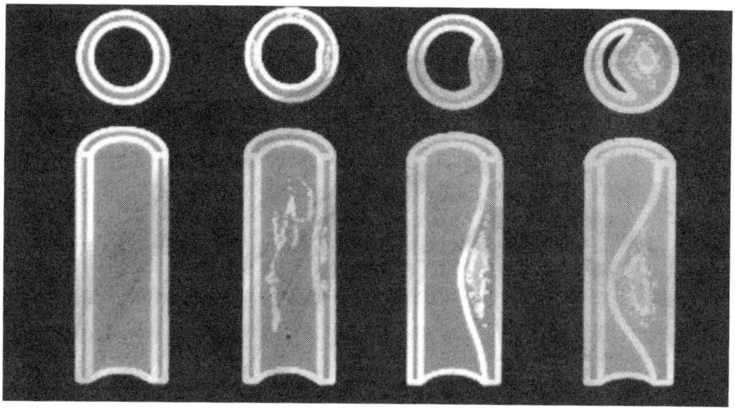

ATHEROSCLEROTIC PROGRESSION

NON-DISEASED VESSEL | PATCHY LIPIDS | FIBROUS PLAQUE | STENOTIC LESION

FIGURE 1.1 Progression of atherosclerosis.

physiologic mechanisms. It has been observed that once a blood clot has started to develop, it extends within minutes into the surrounding blood. Thus the presence of atherosclerotic plaque not only reduces the blood flow to the heart muscle that it nourishes but also is a major predisposing factor in coronary thrombosis.

The methods available for treating the conditions resulting from plaque formation include pharmacologic modalities, such as the administration of nitroglycerin, which dilates the coronary blood vessels to improve flow. In cases too advanced for management by drugs, surgical treatment may be indicated. One surgical technique commonly used is coronary bypass, in which a substitute blood vessel shunts or bypasses blood around the blockage. The bypass operation is effective, but it is expensive and not without substantial risk to the patient.

Percutaneous transluminal balloon catheter angioplasty is an alternative form of treatment. In this method a deflated balloon is inserted into the lumen of an artery partially obstructed by plaque, and the balloon is then inflated in order to enlarge the lumen. The lumen remains expanded after removal of the catheter. The major problems with this technique are restenosis of the narrowed vessel by recurrence of the arterial plaque, and arterial recoil.

1.2.1 Conventional Angioplasty

Angioplasty was first introduced in 1961 by Dotter and Judkins,[4] who used a coaxial catheter system to perform the first transluminal angioplasties. Their system consisted of a wire guide, a size 8 French (8F) catheter, and a size 12F catheter. The atherosclerotic stenosis was dilated by advancing the wire guide through the plaque, then advancing the two catheters on top of one another in succession. The technique was modified by Staple,[5] who used only one tapered catheter: a size 8F catheter tapered to a size 9F. Further modifications were proposed by Gruentzig and Hopff,[1] who developed the first balloon dilation catheter. The balloon can be inflated up to 5 atmospheres (atm) of pressure, exerting—when placed inside the stenotic region—lateral force against the atherosclerotic plaque, thus dilating the vessel.

In 1977 Gruentzig reported encouraging results of his study of patients who had undergone peripherial transluminal angioplasty.[6] His report initiated widespread clinical investigation of the percutaneous transluminal coronary angioplasty (PTCA) technique.

The PTCA technique and the laser technique (thermal) provide the background for the invention of and subsequent research on MBA and are discussed in the following paragraphs.

1.2.1.1 PTCA Technique

Following coronary angiography to determine location and degree of coronary occlusion, the angiographic catheter is replaced by a guiding catheter at the femoral insertion site, and is positioned under fluoroscopic contrast through the site of the occlusion. The dilation catheter is equipped with a steerable wire guide that is advanced through the stenosis distally along the coronary artery. After the wire guide is positioned, the dilation catheter is advanced over the wire guide, positioning the uninflated balloon along the site of the plaque. Contrast material is utilized throughout the procedure (injected through the dilation catheter) to identify the position of the catheter with respect to the stenosis site. The balloon is then inflated for 30–60 s. The inflation procedure can be repeated a few times to ensure dialation and to minimize coronary recoil.

The primary problem associated with PTCA or conventional balloon angioplasty is coronary artery occlusion or restenosis.

Anatomy of the Coronary Arteries An understanding of the anatomy of the coronary arteries, which supply blood to the heart itself, is important to the comprehension of catheter and cable design for angioplasy intervention. Since microwave angioplasty is designed to remodel the coronary artery, this section emphasizes the complexity of coronary circulation, and thus the special requirement for a microwave delivery system that is small, flexible, efficient, and low-loss (i.e., characterized by low power loss).

The continuous heartbeat, and the ability of the heart to change the cardiac output as a function of physical activities, requires the heart muscle to have its own reliable source of oxygen. The left and right coronary arteries provide this oxygen along with other nutrients. It is known that any major reduction of blood supply caused by thrombosis or plaque of coronary vessels leads to rapid muscle necrosis. The anterior aspect of the heart showing the coronary arteries is depicted in Figure 1.2. The left coronary artery arises from the left aortic sinus at the coronary sulcus and divides into a circumflex branch and an anterior descending

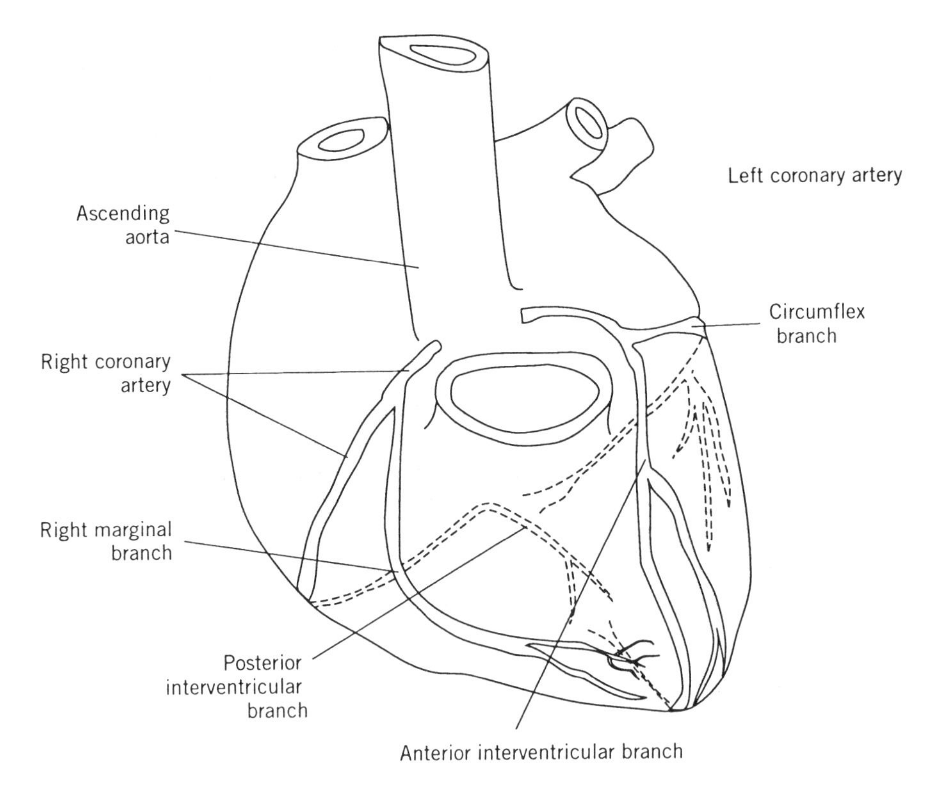

FIGURE 1.2 The anterior aspect of the heart showing the coronary arteries.

branch. As can be seen, branches of the circumflex artery supply the left atrium and the left ventricle.

The right coronary artery arises from the right aortic sinus. It can be seen to pass downward and to the right in the coronary sulcus, which leads to the posterior interventricular groove. Here it anastomoses with the left anterior descending artery and gives off its terminal posterior interventricular branch. The right acute marginal artery runs toward the apex.

Percutaneous Transluminal Balloon Angioplasty Although percutaneous transluminal balloon angioplasty is an important and effective treatment of coronary vascular disease, its efficacy is limited. The frequency of restenosis ranges from 17 to 47% depending on variations in definition of restenosis (clinical, anatomic, physiologic).[7] Abrupt closure at the angioplasty site occurs in 5% of the patients treated with conventional balloon angioplasty. In addition, a low success rate has been reported in patients having totally occluded arteries.

Two morphologic observations have been made following balloon angioplasty that could contribute to the limited success of the procedure:

1. Acute changes less than 30 days after dilation
2. Chronic changes more than 30 days after dilation

ACUTE CHANGES Acute changes such as intimal "cracks," "tears," or "splits" (fracturing of atherosclerotic plaque) in the coronary media (Fig. 1.5), resulting in a dissection, have been reported.[8,9] These results are not unexpected, since the only mechanism involved is a mechanical force applied through the balloon on a plaque that, in most cases, is partially calcified and hence provides high resistance to the inflated balloon.

In addition, abrupt coronary closure has been reported in 5% of the patients treated, although prolonged balloon inflation has reduced the percentage of abrupt closure to about 3%. Following abrupt coronary closure, the patients require surgical intervention; these patients are at increased risk of morbidity and mortality.

CHRONIC CHANGES The most common change in the morphology of a coronary artery following balloon angioplasty 30 days postoperatively is intimal proliferation, which eventually requires a repeat balloon angioplasty procedure. Waller and coworkers reported histologic evidence of restenosis in a coronary artery, 4 months after and on the site of a previous balloon angioplasty, which contained plaque cracks, splits and localized medial dissection.[9]

It is now recognized that balloon angioplasty operates by the following mechanisms:

1. Compression
2. Focal plaque break, fracture, or tear
3. Stretching of plaque-free wall
4. Stretching of artery with minimal plaque compression
5. Focal plaque break, fracture, or tear with localized dissection.

1.2.1.2 Laser Angioplasty

Another technique that has recently received considerable attention is transluminal laser balloon angioplasty. Laser balloon angioplasty utilizes Nd:YAG (neodymium–yttrium aluminum garnet) laser energy, delivered through an optical fiber to a diffusing tip within an otherwise conventional angioplasty balloon, to be radially diffused along the central two-thirds of the length of the balloon to the surrounding arterial wall. This procedure is currently performed only after the most optimal conventional PTCA result is obtained and, therefore, during the final balloon inflation. Histologic studies of vascular tissue after continuous-wave(CW) laser therapy where temperatures as high as 120 °C were utilized have indicated a charred and ragged endothelial surface, which in turn has resulted in thrombosis and vessel perforation. Laser research has now concentrated on ablating tissue by limiting thermal injury.[10] Two methods have attempted to achieve this goal: (1) modification of the tip of the delivery system of CW laser and (2) use of pulsed lasers where the mechanism of ablation appears to be the disruption of molecular bonds, and where no thermal injury occurs because of a rapid highly localized heat effect.[11] At the time of this writing, laser angioplasty research has been reduced significantly, mainly because of the high rate of restenosis.

1.3 APPLICATION OF MICROWAVES TO CANCER TREATMENT, PERTINENT TO DEVELOPMENT OF MBA

1.3.1 Miniature Coaxial Applicator for Treating Deep-Seated Tumors[12–15]

Figure 1.3 is a photograph of a simple miniature coaxial applicator (with integral thermocouple) for treating a variety of deep-seated tumors. The applicator consists of a length of semirigid copper coaxial cable with a radiating antenna that is formed by removing a short length of the outer conductor at the tip of the cable. (Other types of antennas commonly used with miniature coaxial cables include those shown in Fig. 1.4). During hyperthermia treatments, one or more of these miniature applicators are inserted into the tumors. Figure 1.5 illustrates how miniaturized applicators can be used to heat deep-seated brain tumors.

1.3.2 Large Coaxial Applicator for Treating Cancer of the Prostate

Figure 1.6 is a photograph of a large coaxial applicator used for treating cancer of the prostate (as well as benign prostatic hyperplasia). The applicator consists of a section of coaxial cable with an antenna at its tip. An asymmetrically shaped Teflon bulb surrounds the antenna. During hyperthermia treatment, the applicator is inserted into the patient's rectum and the antenna is placed adjacent to the prostate gland. (The asymmetric Teflon bulb helps reduce the heating of the uninvolved parts of the rectum. Water or air cooling is often added to further protect the walls of the rectum.)

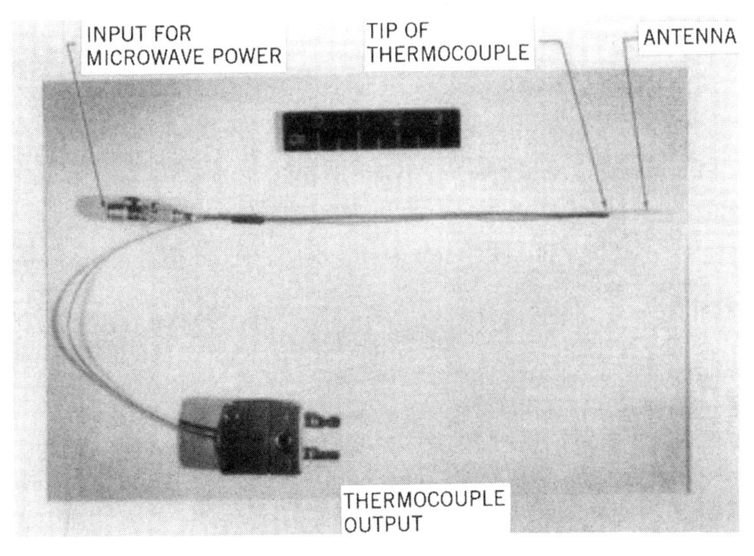

FIGURE 1.3 Miniature coaxial applicator with integral thermocouple.[12]

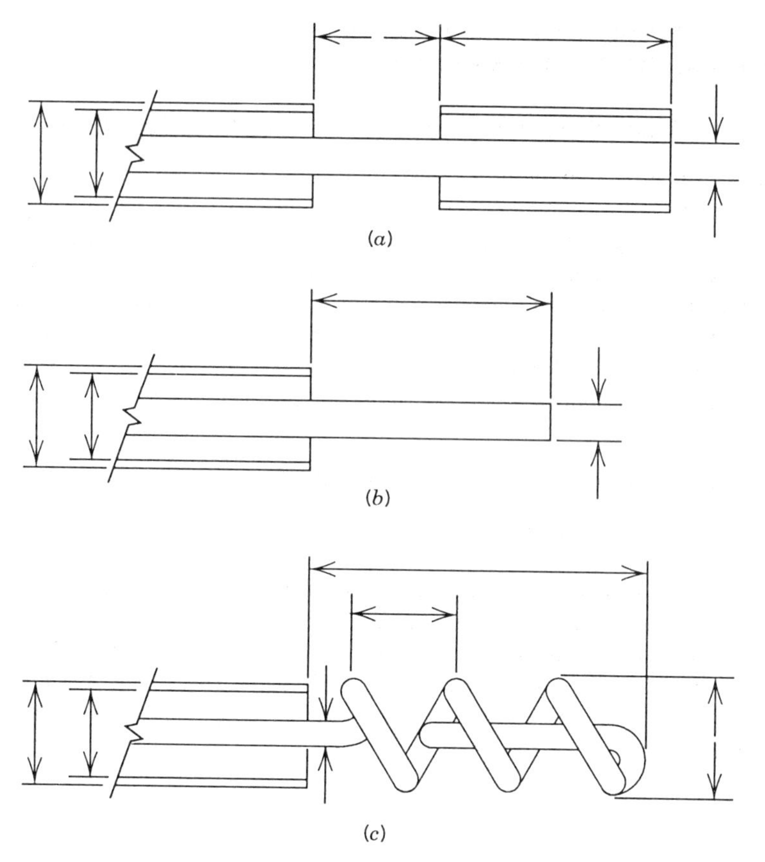

FIGURE 1.4 Antenna configurations: (*a*) gap antenna; (*b*) whip antenna; (*c*) helical antenna.

1.3.3 Microwave Applicator for Transurethral Hyperthermia of Benign Prostatic Hyperplasia (New Approach)[16,17]

Benign prostatic hyperplasia, causing urinary outlet obstruction, is a common disease among aging men. The objective of treating BPH with microwave hyperthermia is to alleviate the symptoms of urinary obstruction by creating necrosis of the BPH tissue and increasing the diameter of the urethra.

An applicator for heating the prostate gland using a transurethral approach is currently being utilized in several centers around the United States (Fig. 1.7). This technique uses a number of microwave antennas attached to the outer surface of a balloon-type urological catheter. The balloon catheter assures the reproducible positioning of the antennas in the prostatic urethra. This new technique might provide a nonsurgical alternative to transurethral prostatectomy for relief from BPH symptoms.

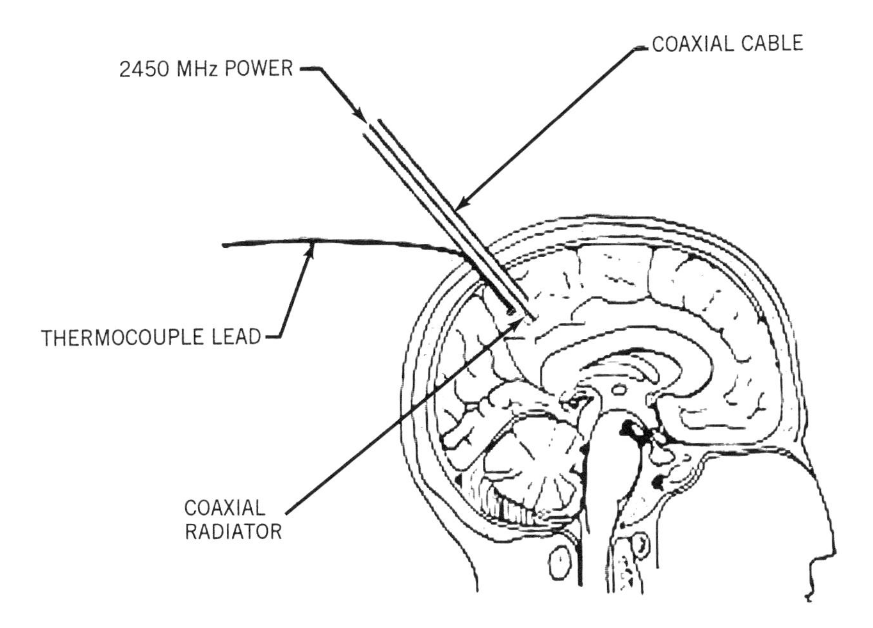

FIGURE 1.5 A sketch of the placement of a miniature coaxial applicator for treating a brain tumor.[13,14]

1.4 ABLATION OF MYOCARDIAL TISSUE UTILIZING A MICROWAVE TECHNOLOGY SIMILAR TO THAT USED FOR MBA

A medical procedure for treatment of tachycardia (rapid heartbeat) or cardiac dysrhythmia uses a catheter that includes a flexible coaxial transmission line (coax) terminated by an antenna. The antenna and the coax are introduced into a chamber of the heart. Then the antenna is brought into contact with a wall of the heart. Action potentials generated by the heart are coupled through the antenna and the coaxial cable to a standard electrocardiograph apparatus for display (Fig. 1.8). Other electrodes placed about the body also produce action potentials, which are displayed by the electrocardiograph. The position of the antenna in the chamber of the heart is adjusted with the aid of the displayed action potentials until the antenna is in contact with the region to be ablated or injured as indicated by its characteristic electrical signature. When the antenna is adjacent to or in contact with the desired location, RF or microwave frequency energy is applied to the proximal end of the coax and through the coax of the antenna. The action potentials may be viewed while the electrical energy is applied. The power of the electrical energy is slowly increased until the desired amount of blockage of the bundle of His or damage to the ectopic focus has been achieved.

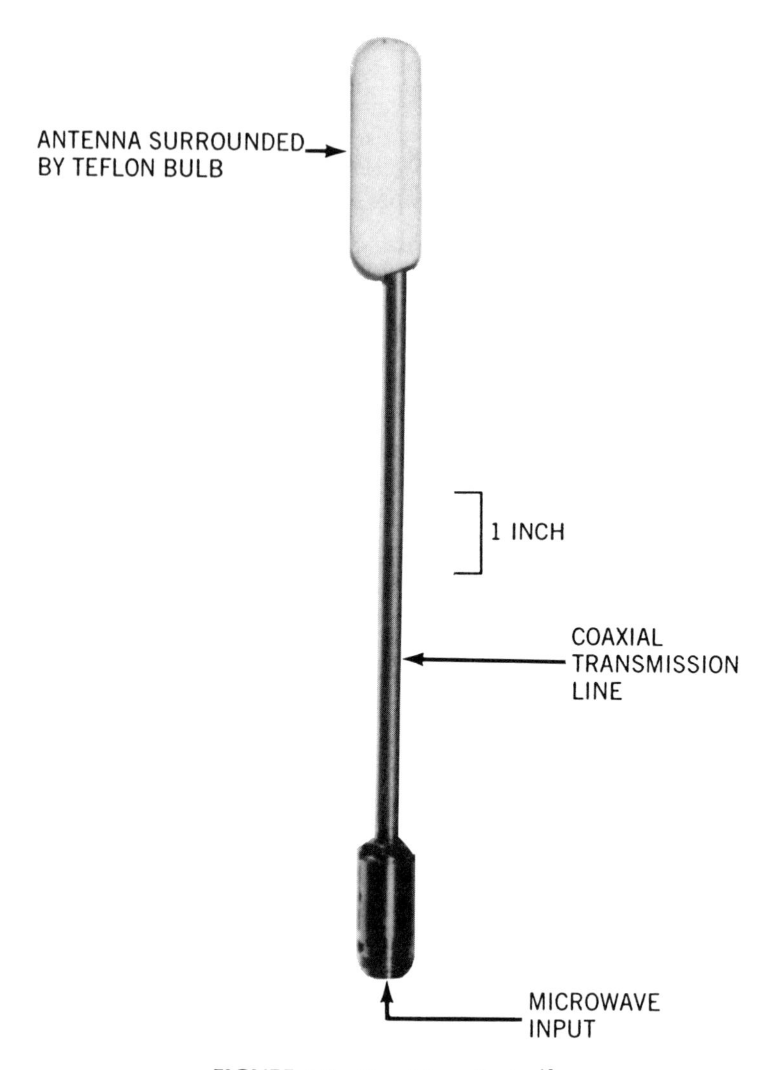

FIGURE 1.6 Coaxial applicator.[15]

1.5 MAJOR MBA COMPONENTS AND CRITICAL ISSUES

This section is concerned with the three primary components of MBA systems: (1) coaxial cable, (2) microwave generator, and (3) antenna.

As a consequence of temperature profile needs for MBA in both radial and coaxial directions and the availability of components, Table 1.1 summarizes the major component requirements (see also Figs. 1.9 and 1.10).

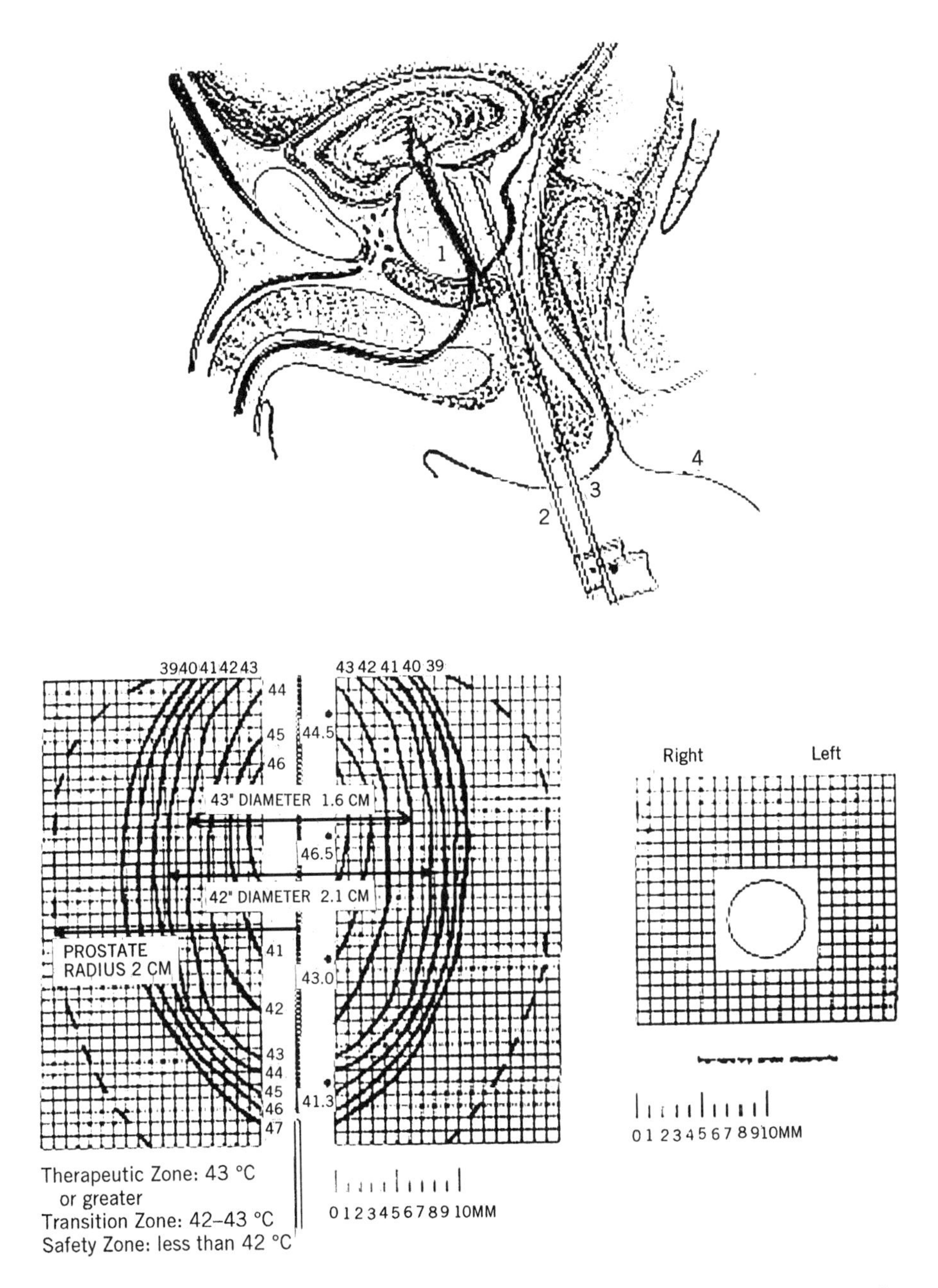

FIGURE 1.7 (*a*) Microwave applicator for transurethral hyperthermia of benign prostatic hyperplasia (BPH)[17]; (*b*) temperature profile after BPH treatment.

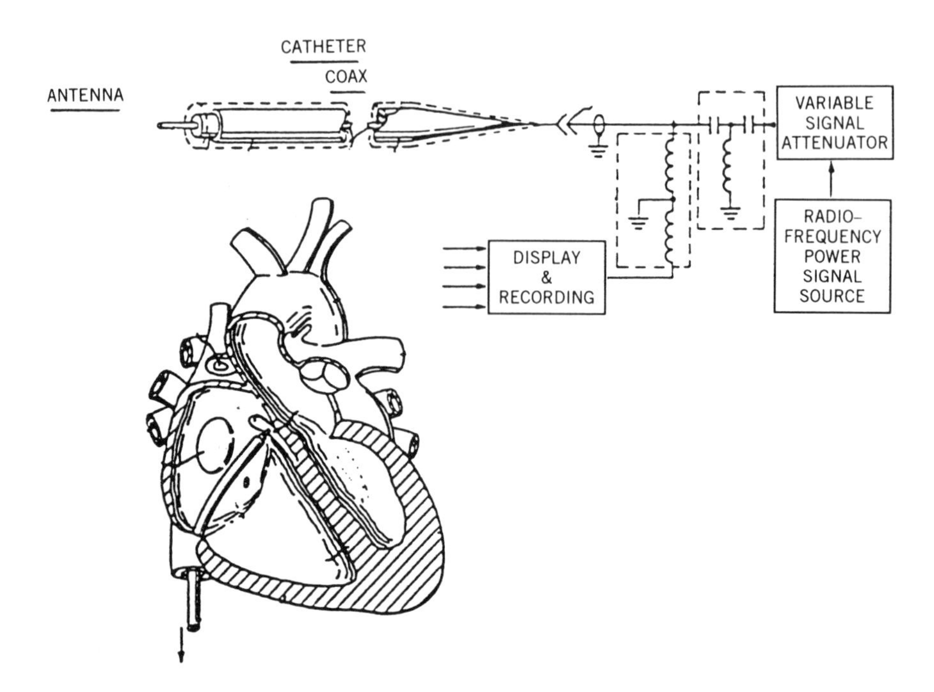

FIGURE 1.8 Microwave system used for myocardial tissue ablation.[18]

TABLE 1.1 Major MBA Component Requirements

Coaxial cable/Antenna
- 50 Ω (ohms)
- Attenuation at 2.45 GHz: 1.0 dB/ft
- Rated power at 2.45 GHz: 80 W (with cooling)
- Flexible outer conductor: copper or stainless steel, coated with Ag (silver) ribbons wound along the dielectric in both directions
- Outer diameter 0.022 in.
- Dielectric constant (between inner and outer conductor) < 1.2
- Dielectric diameter 0.015 in.
- Biocompatibility and sterility

Microwave power generator
- Power 80 W, CW: vacuum-tube technology—magnetron, or solid-state technology—transistors

Antenna configuration
- Whip type: easy to maneuver to the lesion site, and easy to produce (by removing outer conductor)
- Frequency 2.45 GHz.
- Radiating power 10 W.
- Axial heat distribution < 2 cm
- Radial heat distribution 1–3 mm

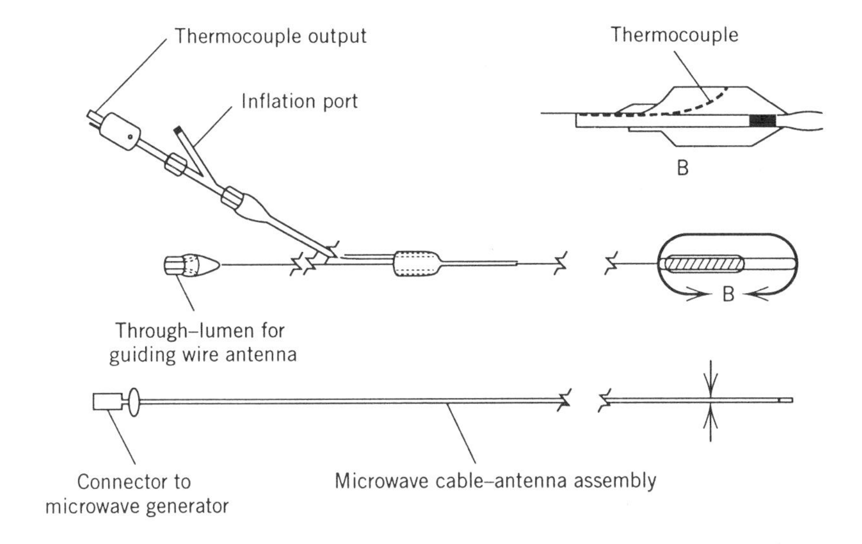

FIGURE 1.9 Microwave balloon angioplasty system.

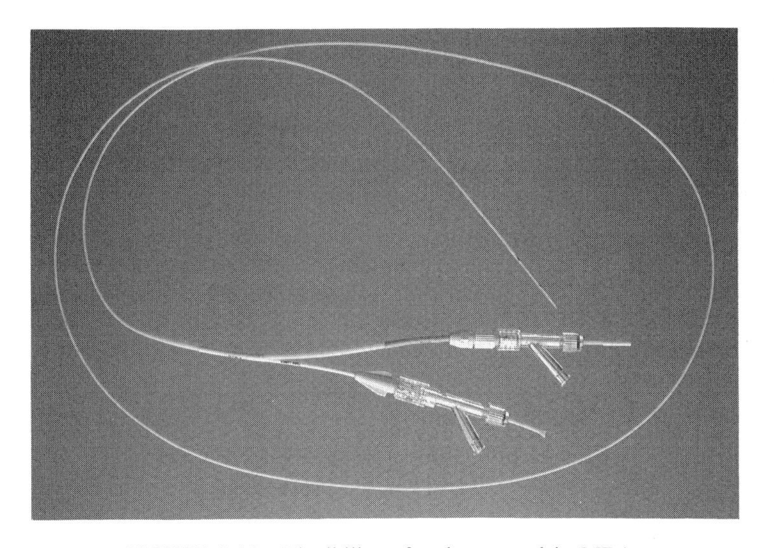

FIGURE 1.10 Flexibility of catheter used in MBA.

1.5.1 Coaxial Cable

The history and the theory of coaxial-cable development are for the most part, trivial to the electrical engineering community. When the cable is introduced into a biologic vessel, however, its complexity changes. Because of characteristics such as very low loss, flexibility, heat capacitance, and material compatibility with tissue, the coaxial cable is a key component in the microwave delivery system and requires special attention.

1.5.1.1 Cable Specifications

A *coaxial cable* is the high-frequency equivalent of a two-wire transmission line necessary to deliver electrical energy from one place to another, and to confine the microwave fields within a specifically delineated structure.

In the MBA application, the role of the coaxial cable is to deliver the microwave power from a generator to an interstitial antenna pushed into an angioplasty balloon that has been positioned in an artery of an experimental animal. The cable must be small enough to slide through the previously positioned catheter, following whatever tortuous path is necessary to reach the treatment site. The small size inherently leads to high loss of the microwave power being transmitted down the length of the cable. The lost microwave power is converted to heat, which raises the temperature of the cable, the catheter, and—to an undetermined extent—the artery and surrounding tissue. Less power is available for radiation from the antenna into the plaque or arterial walls at the treatment site. Another important electrical parameter in a coaxial cable is the characteristic impedance, which is determined by the dimensions and the dielectric material. Imperfections in the cable or connectors or antenna mismatch produce power reflections, which can reduce the radiated power as well as introduce hot spots on the cable itself.

1.5.1.2 Design Considerations

The first design consideration for a coaxial transmission line is the maximum allowable outer conductor diameter, which, for MBA application, is 0.022 in. or less in some circumstances.

Next, the dielectric material must be chosen for dielectric loss and physical properties. The optimum choice (lowest loss) dielectric is air; however, foam dielectric materials that approach the dielectric constant of air are available.

The center conductor diameter is then adjusted to achieve the desired characteristic impedance. An optimum geometry exists for achieving minimum attenuation in a coaxial transmission line. The characteristic impedance of the cable is directly related to the geometry of the conductor diameters.

1.5.1.3 Power Loss

Power loss in a coaxial cable is related to the attenuation constant α. This constant has two components: α_c, due to conductor loss; and α_d, due to dielectric losses; where $\alpha = \alpha_c + \alpha_d$. For those low-loss dielectrics used in microwave cables, the dielectric loss is negligible ($<2\%$) compared to the conductor losses. For copper conductors surrounded by an insulator having a dielectric constant of ε, the losses in decibels per 100 ft can be determined from

$$\alpha_c = \frac{0.214\sqrt{F}}{Z_0}\frac{1}{D}\left(1 + \frac{D}{d}\right) \quad \frac{\text{dB}}{100\,\text{ft}}$$

where F is in megahertz and D is in inches; thus

$$\alpha_d = 2.77F\sqrt{\varepsilon}\tan\delta \qquad \frac{\text{dB}}{100\,\text{ft}}$$

where D is the inner diameter (i.d.) of the outer conductor, d is the outer diameter (o.d.) of the inner conductor in inches, and $\tan\delta$ is the dielectric loss factor.

If the outer diameter is held constant because of mechanical size constraints, as is the case considered here, it can be determined from the above expressions that:[19]

1. The loss has a broad minimum at a D/d ratio of 3.59 for any dielectric at any frequency.
2. For a constant impedance Z_0, the loss decreases as ε approaches unity while the diameter of the inner conductor increases to maintain the constant impedance.

These effects are shown in Table 1.2 for various cables having an outer conductor i.d. of 0.025 in. with various dielectric constant materials (typical loss factor of 0.0001) operating at 2450 MHz.

1.5.1.4 Results

The cable–antenna assemblies consist of lengths of miniature coaxial cable (nominally 45 in. long in animal experiments), terminated in a gap or whip antenna in order to radiate microwave power to treated tissue during balloon angioplasty experiments. The gap is approximately 0.025 in. wide and is located approximately 0.31 in. from the shortcircuited end of the cable. Two cable sizes have been used for most of these assemblies. The basic characteristics of these two commercially available semirigid coaxial cables are tabulated in Table 1.3.

A typical test procedure for the cable–antenna assemblies included the following steps. After assembly, the cable and antenna were first checked for continuity and leakage. Recording the total direct-current (DC) resistance with the gap shorted (short-circuited) is also recommended as a simple check on the

TABLE 1.2 Cable Losses

ε	α_c dB/ft	α_d dB/ft	Z_0	D/d	α_{total}/ft	Dielectric
1.0	0.253	0.068	77	3.59	0.321	Air
1.36	0.29	0.079	66	3.59	0.373	Air-filled Teflon
2.1	0.367	0.0985	53	3.59	0.46	Teflon
1.0	0.279	0.068	50	2.3	0.347	Air
1.36	0.308	0.079	50	2.64	0.387	Air-filled Teflon
2.1	0.368	0.0985	50	3.35	0.468	Teflon

TABLE 1.3 Basic Characteristics of Two Semirigid Coaxial Cables

Manufacturer	Micro-Coax	Precision Tube
Type No.	UT 20-M	CA50020
Outer conductor diameter (in.)	0.023	0.020
Dielectric diameter (in.)	0.015	0.015
Center conductor diameter (in.)	0.0045	0.0044
Attenuation at 2.4 GHz (dB/ft)	1.13	1.33
Rated power at 2.4 GHz	12.5 W at 20 °C	1.8 W at 40 °C
Comments	Ultramalleable, stiff; dielectric core is a loose fit–can slip	Weak, easily kinked; push with extreme caution!

TABLE 1.4 High-Power Test at 2450 MHz

Cable No.	Outer Diameter (in.)	Total Length (in.)	Power (W)		Time (s)	Temperature Rise (°C) at End of 30-s Period
			Incident	Reflected		
A1	0.020	45	30	0.2	30	11.5
A2	0.020	44	30	0.2	30	17.1
A3	0.020	48	30	0.3	30	6.5
B1	0.023	48	30	0.9	30	13.3
B2	0.023	48	30	0.5	30	6.4

potential microwave loss. The cable–antenna assembly was tested under high-power conditions at 2450 MHz to verify its ability to handle the power without breakdown or excessive heating. The cables were surrounded by a test catheter and a simulated arterial environment during this test. For example, the cables were wrapped in wet sponges over their full lengths during the tests. The gap antennas were inserted into a plastic pipette containing deionized water. The pipette was wrapped in a saline-soaked sponge for RF loading purposes. The temperature at the outer pipette surface was monitored as evidence of the RF heating from radiated power. The wetness and contact of the sponges varied considerably, causing the temperature rise to be only qualitative. All units were subjected to an input of 30 W for 30 s, monitored for excessive reflected power, and inspected for damage after removal. Table 1.4 lists the temperature rises measured during one set of postassembly tests. The discrepancy in temperature rise measured in cables B1 and B2 is probably due to the pressure exerted on the phantom.

1.5.2 Microwave Power Requirements

The desired volume of muscle tissue to be heated is defined nominally as a 10 mm long hollow cylinder with an outer diameter of 5.5 mm and an inner diameter of 2.5 mm (1.5-mm-thick wall). The hollow cylinder has a volume of 188 mm^3.

The specific absorption ratio (SAR) is a measure of power deposition in the tissue in which cooling effects are not included.[20–22] Cooling from conduction and circulation will be extremely important in determining the actual temperatures reached during the procedure, but do not impact on the calculated SAR from the following definition:

$$\text{SAR} = k \cdot c \cdot \frac{\Delta T}{\Delta t} \text{ in watts per kilogram}$$

where $k = 4.186\ \text{W} \cdot \text{s/cal}$
$c = 0.83\ \text{cal/}^\circ\text{C} \cdot \text{kg}$ = specific heat of muscle tissue
ΔT = temperature rise in °C
Δt = time in seconds

To produce a 1 °C temperature rise in one second in muscle tissue, SAR = 3.47 mW/mg. Muscle tissue has approximately the same density as water (1 mg/mm^3), and a volume of 188 mm^3 will have a mass of 188 mg. The corresponding required power is

$$\text{Power} = (3.47\ \text{mW/Kg}) \cdot (188\ \text{mg}) = 653\ \text{mW}$$

Since cooling effects are neglected, this amount of power will also raise the tissue temperature 5 °C in 5 s, 10 °C in 10 s, and so on. In actual circumstances, cooling effects will predominate as time progresses, changing the linear relation. With only a 20% efficiency, power into a lossless cable would be

$$653\ \text{mW} \times 5 \approx 3300\ \text{mW}$$

However, in practice, the required rate of temperature rise is 3 °C/s. Therefore, *assuming linearity*, we need 3×3300 mW $\approx$ 10 W of power into a lossless cable, or 10 W of radiated power.

Assuming a radiation efficiency of 80% from the antenna, an antenna input power of 12.5 W would be needed. With all the foregoing assumptions (including no cooling), the required input power to the cable will depend on the total cable loss (α): $P_{\text{in}} = 12.5 \cdot \text{alog}(\alpha/10)$. For a 6-dB cable loss, the input requirement would be 50 W. It is estimated that this is the maximum requirement for the input to the cable. If the foregoing assumptions (volume, temperature rise, heating time) are revised, the power requirements can be changed proportionately.

1.5.2.1 Dimensions

The length of the cable has been defined as 150 cm (approximately 5 ft). If not stepped (i.e., if there is no transition in size), its diameter must not exceed 0.022 in. However, it is recommended that a stepped cable be utilized to minimize cable loss. If stepped, the proximal end will be 25 cm long with an o.d. of 0.030 in.; the central section will be 100 cm long with an o.d. of 0.025 in. and the distal end will be 15 cm long with an o.d. of 0.022 in.

1.5.2.2 Summary and Requirements

The total attenuation for a stepped air-filled Teflon cable could be as low as 3.5 dB at 2450 MHz:

$$[(25\text{ cm} \times 0.62\text{ dB/ft}) + (100\text{ cm} \times 0.74\text{ dB/ft}) + (15\text{ cm} \times 1.03\text{ dB/ft})] \times 0.033\text{ ft/cm} = 3.5\text{ dB}$$

Considering that it may not be feasible or desirable to use the air-filled Teflon over the full length, it is recommended that a maximum attenuation limit of 4.2 dB be specified for the stepped cable.

If a nonstepped separate coaxial cable with an i.d. of 0.022 in. is selected for the clinical cable–antenna assemblies, a maximum attenuation specification of 5.5 dB should be applied. This assumes the use of the lower dielectric-constant insulator material such as air-filled Teflon. If regular Teflon dielectric is used, the loss will be at least 6.5 dB.

1.5.3 Low-Loss Fully Flexible Coaxial Cable

Material other than copper can be utilized in medical applications if the skin effect phenomenon in microwave frequencies is understood. A coaxial cable having a silver-coated stainless-steel center conductor was utilized in our flexible cable–antenna delivery system.

1.5.3.1 Skin Effect

In the design of a coaxial cable at microwave frequencies, the guided wave portion requires special attention. In a coaxial cable, the inner surface of its outer conductor and the outer surface of its inner conductor provide the boundaries for the propagation wave. There could be a sharp transition in the properties of the medium at the surface of the boundaries, specifically, a nonconductive medium followed by a highly conductive surface, such as the center conductor of a coaxial wire that could be hollowed, or made from other material such as silver-coated stainless steel. The metals used in practical guides have finite conductivity; therefore, dynamic fields are propagated to a thickness of δ into the guide walls (skin depth). A simple case is that of a transverse electromagnetic (TEM) wave propagating in a highly conducting medium such as copper and silver, where

$$\delta = \frac{1}{\sqrt{\pi F \mu \sigma}} \text{ meter}$$

where F is the frequency and μ and σ are the permeability and conductivity of the medium, respectively. At the depth δ, the electrical intensity has been attenuated to 37% of the initial value, and the power transported has been attenuated to 13.6% of its initial value.

For all practical purposes, waves can be said to be completely attenuated at depths equal to 5δ. From the immediately preceding expression, it is clear that

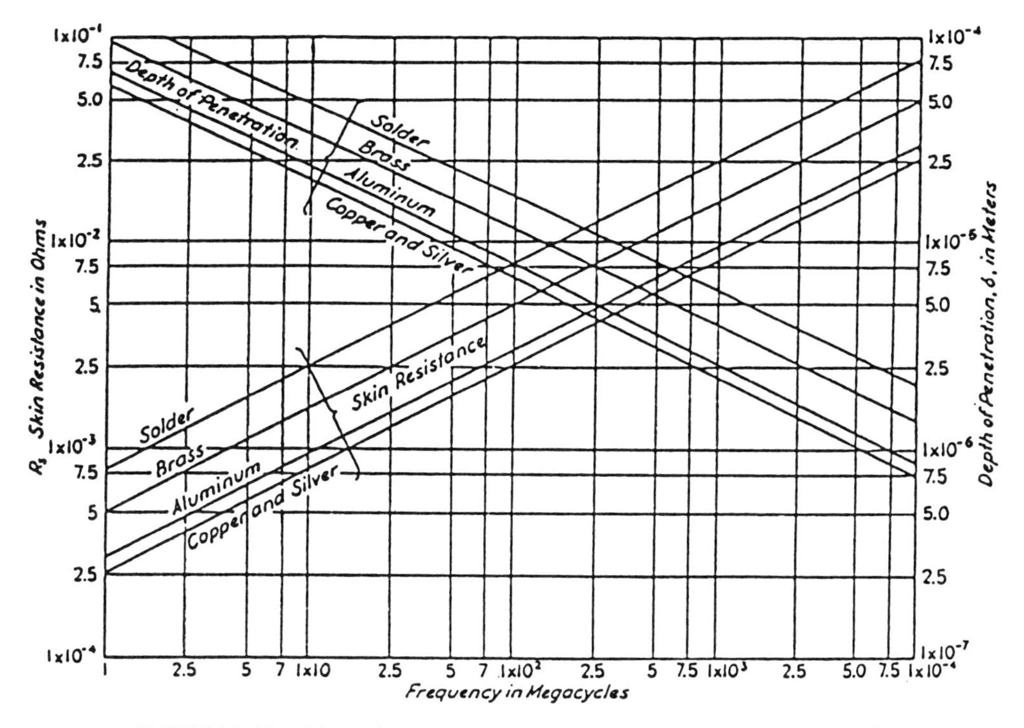

FIGURE 1.11 Skin resistance and depth of penetration for several metals.[23]

a minimum layer of 5δ of a highly conductive material is needed to cover the outside of the inner conductor and inside of the outer conductor for a coaxial cable. Depth of penetration δ (in meters) and skin resistance (in ohms) are plotted in Figure 1.11.

1.5.4 Coaxial Cable for Microwave Balloon Angioplasty

Based on the information above, a microwave cable having the following clinical device specifications was developed:[24]

1. Device requirements
   A. Electrical
      i. Attenuation at 2.45 GHz—6 dB max. (1.0 dB/ft)
      ii. Rated power at 2.45 GHz—80 W
      iii. Distal tip (antenna)
   B. Mechanical: must be flexible enough to track through catheter; around the aortic arch and across the lesion
   C. Functional: must be possible to locate proximal end of antenna under fluoroscopy to line up with thermocouple

2. Device description: coaxial cable with whip or gap antenna
   - A. Center conductor
     - i. Material: Ag-coated stainless steel
     - ii. Dimensions
       - a. SS diameter—0.005 ± 0.0003 in.
       - b. Coated wire diameter 0.006 in. max.
       - c. Coating thickness 0.0003 ± 0.0001 in.
       - d. Length—70.0 in. max.
   - B. Dielectric
     - i. Material: Goretex expanded PTFE (polytetrafluoroethylene)
     - ii. Dimensions: diameter 0.017 in. max. wrapped (0.015 in. under compression of outer conductor)
   - C. Outer conductor
     - i. Material
       - a. Option 1: Ag-coated copper rolled flat wire
       - b. Option 2: Ag-coated SS rolled flat wire
     - ii. Dimensions
       - a. Width 0.007 ± 0.002 in.
       - b. Thickness 0.0015 ± 0.0001 in.
       - c. Ag thickness 0.0003 ± 0.0001 in.
       - d. Wrapped diameter 0.022 in. max.
   - D. Connector
     - i. Material: SS/copper
     - ii. Type: female SMA per MIL-C-39012
   - E. Coating
     - i. Must withstand 100 °C
     - ii. Will cover distal 10 cm of cable
     - iii. Thickness 0.002 in. max.
3. Device design
   - A. Dielectric extruded over center conductor
   - B. Two layers of Ag-coated SS ribbon either overlap–wound over dielectric to form outer conductor, or wound clockwise and counterclockwise
   - C. SS hypotube slid over proximal end of outer conductor coils for support and soldered to coils
   - D. Connector attached to proximal end
   - E. 2.1-cm whip antenna formed at distal end by removing outer conductor coils
   - F. Distal end of cable to be coated with an insulator
   - G. Proximal end of antenna to be radio opaque
4. Biocompatibility and sterility
   - A. Material testing
     - i. All materials to pass USP biologic tests for class VI plastics
       - a. Sensitization assay (does not promote allergic reaction)
       - b. Cytotoxicity test
       - c. Thrombogenicity test

d. Hemolysis (breaking red cells) test
e. Genotoxicity test
ii. Material verification
iii. Heavy metals
iv. Leachables
B. Sterility
i. EtO residuals outgas time;
ii. D Values: sterile cycle
iii. Pyrogen–pyrogen inhibition: detects bacterial endotoxins
iv. Bioburden and spore recovery: natural microbial population of product (quarterly monitoring program)
C. Shelf life: 6 months from date of manufacture

1.6 PROGRAMMABLE HIGH-POWER MICROWAVE SOURCE FOR ANGIOPLASTY APPLICATION

1.6.1 Introduction

The requirement for a programmable microwave source (PMS) delivering up to 80 W CW at 2.45 GHz can be fulfilled by a choice of systems comprised of readily available commercial components.

This section provides a review of two specific technical approaches based on the magnetron and the solid-state technologies and their cost and manufacturability implications.

Generation of 80 W of power at 2.45 GHz can be accomplished in a number of ways, such as[25]

1. Exclusive use of a vacuum-tube technology for power generation and amplification in one embodiment (magnetron, klystron, etc).
2. Exclusive use of a solid-state technology in a distributed manner for power generation, preamplification, and power amplification
3. Hybrid use of a solid-state technology for the oscillator and preamplification followed with a vacuum-tube power amplifier (TWT, etc).

Each approach offers distinct advantages and disadvantages that affect parameters such as the ease of power-level control (probe temperature), system power supplies and cooling requirements (bulkiness, cost), and cost.

Only the magnetron and the solid-state approaches were explored in this section. Each approach warrants a block diagram and an explanation of the system's functionality.

1.6.2 Magnetron-Based Apparatus

A *magnetron* is a vacuum tube providing (1) oscillations at fixed frequency stabilized by internal cavities and (2) an internal amplification. Magnetron tubes originally cost around $1000 each. Fortunately, because of widespread use for microwave-oven applications at 2.45 GHz, the cost of the magnetron has been reduced dramatically. Such magnetrons are typically designed to deliver anywhere from 200 to 600 W at nominal operating voltages. For the desired clinical operation of the apparatus, the output power from the magnetron has to be permanently reduced to 80 W and secured by an active limiter not to exceed this level.

Next, a digital control network is devised to provide a time-controllable output power for levels of 10 W up to 80 W in fixed increments. The block diagram of such a system is provided in Figure 1.12. The magnetron is supplied from a power supply providing the cathode filament heating and the high voltage for the anode. The output power from the magnetron is connected to the four-port high-power isolator.

There are three important reasons for the use of the isolator:

1. The magnetron will be permanently terminated with 50 Ω regardless of the antenna's varying load characteristics. This will assure that no peak output power due to changing load voltage standing-wave ratio (VSWR) exceeds the maximum rating allowed. Furthermore, it will extend the magnetron lifetime and will reduce the dangerous arcing due to load mismatch.

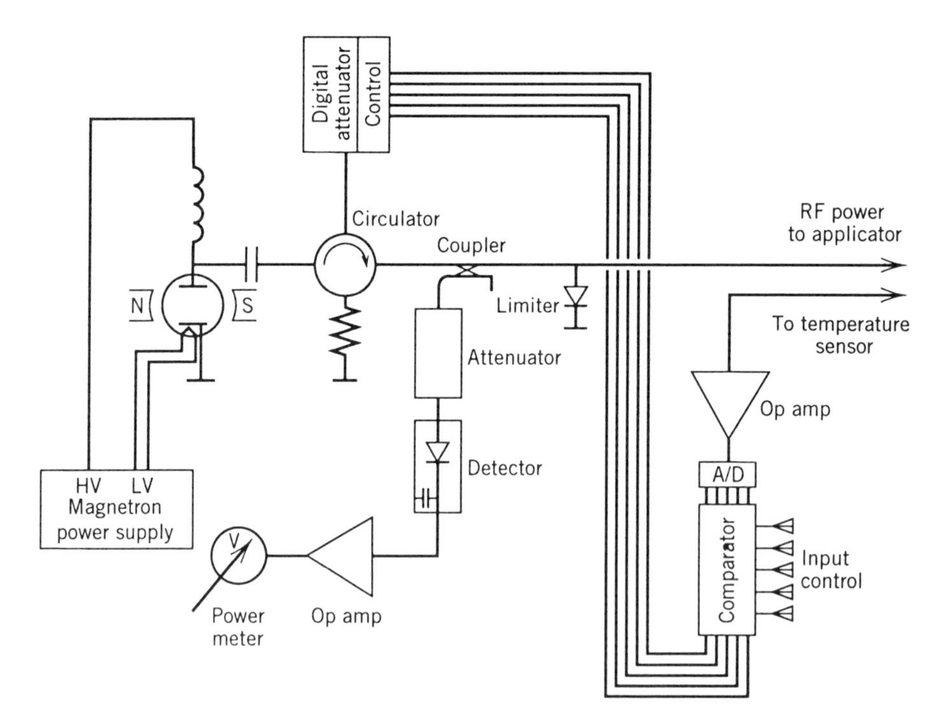

FIGURE 1.12 Magnetron-based microwave generator (Op amp = operational amplifier; HV, LV = high, low voltage; A/D = analog to digital).

2. In case the apparatus is inadvertently activated without the probe connected, or with discontinuity inside the probe, the reflected power is terminated in the isolation resistor of the isolator, thus minimizing the dangers of undesirable radiation.
3. One port of the isolator is connected to the reflective digital attenuator, providing fine control of the output power level.

The coupler at the output along with the detector circuit and a display provide continuous reading at the output power. The limiter diode ensures that the maximum power does not exceed the design limits and, hence, the safety level of the apparatus.

The control network consists of a comparator receiving inputs from the temperature sensor and from the test profile input from the keypad. The output from the comparator drives the attenuator to the position satisfying the control loop's initial conditions.

A generator having most of these capabilities was built and tested under this project.

1.6.3 Solid-State-Based Apparatus

This system is based on an exclusive use of silicon bipolar devices. Since at the present time there is no single, commercially available transistor capable of delivering 80 W CW, a power combining technique can be used.

The conceptual design provided in Figure 1.13 is based on known availability of a power transistor delivering at least 25 W CW at 2.45 GHz.

The power gain block is driven from a dielectric resonator oscillator (DRO) and preamplified by a chain of conventional gain blocks. The digital attenuator is inserted in a low-power segment of the circuit, thus reducing the cost of this component. The coupler on the output line, along with the detector circuit, provides continuous reading of the output power. A limiter diode in the output circuit is not necessary since there is no possibility that the system's output power will exceed 80 W.

The control network consists of a comparator receiving inputs from the temperature sensor and from the test profile input from the keypad. The output from the comparator drives the attenuator to the position satisfying the control loop initial conditions.

1.6.4 Safety Features and Contraindications

Safety features of these microwave devices are as follows:

1. High-temperature shutoff
2. Low-temperature shutoff
3. High microwave reflection shutoff
4. DC block

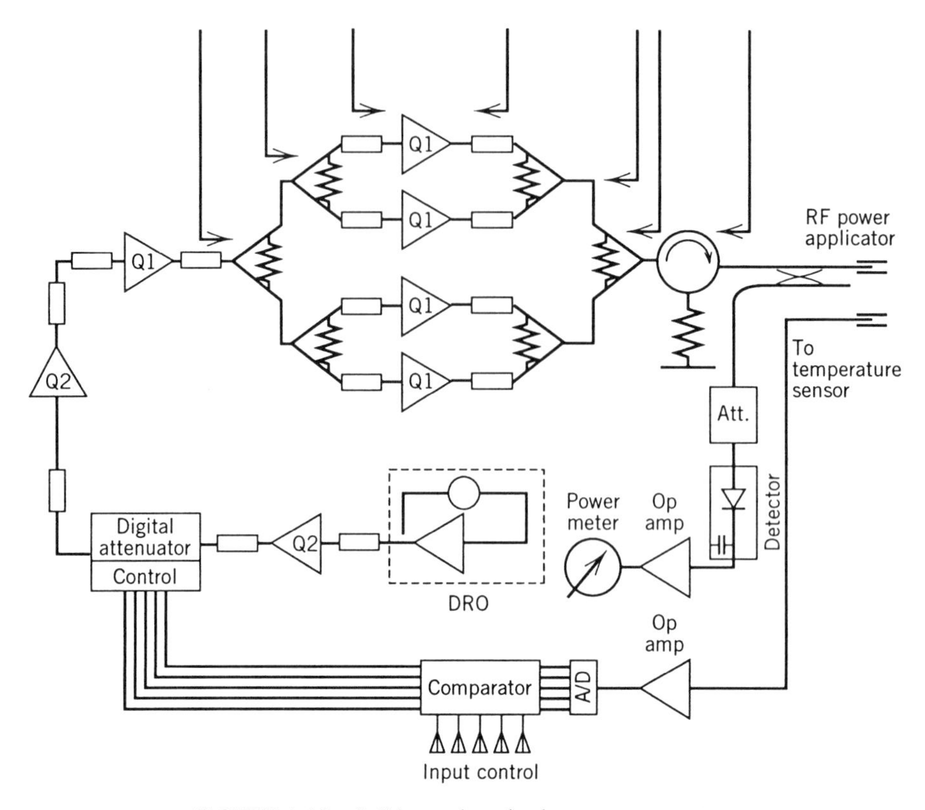

FIGURE 1.13 Solid-state-based microwave generator.

5. Slow temperature rise shutoff
6. Foot switch shutoff

Use of these microwave devices on patients with pacemakers, implanted electrodes, or metal sutures or markers is contraindicated during clinical investigation. Some of these contraindications may not apply to the final product.

1.6.5 Antenna Configuration

To deliver microwave energy to heat myocardial tissue and create histologic changes effectively, it is essential to control the size and location of the radiated field, thus controlling the affected tissue volume. In our studies, three types of microwave antennas at the tip of a coaxial cable were investigated: (1) the gap antenna (Fig. 1.4a), (2) the whip antenna (Fig. 1.4b), and (3) the helical antenna (Fig. 1.4c). Temperature distribution was measured along the tissue cylinder in the direction of the antenna length. A Gaussian shape with peaks was observed adjacent to the gap in the gap antenna and adjacent to the junction in the whip antenna, and there was a temperature drop in both directions axially from the gap–

junction interface. Very little temperature rise, and therefore tissue damage, was observed near the tip of the antennas. Both types of antennas were ideal for the treatment of most cases of atherosclerotic arteries, and were utilized through all our MBA studies. The helical antenna—which has some end-fire characteristics, if designed correctly, thus facilitating heating next to the antenna tip[26]—was not used for MBA because the very small coil size (0.5-mm-diameter) needed for coronary angioplasty failed to show any advantages in preliminary testing in vitro, and is so much more difficult to produce.

1.7 DETAILED INVESTIGATIONAL PROGRAM FOLLOWED DURING THE EVALUATION OF MICROWAVE THERMAL BALLOON ANGIOPLASTY[27–38]

The following sequence was followed during the evaluation and development of microwave thermal balloon angioplasty:

1. Initial bench evaluation of thermal characteristics of the catheter and microwave generator system
2. In vitro studies
   a. In vitro temperature measurements
   b. Myocardial lesion as a consequence of microwave energy at 2.45 GHz
   c. Effect of microwave dissipation on cardiac rhythm
   d. Human aortic plaque in vitro
   e. Biologic stent
3. Following availability of the catheter system, studies performed on (a) normal and (b) atherosclerotic rabbit iliac arteries.
4. Studies performed at various energy levels at 2.45 GHz.

Rabbit studies included both angiography and histology. A catheter was inserted through the carotid or the femoral artery and positioned in the iliac artery. The balloon was inflated and energy delivered. The same procedure was performed on the opposite iliac artery but with a standard balloon dilatation catheter. Repeat angiography was performed at the end of the procedure. The animal was then sacrificed and the vessels sent for histologic analysis.

The preceding studies are designed to assess the capability of producing a microwave thermal balloon angioplasty system and to determine the angiographic and histologic effects of delivery of this energy form in the normal and atherosclerotic rabbit model. Two additional phenomena were studied. These included tissue welding (Boston University)[39] and the effect on thrombi.

1.7.1 In vitro Experiments

1.7.1.1 In vitro Temperature Measurements

The temperature pattern (profile) of the slot antenna used in the MBA experiments (a whip antenna was later used) was measured by placing the antenna between

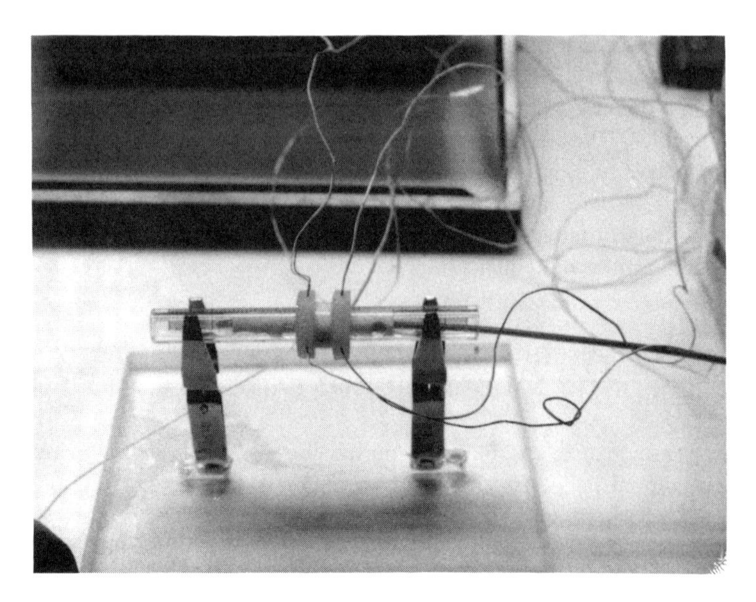

FIGURE 1.14 Temperature measurement apparatus.

two saline-soaked sponges. A liquid-crystal sheet that changes color in the temperature range of 25–30 °C was placed on the heated face of each sponge. The sponges were heated for 30 s with a net input power of 5 W at a frequency of 2450 MHz to measure temperature elevation. The heating pattern of the antenna was bidirectional and reasonably uniform over about 1 cm in length.

In vitro measurements were conducted to establish and verify the in vivo temperature measurements during the MBA procedure and the effect of the thermocouple wire attached to the balloon on the measured temperature. In these experiments, an apparatus was prepared to hold a vessel with an inflated balloon microwave angioplasty catheter having one thermocouple attached to the inside of the balloon, and multiple thermocouples outside of the balloon, in close proximity and at 90° to it. The maximal variation in measured temperature was found to be only 10% when the apparatus was immersed in normal saline solution. Such immersion simulates more closely the heat sink provided by body fluids in the in vivo experiments. The temperature profile was measured again using the same apparatus with multiple thermocouples along the axis of the balloon (Fig. 1.14), which was immersed in saline solution. Uniform temperature measurements along 1 cm were obtained in agreement with the measurements described in the first paragraph of this section.

1.7.1.2 Myocardial Lesion as a Consequence of Microwave Energy at 2.45 GHz

Following determination of the heating properties of the antenna system (Figs. 1.15, 1.16), microwave energy was applied directly to the myocardium of a dog in an open-chest experiment. Figure 1.17 depicts the histologic findings, consisting

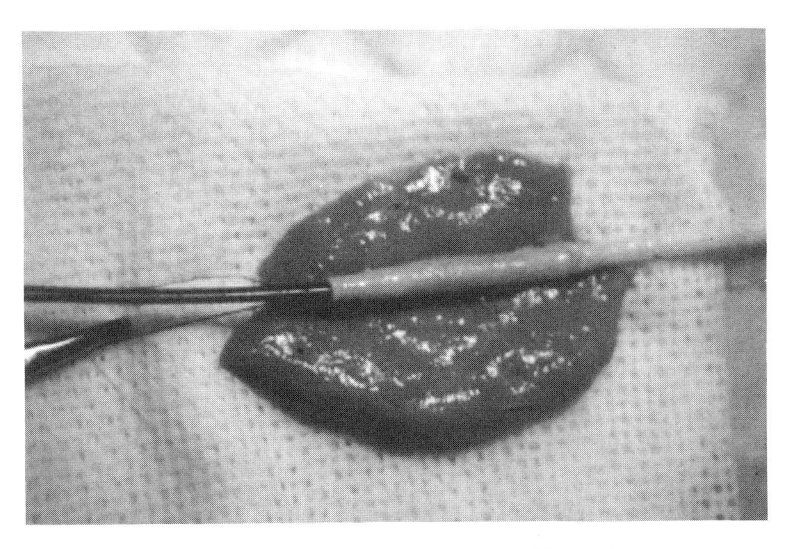

FIGURE 1.15 Antenna applied to a sample of liver tissue in vitro.

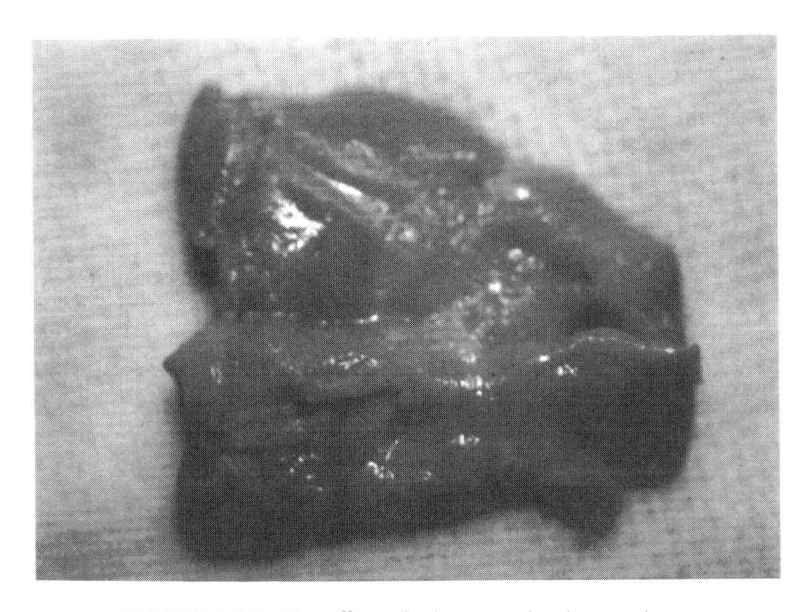

FIGURE 1.16 The effect of microwave heating on tissue.

of a discrete area of tissue disruption, to a depth of 2–3 mm, which followed the application of microwave energy through a coaxial cable and radiating antenna.

Centrally, the lesion contains a zone of severe thermal destruction with resulting coagulation necrosis. The extent of the coagulation necrosis gradually decreases toward the periphery of the lesion, where normal myocardial tissue is

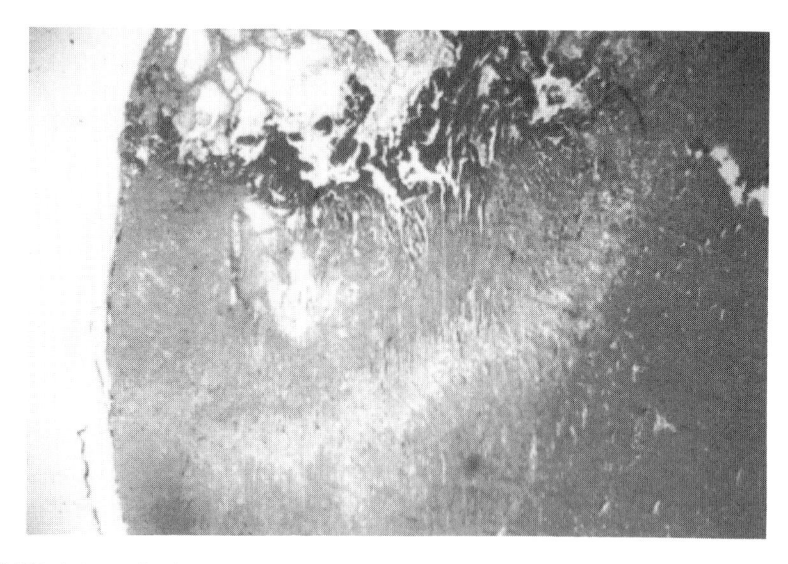

FIGURE 1.17 Histologic section of dog myocardium following thermal disruption by microwave energy.

seen. The outer zones of tissue damage retain some connective-tissue architecture despite the destruction of myocardial tissue, and a few surviving fibroblasts are visible. These experiments, coupled with previous experience with microwave coaxial applicators for use in thermotherapy of brain tumors[12] (Fig. 1.5), indicate that sufficient energy can be introduced into the stenotic lumen of a coronary artery via a special micorwave transmission line–antenna system embedded in a catheter.

1.7.1.3 Effect of Microwave Dissipation on Cardiac Rhythm

The question of the possibility of inducing arrhythmia by applying microwave energy to the heart prompted the following experiments. Microwave energy was applied through a microwave delivery system (cable and antenna) directly on top of the cardiac muscle of an anesthetized and artificially ventilated open-chest dog weighing 50 kg. Various power-levels of 5–35 W, and durations of 1–60 s were applied to various locations in the cardiac muscle (Fig. 1.17 depicts the type of lesion introduced). No arrhythmias were observed in the limits of our experiments, assuring safety of the procedure with regard to arrhythmias.

1.7.1.4 Human Aortic Plaque in vitro

Human plaque varying in plaque consistency from fibrofils (soft) to densely fibrotic to calcified were tested for variation in temperature of heat and pressure. The samples were placed between two glass plates on a heat source. A force of 25 g was exerted on top of the samples, while the hot plate temperature was raised to 50–60 °C. In all cases except the completely calcified sample, we observed

FIGURE 1.18 Human plaque subjected to 60 °C

changes in the geometry of the plaque. At 60 °C the plaque was flat in shape as shown in Fig. 1.18. The difference in temperature observed to obtain the flat configuration in the various plaque samples (soft and hard plaque) was very small, in the order of a few degrees Celsius (even in the case of the hard plaque, which had some degree of calcification but was not completely calcified).

1.7.1.5 Biologic Stent

Normal rabbit arteries were utilized to study the effect of balloon dilation and heat. Figure 1.19 depicts the lateral view of a rabbit artery before (*a*) and after (*b*) MBA. As can be seen, after introducing the combined modalities of pressure and heat, we created a biologic stent that persisted even after immersing the artery in saline solution. Figure 1.20 depicts the luminal view of another rabbit artery before and after in vitro MBA. The preliminary studies described above have indicated that MBA decreases vessel elasticity at the dilation site and volume microwave heating molds the vessel segment to the size and shape of the inflated balloon; thus the creation of a biologic stent.

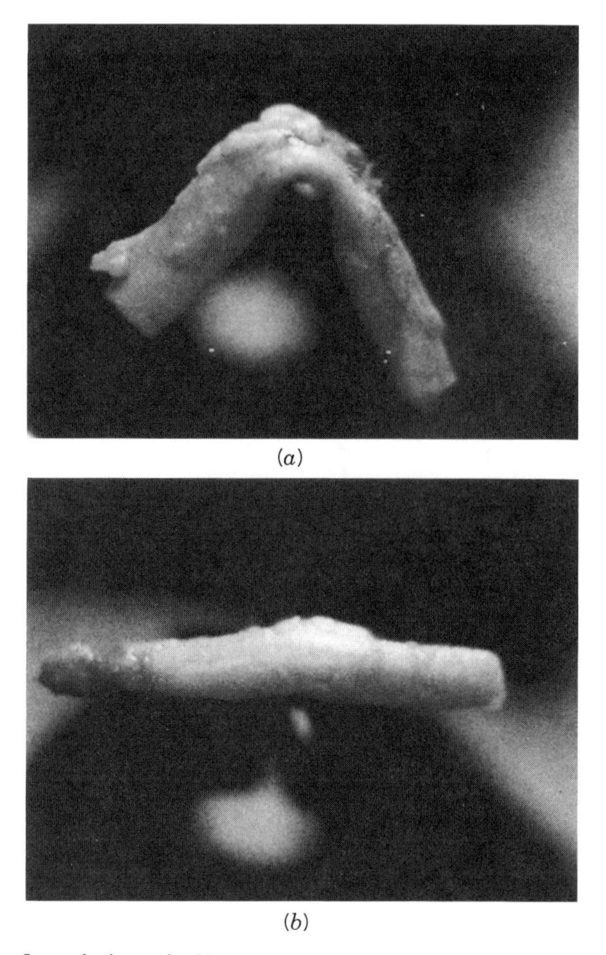

FIGURE 1.19 Lateral view of rabbit artery before (*a*), and after (*b*) in vitro microwave balloon angioplasty.

In summary, these in vitro studies and limited in vivo (open-chest arrhythmia) experiments demonstrate the feasibility of using microwave energy to induce thermal modification of vascular tissue. This modality shows promise as an ancillary means of reducing arterial elastic recoil as evident in the in vitro measurements (Fig. 1.19), forming a biologic stent (Fig. 1.18), and decreasing plaque resistance to dilatation.

1.7.2 MBA on Rabbit Model

1.7.2.1 Method

Our initial studies using this technology were performed in the iliac arteries of normal rabbits. The New Zealand white rabbit was chosen for these studies

FIGURE 1.20 Luminal view of rabbit artery before (center) and after (sides) in vitro microwave balloon angioplasty.

because (1) these rabbits develop atherosclerotic lesions quickly when on a cholesterol-supplemented diet and when the endothelium is removed and (2) the iliac arteries of these rabbits have approximately the same luminal diameter as that of human coronary arteries.

The microwave system used in the experiments consists of a 2450-MHz signal generator (capable of delivering up to 80 W), a directional coupler, and two power meters to measure forward and reflected power. In addition, a thin flexible coaxial cable, 0.023–0.034 in. in diameter, fits within a conventional balloon angioplasty catheter and is terminated by a radiating antenna. The major physical constraint on both the antenna structure and the transmission line is the diameter of the catheter through-lumen. In addition, the radiation pattern from the antenna must be reasonably uniform and confined to the volume around the balloon. One class of microwave-radiating structures that appears to be compatible with these requirements is the insulated asymmetric dipole (gap antenna).

The radiated microwave heating power is generally confined to the region around the gap in the outer conductor of the coaxial line (Fig. 1.21). At any specific frequency of operation, the heating pattern can be optimized by selection of the length of the shorted section and the width of the gap. The dielectric loading effects of the tissue and the balloon inflation fluid are other factors that

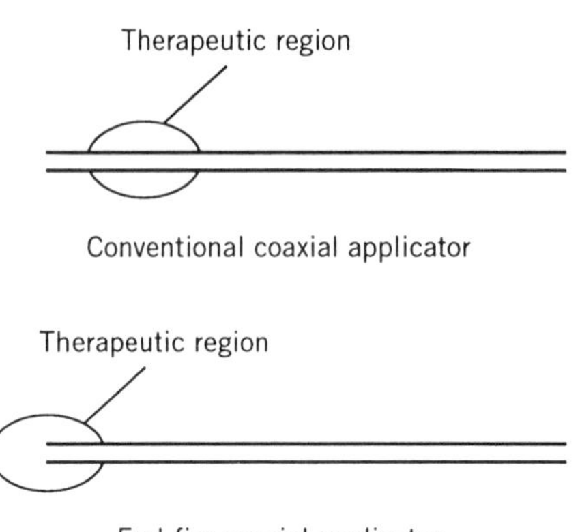

FIGURE 1.21 Therapeutic region for two types of antennas.

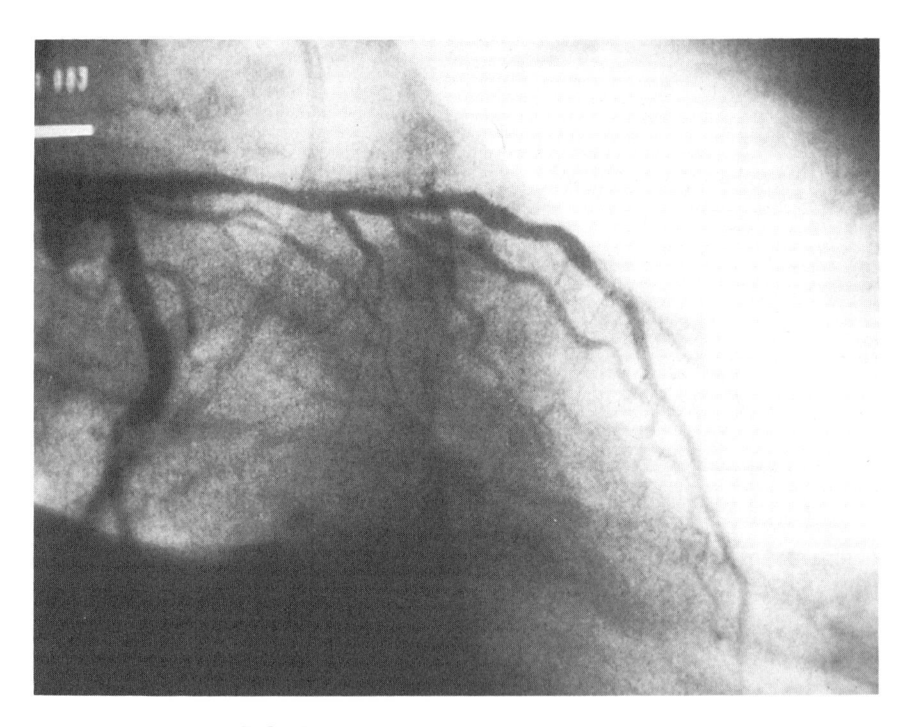

FIGURE 1.22 Coronary arteries in human.

affect the selection of the antenna dimensions, tending to substantially reduce both dimensions as compared to free-space conditions.

The microwave transmission line also must be sufficiently narrow and flexible to fit within, and be advanced through, the catheter (Figs. 1.22, 1.23). As the diameter of a coaxial cable is decreased, the attenuation of the cable

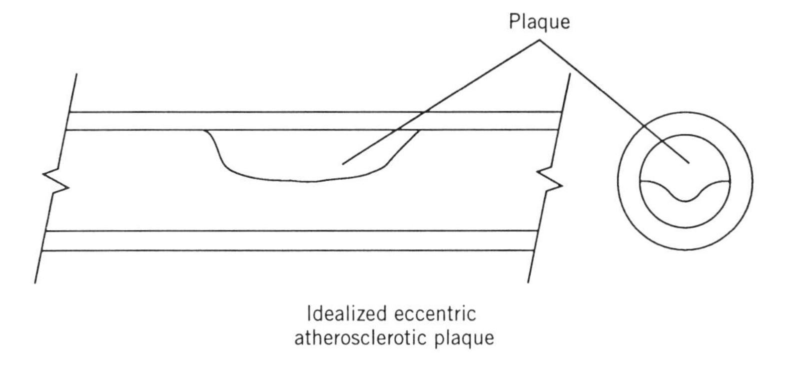

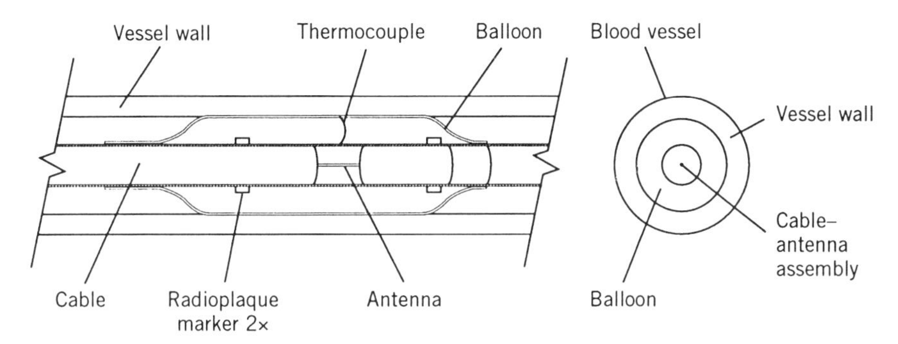

FIGURE 1.23 Position of balloon catheter in relationship to plaque.

increases and the microwave power loss in the cable increases. Although available coaxial cables can be and have been used for experimental work, special cables that optimize loss and mechanical flexibility were developed for practical implementation of the technique. With an experimental 42-in.-long coaxial cable, suitable for installation in a standard angioplasty catheter, and different antenna types, return losses at 915 MHz and 2.45 GHz were measured with the antenna immersed in various media (phantom and deionized water) as shown in Figs. 1.24–1.28. As demonstrated by the small amount of reflected power when the antenna is loaded either by a tissue phantom or by immersion in deionized water, the coupling of microwave power into tissue can be reasonably efficient from this cable–antenna assembly.

A chromel/alumel (K-type) thermocouple is inserted into a separate lumen of the catheter and epoxied to a portion of the balloon as shown in Fig. 1.29a. The coaxial antenna is positioned within the balloon, and microwave energy is broadcast to surrounding arterial tissue. The balloon catheter itself contains three lumens. The through-lumen is the conduit for the guidewire and microwave antenna, while the other lumens are used for inflation of the balloon to a maximum of 3.5 mm and for insertion of the thermocouple.

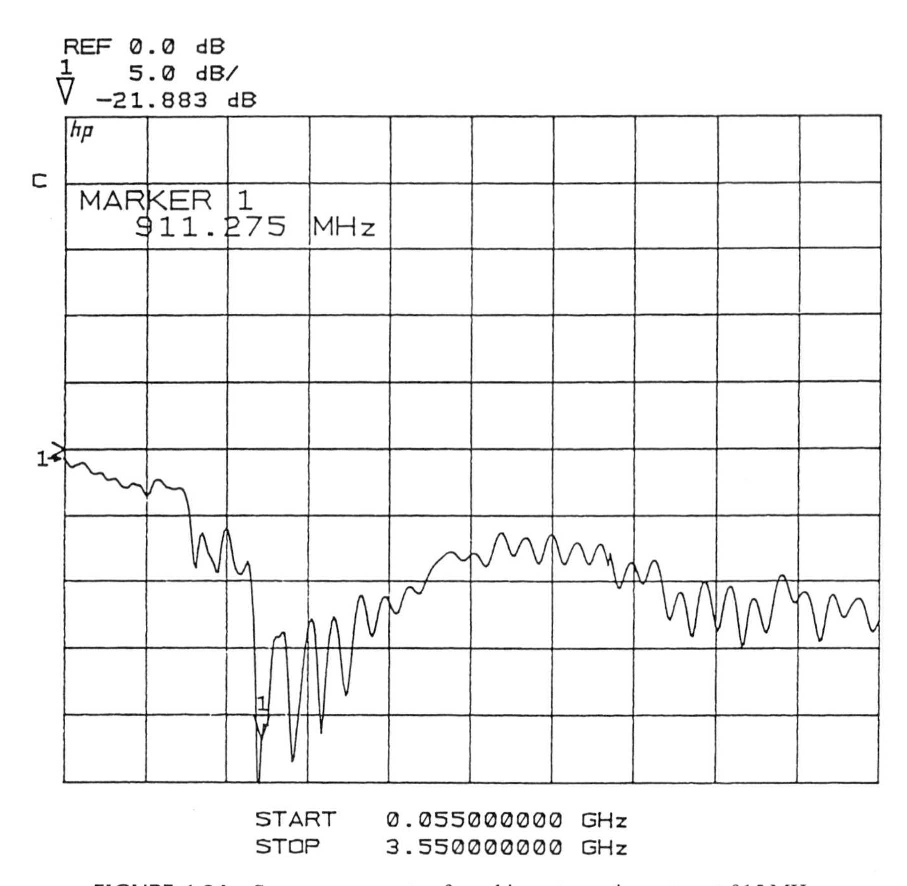

FIGURE 1.24 S_{22} measurements of a whip antenna in water at 915 MHz.

1.7.2.2 Angioplasty Procedure

Angioplasty was performed on 30 normal male and female adult New Zealand white rabbits ranging from 4.5 to 5.5 kg. After induction of general anesthesia with acepromazine maleate, xylazine, and ketamine hydrochloride, both femoral arteries were isolated under sterile conditions. The microwave angioplasty system used in our early experiments was introduced along the guidewire into the carotid artery and positioned in the iliac artery or the distal abdominal aorta (Fig. 1.30). These measurements were performed in part to test the flexibility and integrity of the microwave cable. In later experiments, the microwave angioplasty system was placed in the appropriate femoral artery and advanced retrograde to the external iliac artery under fluoroscopic guidance. Once in place, angioplasty was performed using the specified parameters. Study variables included inflation pressure (1–5 atm), target temperature (50, 70, 85, or 90 °C), and microwave energy duration (30–60 s). Thirty animals were chosen to provide 60 iliac arteries for analysis, at least four vessels for each possible permutation of time, temperature, and pressure,

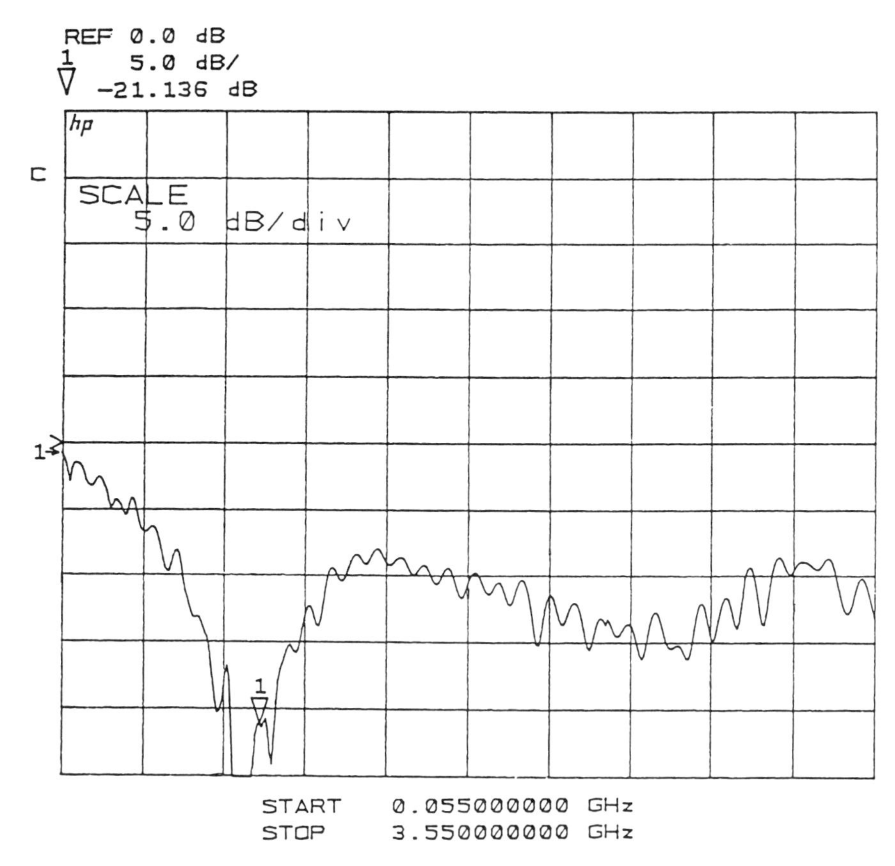

FIGURE 1.25 S_{22} measurements of a helical antenna in phantom at 915 MHz.

as well as eight vessels that underwent conventional angioplasty without heating. Following angioplasty, the catheter was removed and the femoral artery ligated.

1.7.2.3 Histologic Analysis

One week after angioplasty, the animals were again anesthetized and the inferior vena cava and abdominal aorta were isolated. After heparinization, animals were sacrificed with an overdose of pentobarbital and the aorta was perfused with 10% formalin at 100 mmHg for 30 min. Following this, the external iliac arteries were removed and sent for histologic analysis.

Each vessel was embedded in paraffin and sectioned at 4-mm intervals (Fig. 1.31), then stained with hematoxylin–eosin, trichrome, and Verhof–Van Giesen stains. Each vessel was examined by a pathologist who was blinded to the conditions of each angioplasty. Medial injury for each section was qualitatively graded on a scale of 1–5 with 1 representing medial reaction and 5 representing full-thickness medial necrosis. In addition, medial injury was expressed as the

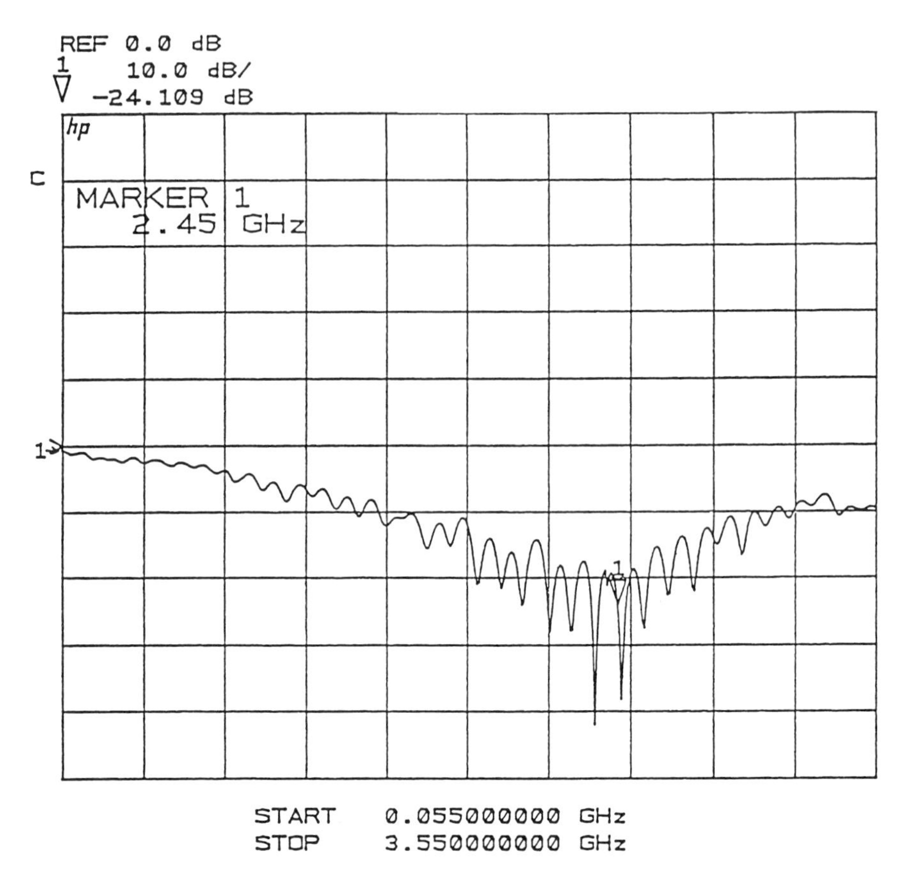

FIGURE 1.26 S_{22} measurements of a helical antenna covered with diamond paste at 2.45 GHz.

percent of the total vessel circumference involved (Fig. 1.32). The product of the depth of injury and percent circumference ($\times 100$) was defined as the medial injury index (MII). For each vessel, the MII for all vessel sections was averaged to yield the mean MII. Intimal proliferation was graded on a scale of 1–4, and also expressed as a percent of vessel circumference. The intimal proliferation index (IPI) was defined as the product of the two ($\times 100$). The mean IPI was computed by averaging the IPI of all vessel sections for a given vessel.

Quantification of Vessel Injury The quantification of vessel injury is described in the following paragraphs.

Each vessel is sectioned at 2- or 4-mm intervals. The medial depth of injury for each section was graded on a scale of 1–5, with 1 representing medial reaction and full-thickness medial necrosis, as follows:

Grade 1 = occasional pyknotic nuclei with an occasional inflammatory cell

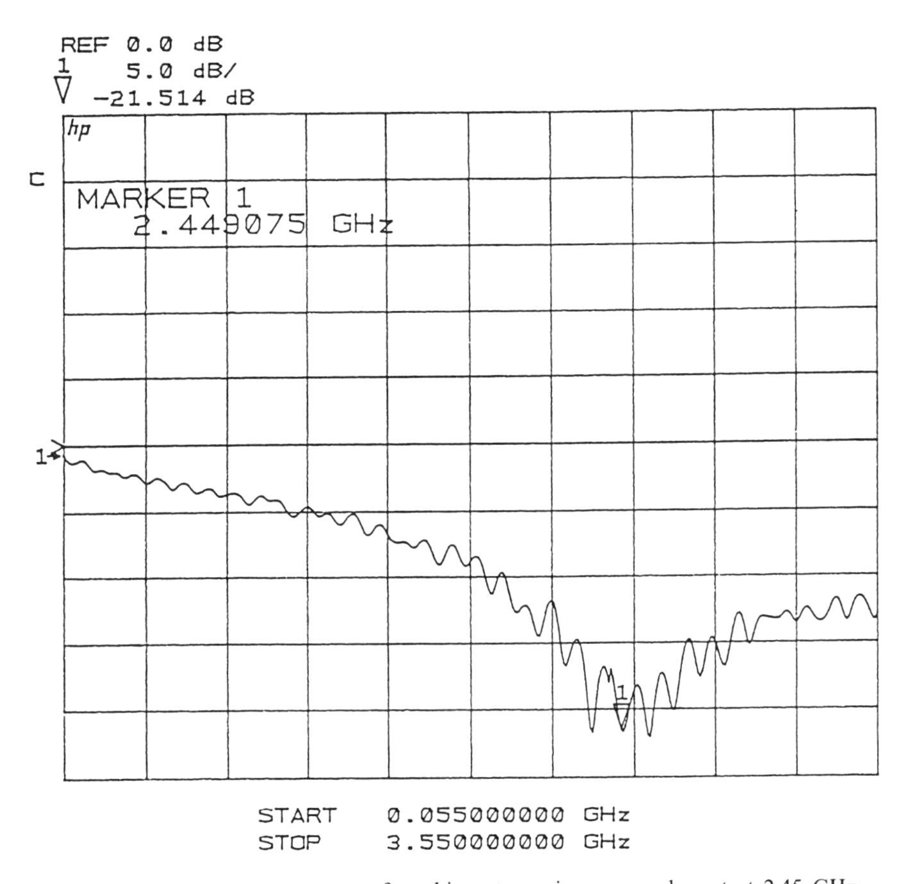

FIGURE 1.27 Measurements of a whip antenna in processed meat at 2.45 GHz.

Grade 2 = obvious inflammatory infiltrate (polymorphonuclear cells) and occasional necrotic smooth-muscle cells with isolated foci of fibrosis

Grade 3 = moderate fibrosis with moderate necrotic smooth-muscle cells

Grade 4 = extensive cell muscle necrosis, extending at some point to the adventitia, and prominent inflammatory infiltrate with prominent fibrosis

Grade 5 = complete circumferential necrosis or fibrosis with adventitial inflammatory infiltrate.

A visual estimate of the medial circumference involved by the injury was expressed as the percent of the total circumference. For each section, the product of the grade of injury and percent circumferential involvement ($\times 100$) was defined as the medial injury index. Since each artery was studied in at least

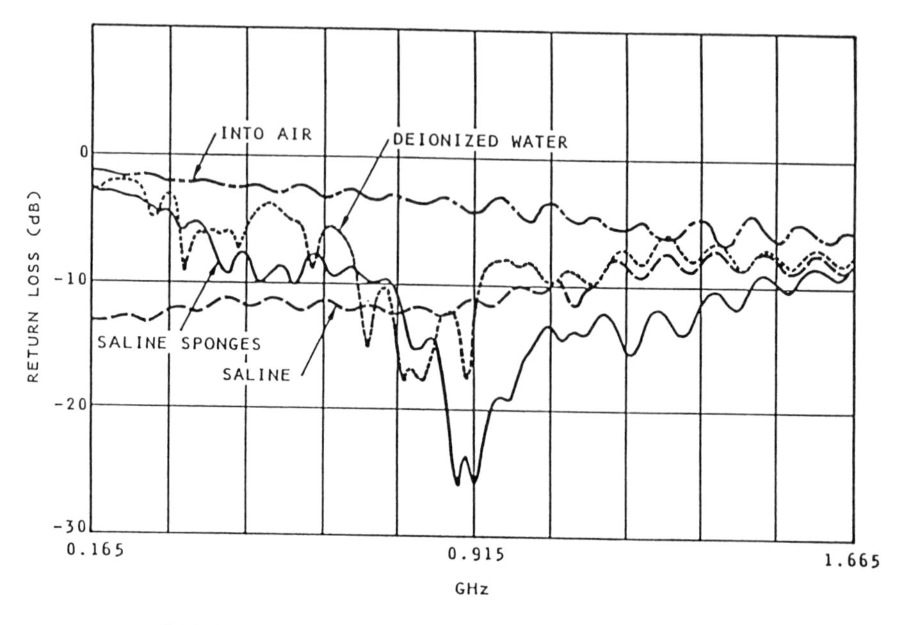

FIGURE 1.28 Return loss measurements of a gap antenna.

five sequential sections, the medial injury index was averaged for all the sections of a given artery. This average was defined as the mean MII.

Intimal proliferation was graded on a scale of 1–4, with 4 representing severe proliferation. Intimal proliferation was also defined as a percent of vessel circumference. Histologic criteria for intimal proliferation were as follows:

Grade 1 = occasional regions of intimal thickening of more than two cells

Grade 2 = occasional regions of intimal thickening by a layer of more than three cells

Grade 3 = 25–50% of the intimal circumference covered by a layer of more than three cells

Grade 4 = 75% of the intimal circumference covered by a layer of more than four cells.

The IPI was defined as the product of the intimal proliferation score and the percentage of the circumference that was involved ($\times 100$). The mean IPI was obtained by averaging the intimal proliferation index of all the sequential sections of each artery.

Injury circumference was expressed as a percentage of total circumference.

$$\text{Mean MII} = \sum \frac{(\text{Medial depth of injury} \times \text{injury circumference})}{\text{number of sections showing medial injury}}$$

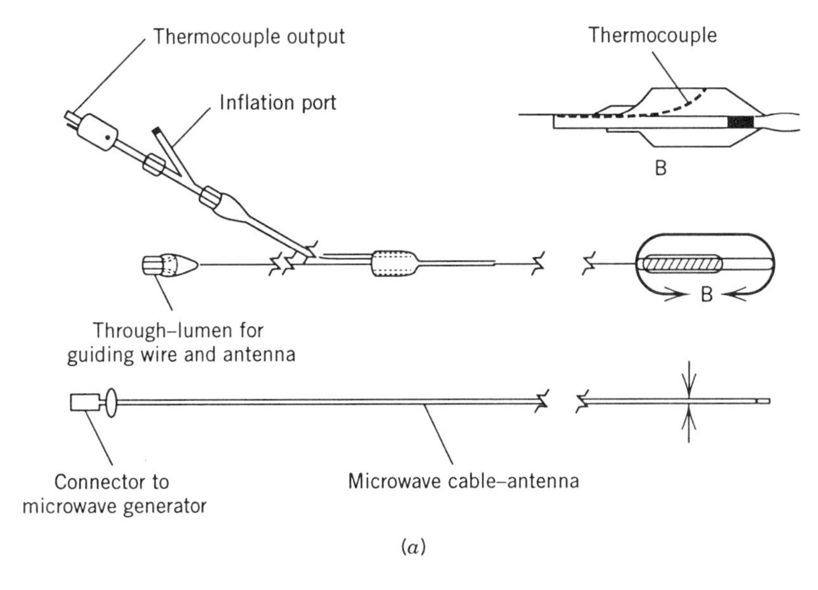

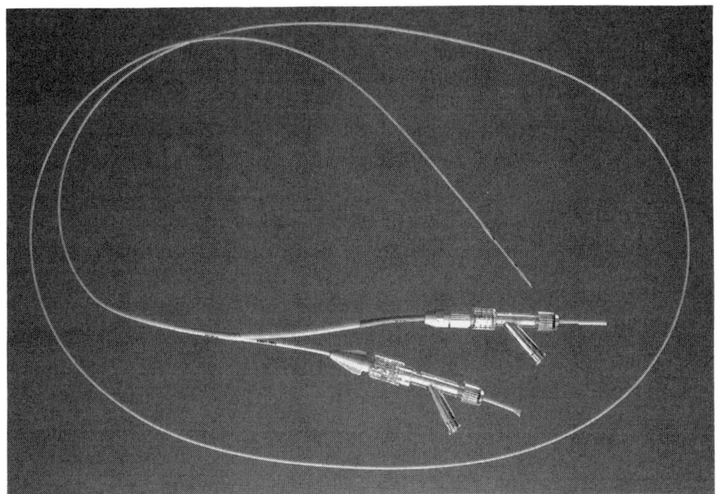

FIGURE 1.29 (*a*) Balloon catheter; (*b*) microwave delivery system.

$$\text{Mean IPI} = \sum \frac{(\text{IPI score} \times \text{injury circumference})}{\text{number of sections showing medial injury}}$$

Peak MII = medial depth of injury × injury circumference for the vessel section showing the greatest injury

Results Of 30 animals, 1 died following the administration of anesthetics, and 1 animal underwent angioplasty of the right iliac artery only as the result of

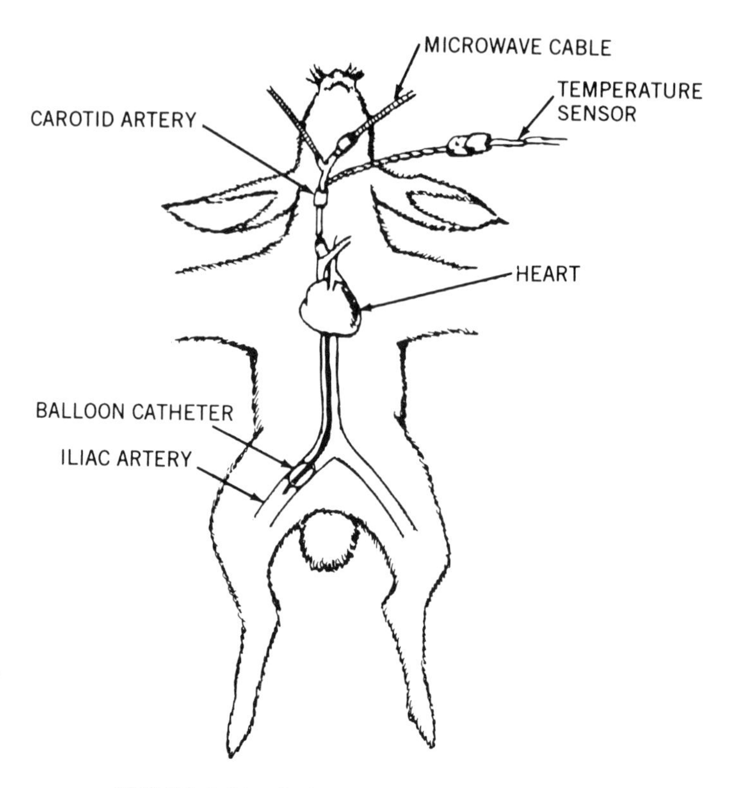

FIGURE 1.30 Catheter inserted in rabbit model.

laceration of the left femoral artery during catheter insertion. The iliac arteries of two animals were lost during histologic processing, leaving 53 iliac arteries for analysis. Representative histologic sections are shown in Figures 1.33 and 1.34. Arteries exposed to low temperature (50–60 °C) and arteries exposed to angioplasty without microwave energy had prominent proliferation without any noticeable medial changes (Fig. 1.35). Arteries exposed to higher temperatures (70–80 °C) had mild intimal fibrosis and were covered by apparently normal endothelial tissue. Extensive medial fibrosis as a consequence of previous medial necrosis was also noted (Fig. 1.36).

Target temperature was defined as a discrete variable (50, 70, and 90 °C). Validation of the thermocouple was performed with simultaneous temperature recordings on the internal and external surfaces of a rabbit carotid artery in vitro, as discussed before. There was parallel tracking of temperature over a variety of temperatures, although the thermocouple within the balloon read 8–10 degrees lower than that recorded on the adventitial surface. This discrepancy is likely an artifact caused by the orientation of the thermocouple to the microwave antenna and has subsequently been eliminated. As a result, recorded temperatures may be eight to ten degrees higher than actual temperatures.

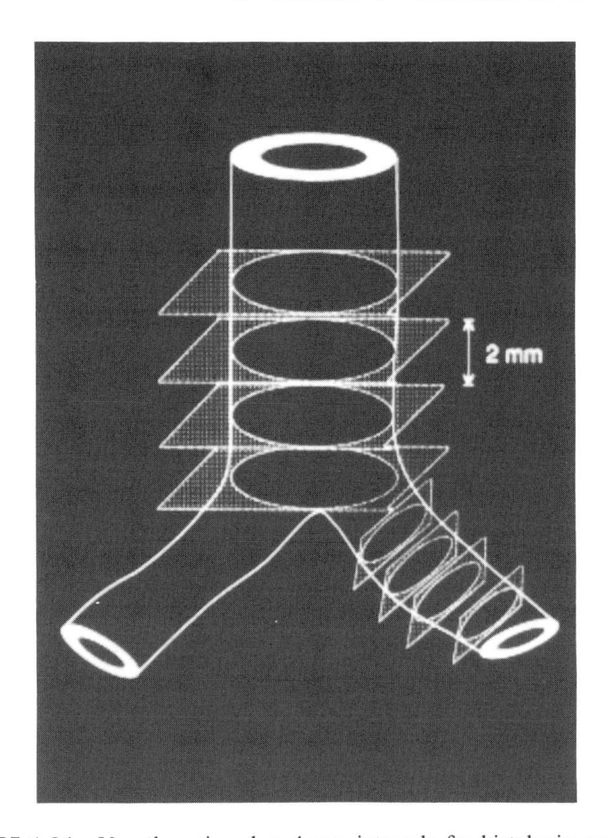

FIGURE 1.31 Vessel sectioned at 4-mm intervals for histologic analysis.

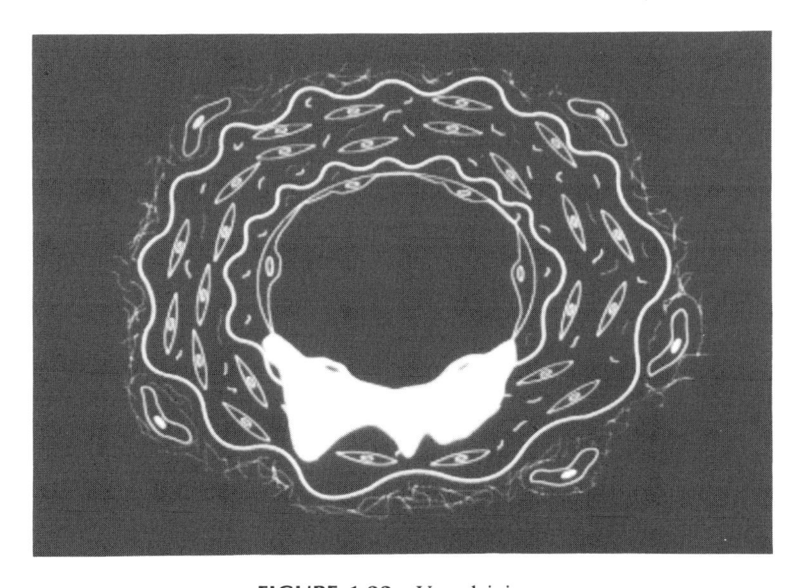

FIGURE 1.32 Vessel injury.

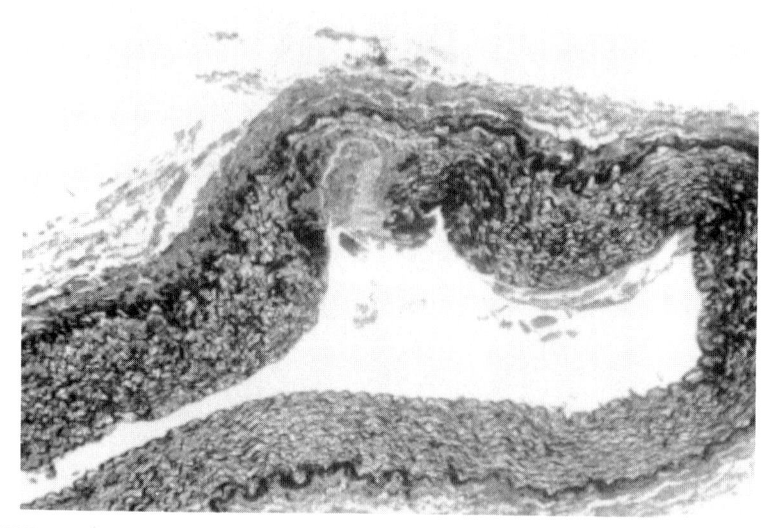

FIGURE 1.33 Histologic section of rabbit iliac artery following in vivo microwave balloon angioplasty.

FIGURE 1.34 Histologic section following in vivo MBA.

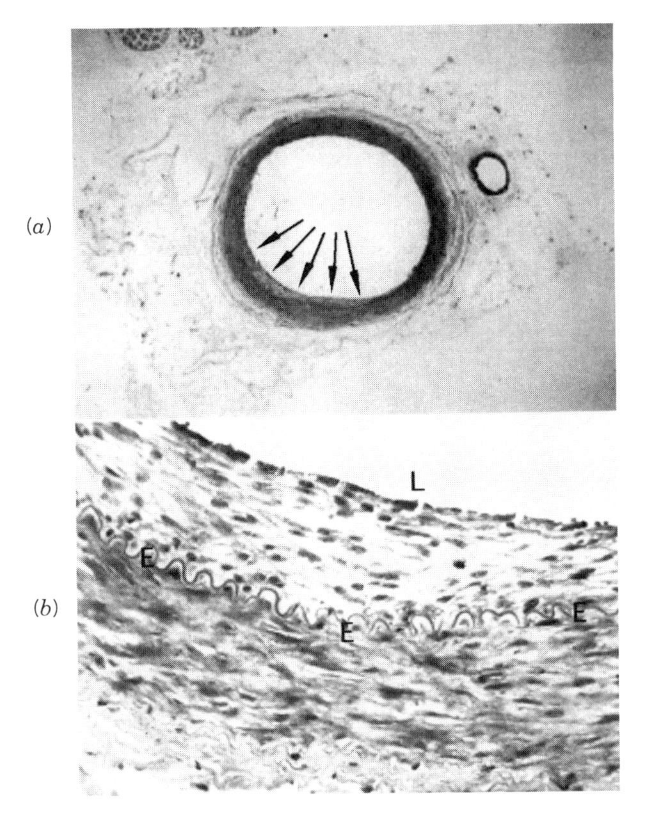

FIGURE 1.35 (*a*) Artery exposed to 50 °C for 30 s. There is prominent intimal thickening (*arrows*). (Trichrome stain; original magnification ×20). (*b*) Detail of the intimal reaction. The intima is thicker and the change is more extensive than in the control artery. E, internal elastic lamella; L, lumen. (trichrome stain; original magnification ×20).

A linear relation was seen between mean MII and peak temperature (Fig. 1.37). It should be noted that below 60 °C, no more medial injury was observed than was seen in conventional angioplasty. Not surprisingly, the same relationship holds for mean temperature and mean MII (Fig. 1.38).

There is no clear relation between intimal proliferation and temperature (Figs. 1.39, 1.40). However, a plot of mean IPI and mean MII suggests an inverse relationship between intimal proliferation and medial injury (Fig. 1.41). The longitudinal distribution of medial injury is shown in Figure 1.42. Medial injury falls off to close to control at a distance of 1.2 cm from the point of peak medial injury.

Microwave duration and inflation pressure had no effect on intimal proliferation or medial injury.

Discussion In this study we have demonstrated that RF energy, specifically microwaves delivered through a coaxial cable terminated by an antenna, can be

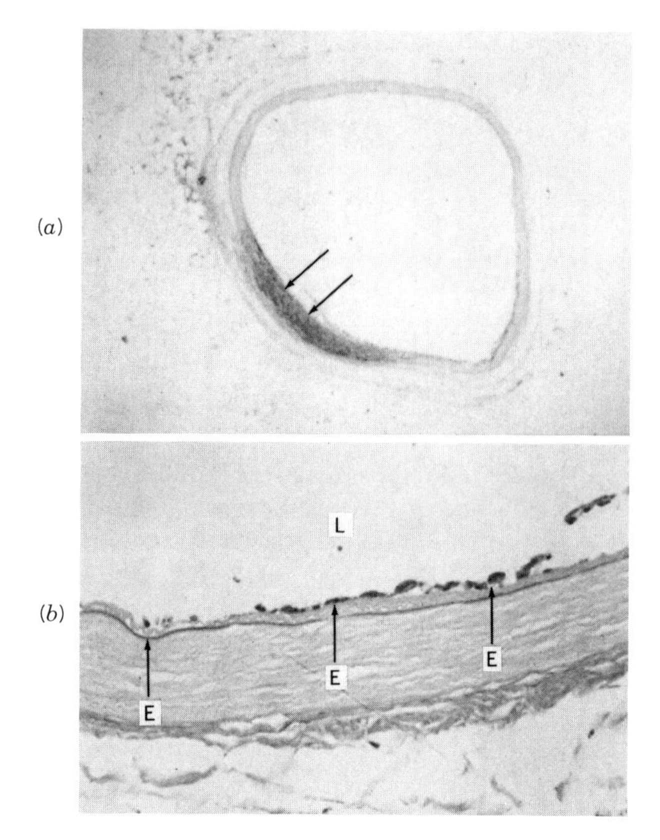

FIGURE 1.36 (*a*) Artery exposed to 80 °C for 30 s. There is extensive medial fibrosis with only one area containing viable smooth muscles (*arrows*). (Trichrome stain; original magnification ×200). (*b*) Details of the medial fibrosis. In this segment smooth muscle cells are absent. The intima is composed of a single layer of endothelial cells. E, internal elastic lamella; L, lumen. (Trichrome stain; original magnification ×200).

used to deliver intravascular thermal energy during balloon angioplasty. The use of a thermocouple in the balloon coupled with a feedback control allows on-line monitoring and control of the resultant temperature rise to any predetermined point.

We observed a linear correlation between the peak temperature achieved during angioplasty and the degree of medial injury observed at 1 week. A threshold effect was noted, with medial injury occurring predominantly at temperatures greater than 60 °C. At temperatures greater than 85 °C medial injury was of uniformly full thickness. The axial extent of injury utilizing the current catheter delivery system was approximately 1.6 cm.

Although multiple mechanisms may affect restenosis, myointimal proliferation from the media is thought to be one of the major determinants. In this regard,

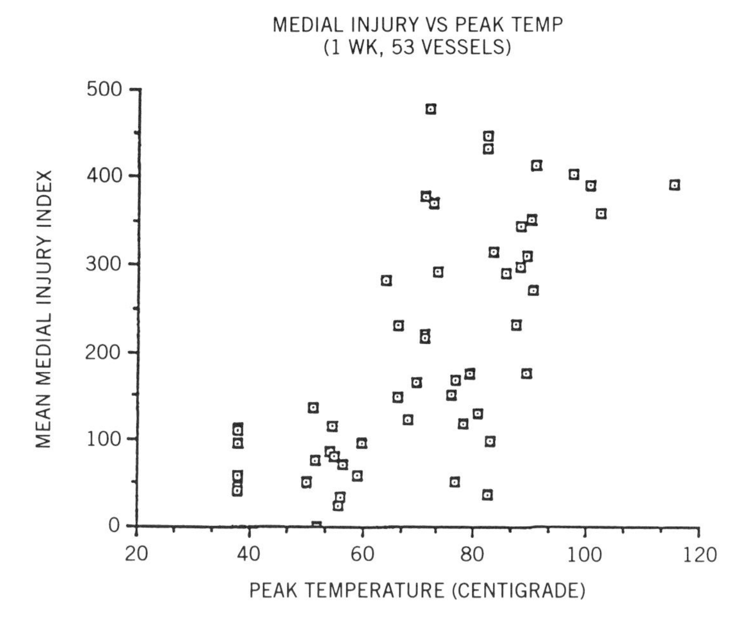

FIGURE 1.37 Medial injury versus peak temperature (1 week, 53 vessels).

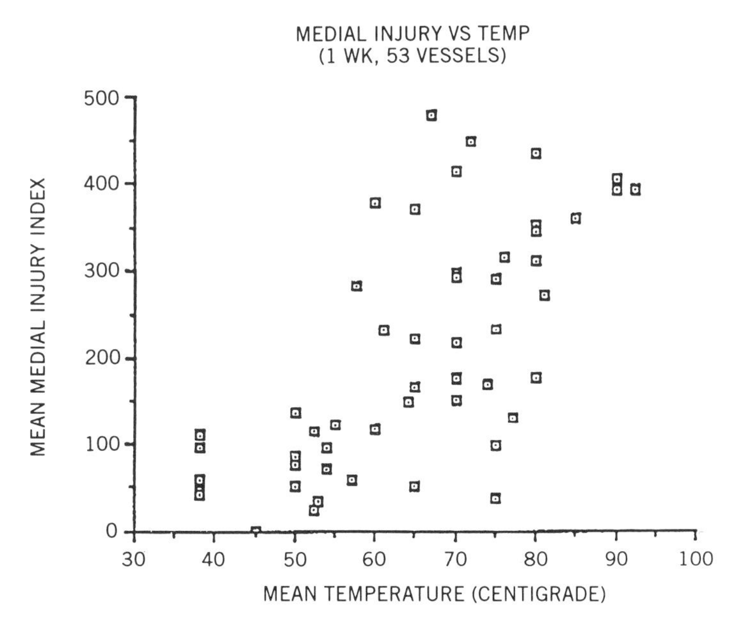

FIGURE 1.38 Medial injury versus mean temperature (1 week, 53 vessels).

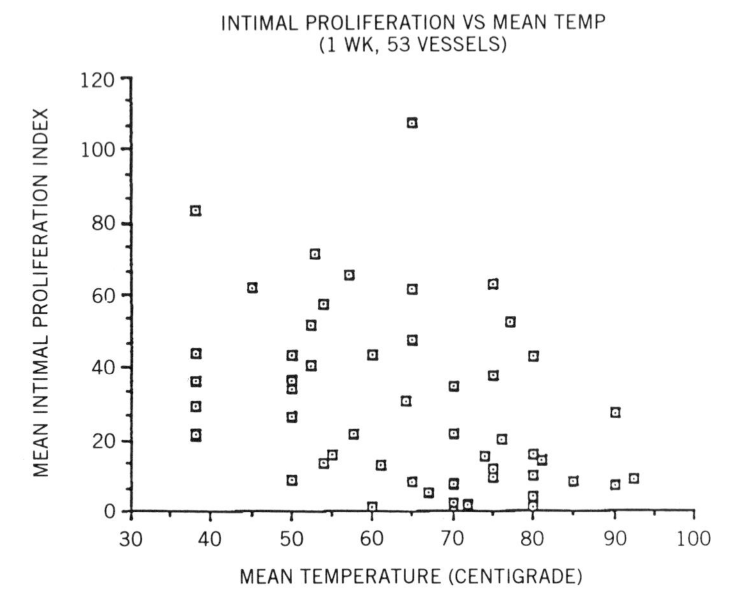

FIGURE 1.39 Intimal proliferation versus mean temperature (1 week, 53 vessels).

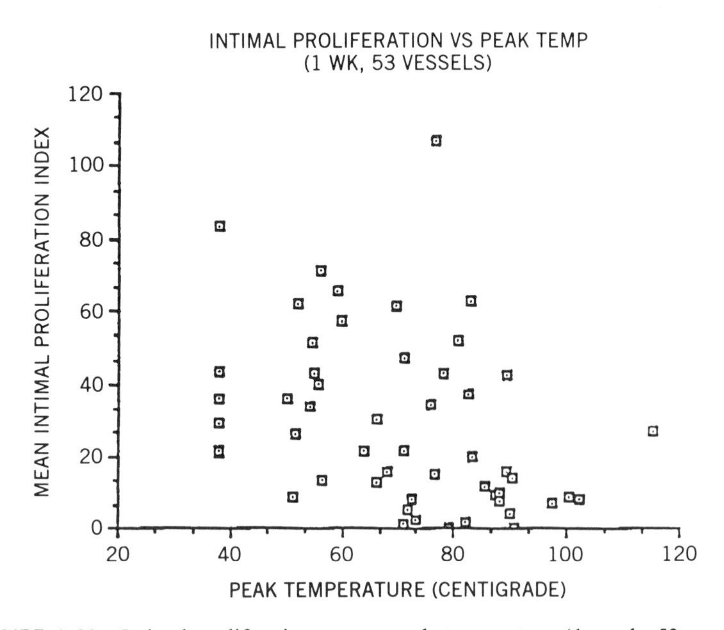

FIGURE 1.40 Intimal proliferation versus peak temperature (1 week, 53 vessels).

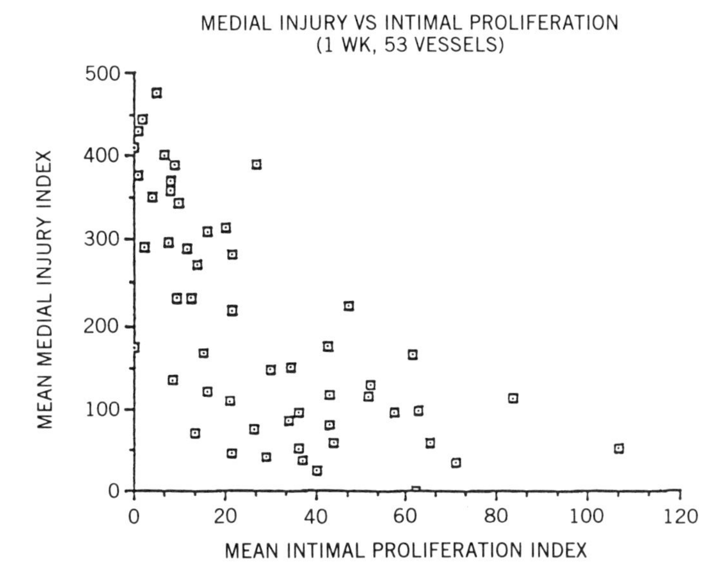

FIGURE 1.41 Medial injury versus intimal proliferation (1 week, 53 vessels).

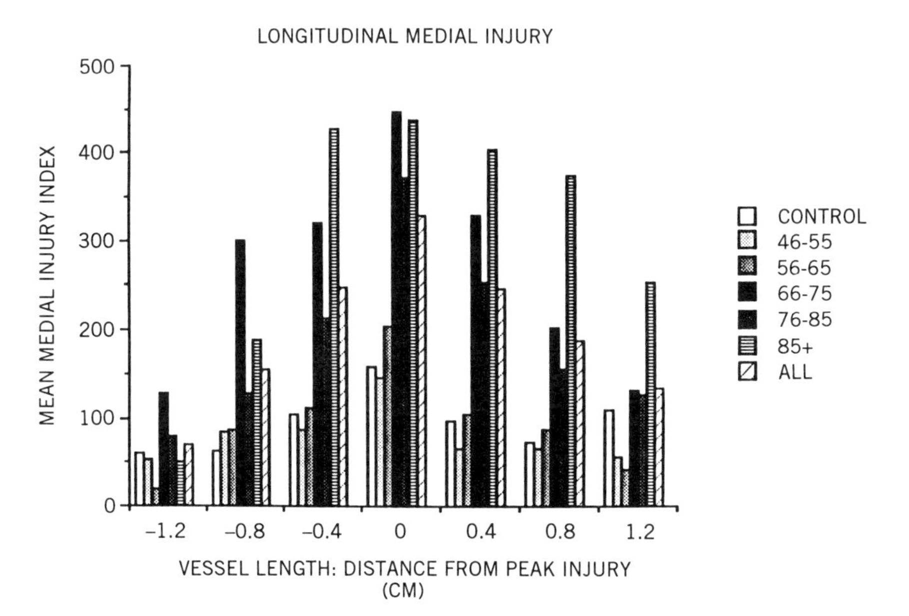

FIGURE 1.42 Longitudinal medial injury.

the inverse relationship between the MII and IPI was of interest. Perhaps a selective injury to the smooth-muscle cells in the media may be of value in reducing the extent of myointimal hyperplasia. Conceivably, such injury alone, or in combination with a pharmacologic agent, could retard the process of restenosis.

Further evaluation is required since the current histologic samples were obtained at 1 week after angioplasty.

Microwaves provide a convenient means of delivering thermal energy. The physical characteristics of the delivery system are tolerant of the demands for a cable small and flexible enough to be positioned in the coronary tree. The distal portion of the cable, the antenna, is the site of energy radiation. The antenna can be modified to vary the pattern of energy radiation to meet individual requirements. Energy input can be modified to vary the resultant tissue temperature.

It should be stressea that this study was a first attempt to define the biologic effects of intravascular microwave energy and determine optimal angioplasty parameters for the following study on the atherosclerotic rabbit model.

1.7.3 Atherosclerotic Rabbit Model

Our next study was performed on an atherosclerotic rabbit model. Eight animals were administered a high-fat diet. After 2 weeks on this diet, endothelial denudation of the external iliac arteries was performed by positioning a balloon catheter in the external iliac artery, and then pulling it back into the distal aorta. This procedure predictably resulted in an atherosclerotic lesion at the site of denudation. Four weeks later these animals were brought back to the laboratory and bilateral femoral artery cutdowns were performed. The balloon catheter–cable system was then positioned in the atherosclerotic external iliac artery. Animals had microwave balloon angioplasty (MBA) performed on one external iliac artery and conventional balloon angioplasty (CBA) performed on the contralateral artery. Each animal thus served as its own control. Angioplasty was performed with a 3 mm balloon inflated to 5 atm for 1 min. In the animals treated with MBA a similar procedure was performed with the addition of microwave energy for 30 s during the inflation. Energy was delivered to raise the temperature, as measured on the balloon surface, to 70 or 85 °C. After treatment of both vessels, angiography was performed and the femoral arteries were ligated.

Four weeks after the angioplasty procedure, the animals were brought back to the laboratory. Angiography of the iliac arteries was repeated and the animals were sacrificed. Following perfusion fixation of the iliac arteries, the vessels were excised, fixed, sectioned, and stained in the same manner as described above. Histologic analysis and quantitative angiography of the vessels were performed.

Histologic analysis revealed that on the side treated with MBA there was loss of lipid-laden cells and replacement in the intima, and in some cases in the media, with a hypocellular fibrotic matrix. Quantitative angiography of the iliac arteries revealed a significant improvement in the diameter of the arteries treated with MBA at 85 °C both immediately and at 4 weeks (Figs. 1.43–1.45). The increase in luminal diameter with MBA at 85 °C was greater ($p < .05$) than that with CBA immediately postangioplasty (Table 1.5). A trend toward enhanced benefit was also noted at 4 weeks. Although there was a trend for benefit immediately post procedure in the animals treated with MBA at 70 °C

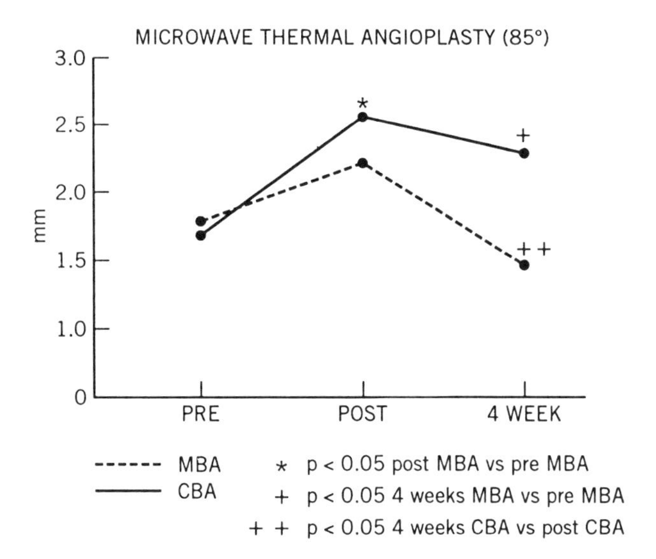

FIGURE 1.43 Microwave thermal angioplasty (85 °C).

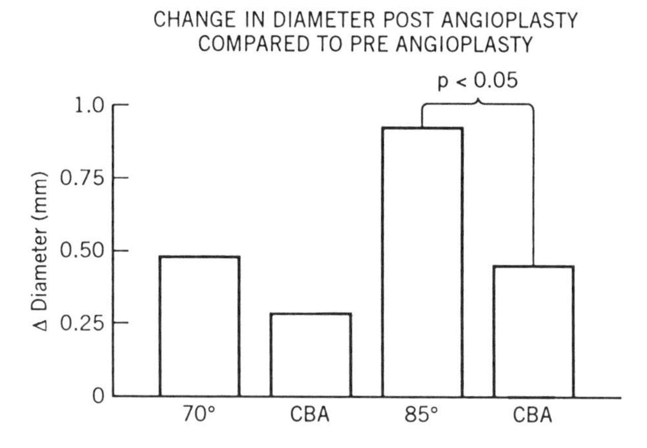

FIGURE 1.44 Change in arterial diameter postangioplasty compared to preangioplasty.

(Figs. 1.44–1.46), this was not statistically significant (Table 1.6). Furthermore, at 4 weeks there was loss of the immediate benefit in the 70 °C MBA vessels and no difference compared to the control arteries treated with conventional angioplasty. In Fig. 1.47*a–c*, angiography of the iliac arteries in a rabbit demonstrates bilateral atherosclerotic lesions. Microwave balloon angioplasty was subsequently performed on the right and conventional balloon angioplasty on the left iliac artery. Immediately postangioplasty a dissection was noted on the left (Fig. 1.47*b*). A larger lumen without dissection is noted on the right (Fig. 1.47*b*).

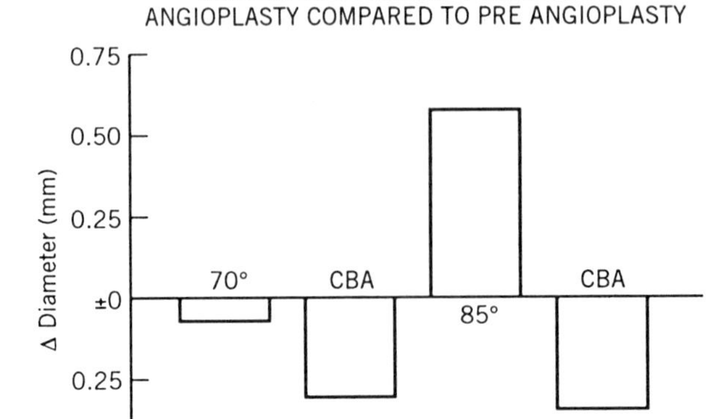

FIGURE 1.45 Change in arterial diameter 4 weeks postangioplasty compared to preangioplasty.

TABLE 1.5 Microwave Thermal Angioplasty (85 °C)

	MBA	CBA
Preangioplasty	1.66 ± 0.32	1.77 ± 0.31
Postangioplasty		
Immediately	2.54 ± 0.23^{a}	2.24 ± 0.21
After 4 weeks	2.29 ± 0.74^{b}	1.42 ± 1.02^{c}

[a] $p < .05$ post-MBA versus pre-MBA.
[b] $p < .05$ 4 weeks MBA versus pre-MBA.
[c] $p < .05$ 4 weeks CBA versus post-CBA.

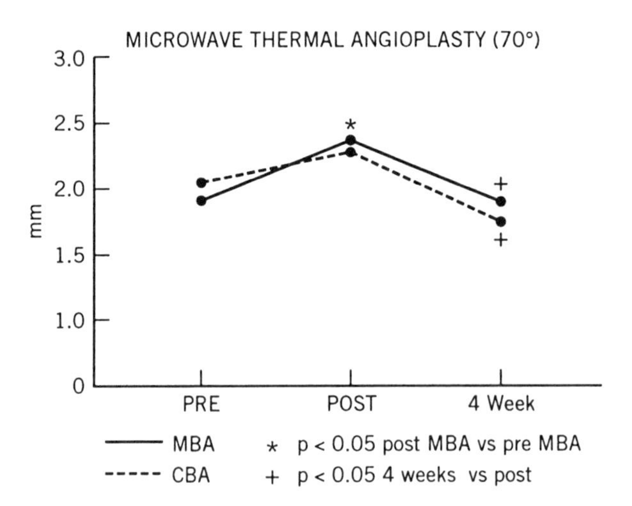

FIGURE 1.46 Microwave thermal angioplasty (70 °C).

TABLE 1.6 Microwave Thermal Angioplasty (70 °C)

	MBA	CBA
Preangioplasty	1.92 ± 0.37	2.03 ± 0.36
Postangioplasty		
Immediately	2.40 ± 0.22^a	2.32 ± 0.40
After 4 weeks	1.68 ± 0.28^b	1.73 ± 0.62^c

[a] $p < .05$ post-MBA versus pre-MBA.
[b] $p < .05$ 4 weeks MBA versus post-MBA.
[c] $p < .05$ 4 weeks CBA versus post-CBA.

At 4 weeks postangioplasty there is a sustained benefit on the right. Restenosis is noted on the left (Fig. 1.47*c*).

These studies demonstrate that MBA is technically feasible, and that the initial results of angioplasty can be enhanced with MBA (Table 1.5). The finding of sustained benefit at four weeks has encouraged us to explore further the potential of this modality in reducing the incidence of restenosis. At the current time, a target temperature of 85 °C seems optimal for this effect. Studies are also in progress to evaluate the potential of this technique to treat additional mechanisms that may decrease coronary flow and lead to unsuccessful angioplasty. In one such study, the ability to seal dissections is being evaluated. Initial observations suggest that such tears in the vessel wall can be sealed by microwave thermal energy.

1.7.4 The Effect of MBA on Vascular Thrombosis

Angioplasty of the coronary arteries and arterial supply to the lower extremities and kidneys is now an established technique for the treatment of arterial stenosis. However, one major complication with this procedure is the presence or development of thrombi. A thrombus can result in complete occlusion of the vessel despite adequate dilatation of the underlying vessel. Such thrombi may be present before an angioplasty procedure begins, or may form as a consequence of the procedure. It is possible to treat the thrombus through the installation of thrombolytic therapy such as streptokinase or tissue plasminogen activator. Mechanical means of treating such thrombus formation were also investigated previously. In the current study we evaluated the feasibility of utilizing microwave thermal energy to more effectively resolve coronary thrombus than CBA.

In 13 mongrel dogs anesthesia was induced with pentobarbital. The chest was then opened and the heart exposed and suspended in a pericardial cradle. The left anterior descending coronary artery was isolated. The artery was occluded, and thrombin was distilled into the artery to induce a thrombus. The presence of a stable thrombus was verified by coronary angiography. Following the formation of a stable thrombus, the animals were observed for 30 min of stability. The artery was then traversed by a balloon angioplasty system. The arteries were

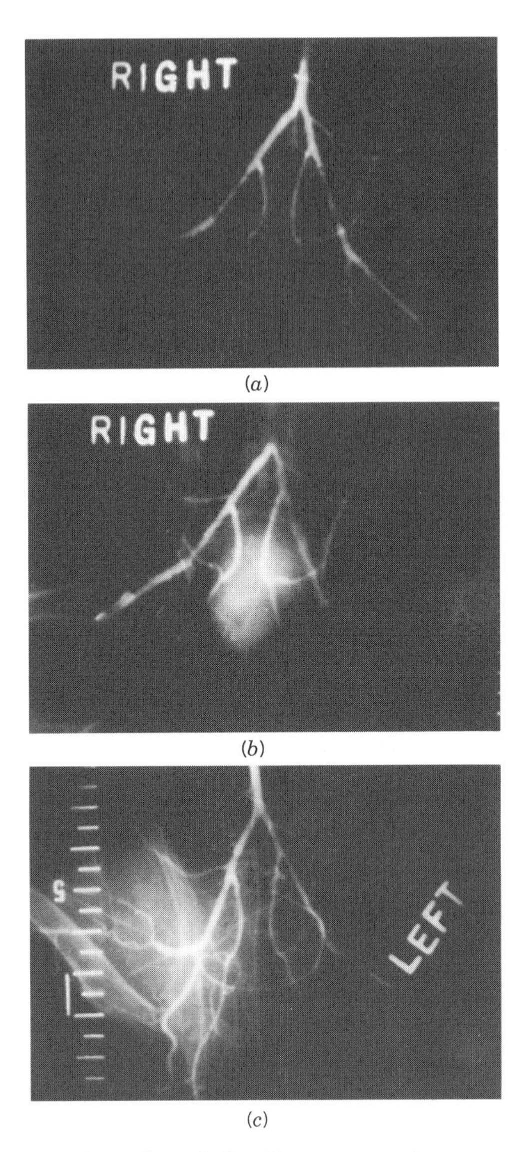

FIGURE 1.47 (*a*) Angiography of the iliac artery prior to angioplasty. Bilateral atherosclerotic lesions are seen. (*b*) Angiography of the iliac arteries following angioplasty. Microwave balloon angioplasty was performed on the right, and conventional balloon angioplasty was performed on the left. Note the dissection on the left (*arrow*). (*c*) Angiography of the atherosclerotic iliac arteries 4 weeks following angioplasty. Restenosis is noted on the left.

treated with either CBA or MBA. The balloon inflation time was 1 min. With MBA, the balloon was inflated for 1 min with microwave energy delivery for 30 s. Microwave energy was delivered at 2450 MHz. The energy input was regulated by feedback from a thermocouple on the surface of the balloon so that the temperature as monitored on the internal balloon surface was 85 °C. Immediately after angioplasty the animals were administered 3000 units of intravenous heparin.

Angiography was performed before angioplasty, immediately after angioplasty, and 30 min after angioplasty. Following the 30-min angiography, the animals were sacrificed and the treated segment of vessel was removed for histologic examination.

Prior to angioplasty all animals demonstrated complete occlusion of the left anterior descending artery. On angiographic evaluation two out of six animals treated with MBA manifested thrombus formation. Five out of six animals treated with CBA showed evidence of thrombi on angiographic evaluation. The angiographic findings varied from filling defects in the lumen to complete occlusion.

Histologic evaluation revealed complete occlusion of the vessel in four out of seven animals treated with conventional balloon angioplasty (Fig. 1.48). In addition, two out of seven animals treated with CBA exhibited a thrombus in up to 50% of the vessel lumen. In contrast, in one out of six animals treated with MBA complete occlusion of the vessel was noted. In addition, two out of six animals treated with MBA had a thrombus occluding up to 50% of the vessel lumen. Histologic examination of the thrombus following MBA revealed

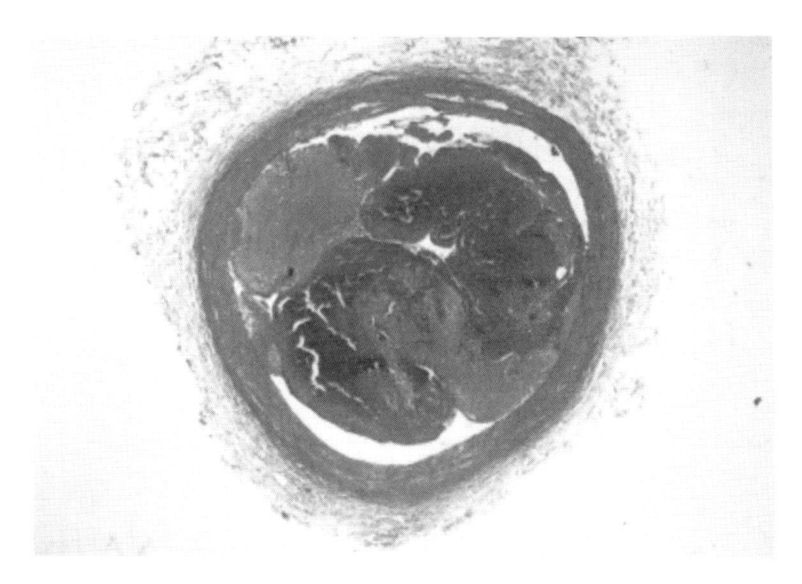

FIGURE 1.48 Canine left anterior descending artery following occlusion with a thrombus, and conventional balloon angioplasty. The vessel remains occluded with the thrombus despite the previous application of balloon angioplasty.

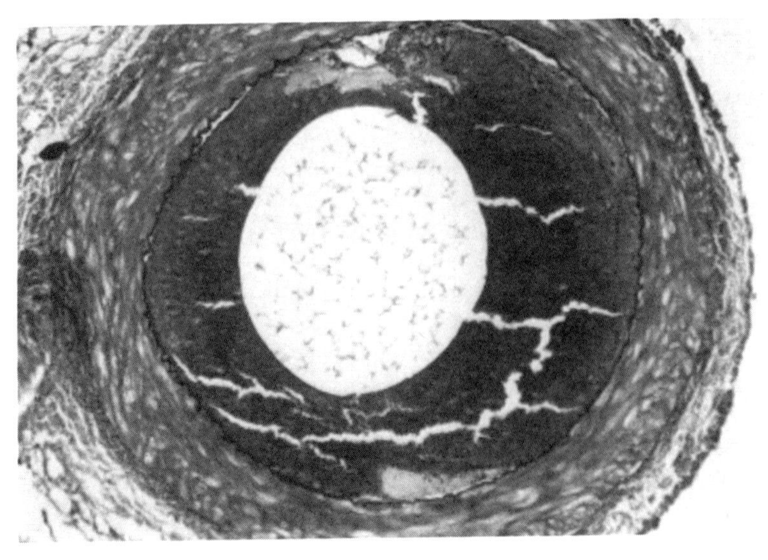

FIGURE 1.49 Canine left anterior descending artery. The artery was previously occluded with a thrombus and has been treated with balloon angioplasty in conjunction with microwave energy thermal delivery. Note the peripheral lamination of the thrombus. The thrombus has been coagulated and remains stable in a peripheral position. A central channel is patent and provides an adequate lumen for blood flow.

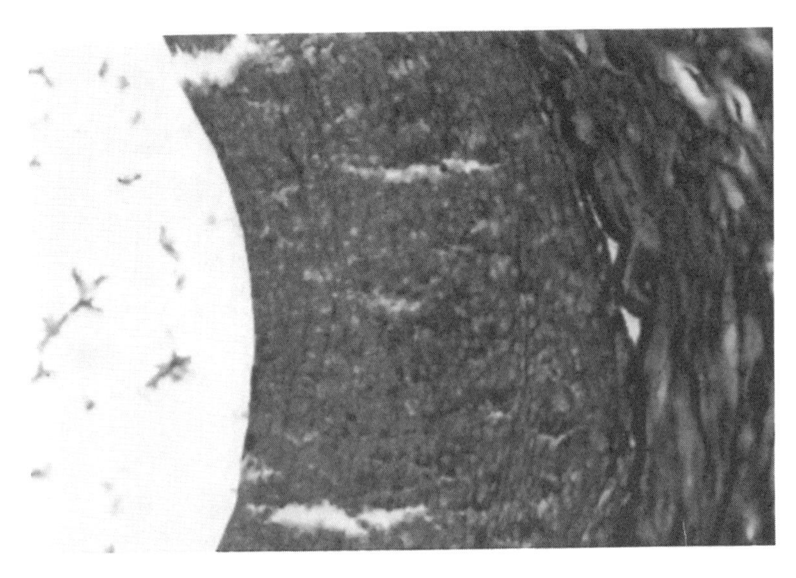

FIGURE 1.50 Magnified section of the coagulated thrombus shown in Figure 1.49.

coagulated thrombus with peripheral lamination of the thrombus (Figs. 1.49, 1.50). A patent central lumen was noted in the center of the laminated thrombus.

We have demonstrated that thermal angioplasty using microwave energy is an effective means, in conjunction with balloon angioplasty, in the treatment of intravascular thrombi. The combination of balloon angioplasty in which a thrombus is displaced to the periphery of the vessel, along with thermal energy delivery to coagulate and stabilize the thrombus, therefore provides promise of application in a variety of clinical syndromes in which a thrombus may complicate an angioplasty procedure and occlude a vessel. Such syndromes include acute myocardial infarction and unstable angina pectoris.

1.8 CONCLUSIONS

The popularity of CBA and its less-than-optimal results have prompted the early investigation into MBA. As a consequence of in vitro success of thermal therapy – and, more specifically, volume thermal therapy – a patent entitled "Percutaneous transluminal microwave catheter angioplasty," by Arye Rosen and Paul Walinsky, was filed in October 1985, and was granted as U.S. Patent Number 4,643,186 on February 17, 1987. Through the combination of two modalities—pressure (balloon) and volume heating (microwave)—the problems now facing a large number of patients treated with CBA were shown to have been solved in animals. This thesis discusses in detail the contributions that have been made in an attempt to solve the key problems. In view of the novelty of the approach, only limited prior art was available, and the research had to include the microwave system, the in vitro research, and the in vivo investigation.

The problems encountered as a consequence of CBA are as follows:

- Restenosis in 30% of patients treated in the first 6 months following CBA
- Dissection following CBA application
- Occurrences of thrombi as a consequence of dissection and plaque cracking
- Elastic recoil

Attempts to solve the problems in an animal model were divided into several steps, as follows:

1. An effort to understand the effects of microwaves on tissue, which relied on previously reported studies and adapted to the specific cases treated with MBA
2. The development of a microwave system that required a new coaxial cable having parameters that were not available prior to the project, and the integration of the cable–antenna assembly with a balloon catheter
3. In vitro investigation of the validity of the microwave therapy

4. Extensive in vitro and in vivo experiments to validate antenna–cable assembly effectiveness, as related to
   a. The delivery of the thermotherapy
   b. The maneuverability of the cable and/or catheter to the site of interest (stiffness and flexibility)
5. In vivo investigation to verify the validity and efficacy of MBA versus CBA

Our understanding of the mechanisms and therapeutic applications of intravascular thermal energy (i.e., heat by microwave or by another energy form) is in an early stage. Complex changes occur in all layers of the vessel, and different thermal dosimetry profiles may be required to optimize therapeutic benefit. A high temperature ($\sim 80\,^{\circ}\mathrm{C}$) in treated tissue may be optimal for tissue welding in the therapy of dissection. Maintenance of vascular distention may be accomplished with a lower temperature. Modification of the injury response of the media may be best accomplished with yet another temperature profile. Coagulation and stabilization of thrombi is still another therapeutic goal requiring its own thermal characterization.

Clarification of the therapeutic potential of heating will be aided by a further understanding of such thermal dosimetry. Our studies thus far have led us to the conclusion that MBA is effective in enhancing the primary result of angioplasty. The addition of microwave energy is effective in increasing the diameter of the vessel postangioplasty and in sealing dissections, and seems to be effective in aiding in the stabilization of thrombi. The role of microwave angioplasty in reducing the incidence of restenosis is a complex question. The current results are too preliminary to offer definitive answers.

The MBA system is relatively easy to deploy, and the system is cost-effective compared to alternative heating modalities, such as lasers, since the microwave generator at 2.45 GHz is essentially identical to the one used in microwave ovens.

As an extention of the research carried out thus far, a limited investigation in animals regarding the possible cause of restenosis is recommended. It is suspected that the mere handling (and probable contamination) of the deionized water (used for balloon inflation) might change the characteristics of the water by lowering its resistance and making it more absorbant to microwave radiation. The heat generation (microwave absorption) in the balloon fluid could cause excessive heating, and thus excessive damage to the endothelium—a possible trigger to restenosis. It is suggested—for investigational purpose only (not for clinical use)—that the fluid in the balloon be eliminated, thus reducing the heat effect through the fluid, and relying only on microwave volume heating (realizing a change in microwave matching conditions). Possibly, this would allow the radiation of RF power to affect the media rather than the endothelium. Should results indicate it, adjustments could then be made in the type of fluid used and/or in the manner of its handling.

1.9 RECOMMENDATIONS FOR FUTURE RESEARCH

The utilization of integrated circuits at microwave frequencies, and more specifically the use of microwave monolithic integrated circuits (MMICs) in many microwave systems, has afforded opportunities for the use of ICs in medicine. In this chapter on the future use of microwaves in cardiology, the use of a GaAs (gallium arsenide) monolithic microwave circuit, in conjunction with a GaAs laser, is envisioned. Through the chapter, the subject of a microwave-aided balloon angioplasty system was treated. In the arrangement described, a catheter including a microwave transmission line and terminated, at its distal end, with an antenna is detailed. During angioplasty, the catheter is introduced into a blood vessel, and the distal end with the balloon and the antenna is manipulated to a point adjacent to the plaque. Microwave power is applied to the proximal end of the catheter to reach the antenna through the microwave transmission line. The very small size of the coaxial cable (22–23 mils) tends to result in relatively large losses. These losses may be so great that up to 75% of the microwave power applied to the proximal end of the transmission line may not reach the distal end for application to the tissue. Instead, this power is dissipated in the form of heat, which tends to be greatest at the proximal end. Since the heat is concentrated within the relatively narrow transmission line rather than being distributed throughout a larger volume, the transmission line may become hot enough, if not cooled, to cause burns at the point at which the catheter enters the body. To alleviate heat dissipation, *a catheter with a distally located MMIC circuit* at 2.45 GHz, and a distally located semiconductor laser can be utilized. The MMIC includes an antenna to aid in coupling radiation from a semiconductor oscillator–amplifier circuit to the surrounding tissue. The catheter may also include an axial aperture adapted for use with a guide filament. In this configuration, only very small DC conductors occupy the catheter, thus permitting the addition of other components such as fiber optic scopes and the like.

The future also holds the possibility of treatment that combines the use of lipid-lowering drugs and MBA. Combinations of drugs that reduce lipids in the blood could slow the progress of restenosis even further following MBA.

In conclusion, the reduction of the rate of restenosis as a consequence of MBA as depicted in Fig. 1.47, may result in a substantial reduction of repeat angioplasty procedures if combined with the administration of lipid-lowering drugs.

Acknowledgements

We wish to recognize the assistance and advice of many who, since 1987, have participated in research in the area of microwave balloon angioplasty: from Jefferson Medical College, Drs. Donald Nardone, Barry Bravette, Yi Shi, Antonio Martinez-Hernandez, Andrew Zalewski, Richard M. Rosenbaum, Arnold J. Greenspon, Steve Hsu, and Michael Smith; from Baxter Edwards LIS Division, Dr. Jin-Son Chou; from MMTC, Dr. Fred Sterzer, Mr. Dan Mawhinney, and Mr.

Adolph Presser; from Sinergetics, Mr. Zygmond Turski; from Gore, Mr. Mike Brown; and from UMDNJ-Robert Wood Johnson Medical School, Dr. Harel D. Rosen.

Gratitude is also due to Professor Peter R. Herczfeld for his professional insight and continuous encouragement, and to Mr. Walter Janton for his technical skills and invaluable support; and finally, to Daniella Rosen, for typing and revising this manuscript again and again.

REFERENCES

1. Gruentzig A, Hopff H: Perkutane-Rekanalisation chronischer arterieller Verschlusse mit einem neuen Dilatationskatheter: Modifikation der Dotter-Technik. *Dtsch Med Wochenschr* 99:2502, 2511, 1974.
2. Gruentzig AR, Senning A, Seigenthaler WE: Nonoperative dilatation of coronary-artery stenosis: Percutaneous transluminal coronary angioplasty. *N Engl J Med* 301:61, 1979.
3. Rosen A, Walinsky P: Percutaneous transluminal microwave catheter angioplasty. U.S. Pat. 4,643,186, Feb. 17, 1987.
4. Dotter CT, Judkins MP: Transluminal treatment of arteriosclerotic obstruction: Description of a new technique and a preliminary report of its application. *Circulation* 30:654, 1964.
5. Staple TW: Modified catheter for percutaneous transluminal treatment of arteriosclerotic obstructions. *Radiology* 91:1041, 1968.
6. Gruentzig A: *Die Perkutane Transluminale Rekanalisation Chronischer Arterienverschlusse Mit Einer Neuen Dilatationstechnik.* Witzstrock, Baden-Baden, 1977.
7. Vlietstra RE, Holmes Jr, DR: *PTCA, Percutaneous Transluminal Coronary Angioplasty*. FA Davis, Philadelphia, 1987.
8. Waller BF: Early and late morphologic changes in human coronary arteries after percutaneous transluminal coronary angioplasty. *Clin Cardiol* 6:363, 1983.
9. Waller BF, Girod DA, Dillon JC: Transverse aortic wall tears in infants after balloon angioplasty for aortic valve stenosis: Relation of aortic wall damage to diameter of inflated angioplasty balloon and aortic lumen in seven necropsy cases. *J Am Coll Cardiol* 4:1235, 1984.
10. Grundfest WS, Litvack F, Forrester JS, et al: Laser ablation of atherosclerotic plaque without adjacent tissue injury. *J Am Coll Cardiol* 5:929, 1985.
11. Abela GS, Normann S, Cohen DM, et al: Laser recanalization of occluded atherosclerotic arteries in vivo and in vitro. *Circulation* 71:403, 1985.
12. Winter A, Laing J, Paglione R, et al: Microwave thermotherapy for the treatment of human brain cancer. *IEEE MTT-S Digest*, 1983, p. 180.
13. Winter A, Laing J, Paglione R, et al: Microwave hyperthermia for brain tumors. *Neurosurgery* 17:387, 1985.
14. Paglione RW: Miniature microwave antennas for inducing localized hyperthermia in human malignancies. *RCA Rev* 44:611, 1983.

15. Mendecki J, Friedenthal E, Botstein C, et al: Microwave applicators for localized hyperthermia treatment of cancer of the prostate. *Internatl J Radiat Oncol Biol Phys* 6:1583, 1980.

16. Astrahan MA, Sapozink MD, Cohen D, et al: Microwave applicator for transurethral hyperthermia of benign prostatic hyperplasia. *Internatl J Hyperthermia* 5:283, 1989.

17. Arastu HJ, Hightower M, Gisberg P, et al: The efficacy of transurethral microwave hyperthermia in the management of benign prostatic hyperplasia. International Scientific Meeting on Microwaves in Medicine, Belgrade, Yugoslavia, April 8–11, 1991, *Digest of Papers*, pp. 53–58.

18. Walinsky P, Rosen A, Greenspon AJ: Method and apparatus for high frequency catheter ablation. U.S. Pat. 4,641,649, Feb. 10, 1987.

19. Franke E: Minimum attenuation geometry for coaxial transmission line. *RF Design* 58, May 1989.

20. Emery AF, et al: The numerical thermal simulation of the human body when absorbing non-ionizing microwave irradiation with emphasis on the effect of different sweat models. *HEW Publ.* (FDA) 77-8011, 1976, pp. 96–118.

21. Cleary SF: Uncertainties in the evaluation of the biological effects of microwave and radiofrequency radiation. *Health Phys* 25:387, 1973.

22. Biological effects of electromagnetic fields. Paper presented at Swedish Academy of Engineering Sciences, Stockholm, 1976.

23. Reich HJ, Ordung PF, Krauss HL, et al (eds): *Microwave Theory and Techniques.* The Van Nostrand Series in Electronics and Communications, Van Nostrand, New York, 1953.

24. Chou J-S: Private communications.

25. Turski Z: Private communications.

26. Strohbehn JW, Bowers E, Walsh J, et al: An invasive microwave antenna for locally induced hyperthermia for cancer therapy. *J Microwave Power* 14:181, 1979.

27. Rosen A, Walinsky P, Smith D, Shi Y, Kosman Z, Martinez A, Rosen H, Sterzer F, et al: Percutaneous transluminal microwave angioplasty catheter. *IEEE MTT-S Digest*, 1989, p. 167.

28. Rosen A, Walinsky P, Smith D, et al: Advances in microwave thermal balloon angioplasty in animal model. The 3rd Asia–Pacific Microwave Conference Proceedings, Tokyo, Japan, September 18–21, 1990, pp. 849–852.

29. Rosen A, Walinsky P, Herczfeld PR, Greenspon AJ: The utilization of rf/microwaves in the treatment of cardiac dysfunction. 1993 SBMO International Microwave Conference Proceedings, Sao Paulo, Brazil, August 2–5, 1993.

30. Rosen A, Walinsky P, Smith D, et al: Studies of microwave thermal balloon angioplasty in rabbits. *IEEE MTT-S Digest*, 1990, p. 537.

31. Rosen A, Walinsky P, Nardone D, et al: Microwave thermal angioplasty in the normal and atherosclerotic rabbit model. *IEEE Microwave and Guided Wave Letters*, Vol. 1, 4:73, April 1991.

32. Walinsky P, Rosen A, Smith D, et al: Microwave balloon angioplasty. *IEEE MTT-S Digest*, 1991.

33. Walinsky P, Rosen A, Martinez-Hernandez A, et al: Microwave balloon angioplasty. *J Invasive Cardiol*, Vol. 3, 3:152, May–June 1991.

34. Rosen A: Microwave applications in medicine in the U.S.A.—A short overview. The 21st European Microwave Conference Proceedings, Stuttgart, Germany, September 9–12, 1991, pp. 139–149.

35. Rosen A, Walinsky P, Nardone D, et al: Treatment of intracoronary thrombus with microwave thermal balloon angioplasty. *IEEE MTT-S International Microwave Symposium Digest*, 1992.

36. Rosen A, Walinsky P, Herczfeld P: Microwaves in medical applications: Microwave balloon angioplasty. MM '92 International Conference Proceedings, October 13–15, Brighton, England.

37. Rosen A, Walinsky P, Herczfeld P: Microwave in medical applications: Microwave balloon angioplasty. Electro International 1993 Proceedings, Edison, NJ, April 27–29, 1993.

38. Walinsky P, Rosen A, Martinez-Hernandez A, et al: Microwave balloon angioplasty. In Vogel JHK, King SB III (eds): *The Practice of Interventional Cardiology*, 2 ed. Mosby Year Book, 1993, Vol. 27, pp. 281–285.

39. Landau C, Currier JW, Haudenschild CC, et al: Microwave balloon angioplasty effectively seals arterial dissections in an atherosclerotic rabbit model. *J Am Coll Cardiol* 23:1700, 1994.

CHAPTER TWO

Catheter Ablation for the Treatment of Cardiac Arrhythmias

ARNOLD J. GREENSPON, M.D., AND PAUL WALINSKY, M.D.,
Division of Cardiology, Department of Medicine,
Jefferson Medical College, Philadelphia, PA
ARYE ROSEN, Ph.D., *Department of Electrical and Computer Engineering,*
Drexel University, Philadelphia, PA

2.1 INTRODUCTION

Catheter techniques have been developed for the treatment of cardiac arrhythmias. The technique for catheter-based treatment of cardiac arrhythmias rests on the concept that arrhythmias arise from regions in the heart of abnormal impulse conduction or formation. Ablation of the arrhythmia is possible by identifying and targeting these sites. Various energy forms have been developed in an attempt to specifically injure or destroy the cardiac tissue involved in the cardiac arrhythmia. The goal of effective catheter ablation is to create localized injury that will effectively remove the arrhythmic focus without injuring the surrounding normal heart. This chapter reviews three energy forms that have been investigated for catheter ablation: high-energy direct current, radiofrequency energy, and microwave energy (Table 2.1). Catheter ablation of cardiac arrhythmias is a promising technique for the nonsurgical cure of symptomatic arrhythmias. It is

New Frontiers in Medical Device Technology, Edited by Arye Rosen and Harel Rosen
ISBN 0-471-59189-0

TABLE 2.1 Energy Sources for Catheter Ablation

	Direct Current	Radiofrequency	Microwave
Waveform	Monophasic, damped sinusoidal	Continuous unmodulated sinusoidal	N/A[a]
Frequency	DC	550–750 kHz	915, 2450 MHz
Voltage V	2000–3000 V	< 100 V	N/A
Mechanism of injury	Passive heating, baro trauma, electric field effects	Resistive heating	Radiant heating
Sparking, barotrauma	Yes	No	No
General anesthesia	Yes	No	No
Lesion size	Moderate	Small	Unknown
Control of injury	Low	High	High

[a]N/A = data not available.

the hope that many symptomatic patients who now require lifelong drug treatment of their arrhythmias will be treated effectively in the future with this technique.

2.2 DIRECT CURRENT FOR CATHETER ABLATION

2.2.1 Biophysics of Direct Current for Cardiac Ablation

High-energy direct-current (DC) energy delivered through intracardiac electrode catheters causes selective cardiac injury suitable for catheter ablation.[1–4] The delivery of 200–400 J (joules) of stored energy from a standard cardiac defibrillator through an electrode catheter results in 2000–3000-V potentials at the electrode surface.[5–7] When this discharge is placed in a saline bath, an explosive flash results. When this discharge is placed into a cardiac chamber, there is direct myocardial damage. The mechanism of myocardial damage resulting from high-energy direct current is due to a combination of three factors: (1) the concussive shock wave that causes barotrauma; (2) heat developing at the electrode surface during the discharge, causing thermal injury to tissue; and (3) disruption of cellular membranes due to electric field effects.

Myocardial injury results from the sequence of events that occurs when a DC pulse from a standard defibillator is passed through a multielectrode catheter[5–7] (Fig. 2.1). Standard defibrillators in clinical use deliver capacitor discharges over 5–10 ms, with peak voltages developing within 1–2 ms. The delivery of high-energy to the tip of an electrode catheter causes electrolysis of water at the electrode tip with formation of oxygen and hydrogen gas. Although the quantity of gas formed is small, a bubble develops at the tip of the electrode. This bubble insulates the surrounding blood from the electrode, thereby impeding the flow of current from the electrode into the blood. Current continues to flow from the defibrillator to the electrode tip despite the increase in impedance. There is a rise

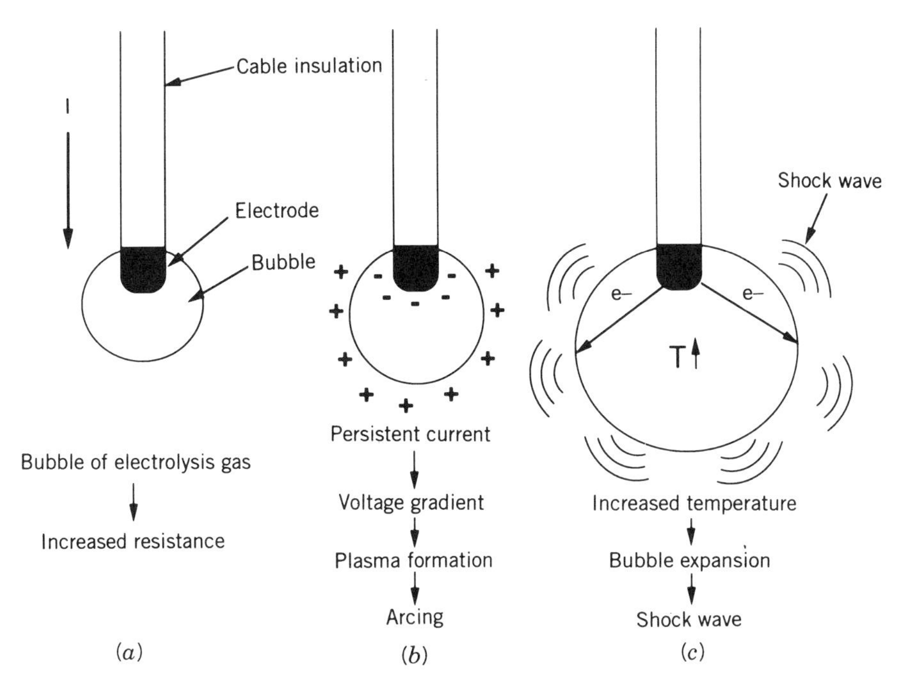

FIGURE 2.1 Mechanism of DC ablation. When direct current from a standard defibrillator is delivered to the tip of an electrode catheter, a series of events occurs, resulting in the generation of a shock wave. The charge delivered to the electrode tip first results in the electrolysis of water into hydrogen and oxygen gas. An insulating bubble develops resulting in an increased impedance (*a*). Current continues to flow to the tip of the catheter despite the rise in impedance, which results in a voltage gradient across the bubble. Arcing occurs. (*b*) Once arcing occurs, there is a tremendous rise in temperature, causing the bubble to expand generating a shock wave (e^-= flow of electrons). (*c*) (Adapted from *Clinical Cardiology*, 1990.[7])

in voltage measured between the electrode tip and the blood. This voltage gradient is responsible for disrupting cell membranes and tissue damage. When this voltage gradient reaches a critical value, a flash or arc develops. The temperature on the electrode surface dramatically rises to as high as 6000 K as the bubble rapidly expands. When the bubble expands a shock wave is generated with pressures as high as 50,000 atms, although they usually range from 10 to 20 atms. The shock wave then causes barotrauma to adjacent cardiac tissue. If the electrode catheter is not properly positioned, unwanted damage to cardiac structures or cardiac perforation could occur.

Direct-current pulses are generally passed through standard multielectrode catheters. Delivery of high-energy to the tip of such an electrode catheter may produce disruption of the catheter material. Insulation breakdown and shunting of energy are possible. These factors obviously limit the safety and efficacy of this technique.

Direct-current shocks through an electrode catheter produce hemispheric lesions 1–2 cm in depth.[8,9] Acutely, these lesions show hemorrhagic necrosis with a homogeneous loss of cellular detail, pyknotic nuclei, and cellular deformation. In addition, there is interstitial hemmorhage, edema, and inflammatory cell infiltration. When these lesions are examined chronically 5–7 days later, there is intense fibrosis, fatty infiltration, and fibroblastic cell proliferation at the margin of the lesion.

Unfortunately, proarrhythmic effects of these shocks occur immediately and up to 12–48 hs later. Jones and coworkers described the effects of countershock electric fields on cultured myocytes.[10,11] As stimuli increased in intensity from 1 to 80 times threshold value, cellular electrical stability decreased. At 24 times threshold value there were transient tachyarrhythmias. When the stimulus was increased to 42 times threshold value there was arrest of rhythmic activity. At 80 times the threshold value asynchronous contraction of sarcomeres or *cellular fibrillation* was observed. In animal studies reported by Westveer, 4 of 15 animals died immediately following the delivery of high-energy DC current to the left ventricular chamber.[12] Similarly, Lerman and coworkers reported that 8 of 11 animals subjected to DC electrode shocks developed lethal arrhythmias 24 hs postprocedure.[13] These data suggest that high-energy DC catheter shocks may promote deleterious cardiac arrhythmias.

Several investigators have modified DC techniques in the hope of minimizing the untoward effects of high-energy shocks. In a series of experiments, Bardy and coworkers demonstrated that when energy was held constant, it was voltage and not charge that was most responsible for cell injury.[14] Therefore, a high-voltage, low-capacitance, low-energy waveform is capable of producing tissue injury without the barotrauma and marked rise in electrode temperature observed during high-energy discharges. Cunningham and coworkers developed a nonarcing DC system that did not produce a flash or explosive force.[15,16] A defibrillator was modified to deliver 0.6 J instead of 200–400 J during a period of 10 μs instead of 2–4 ms. These high-intensity shocks delivered 12–25 A (amperes) and 1.8–2.8 kV (kilovolts) without arcing or increased pressure at the electrode surface. The catheter electrode was also modified to obtain a more uniform current density at the electrode surface and therefore a higher threshold for arcing. Such modifications in the design of DC systems decrease but do not eliminate the safety concerns of direct current as an energy source for catheter ablation.

2.2.2 Clinical Studies

Ablation of conduction at the AV junction was the first application of DC catheter ablation. In 1982, Gallagher and Scheinman independently reported a technique for ablating the AV conduction system.[1,2] An electrode catheter was placed across the tricuspid valve in close proximity to the bundle of His (Fig. 2.2). A DC shock was delivered to the distal electrode at a point recording the maximal unipolar His bundle deflection. Following the shock there was complete loss of AV conduction, the desired effect of the procedure. In 1988, the results of the

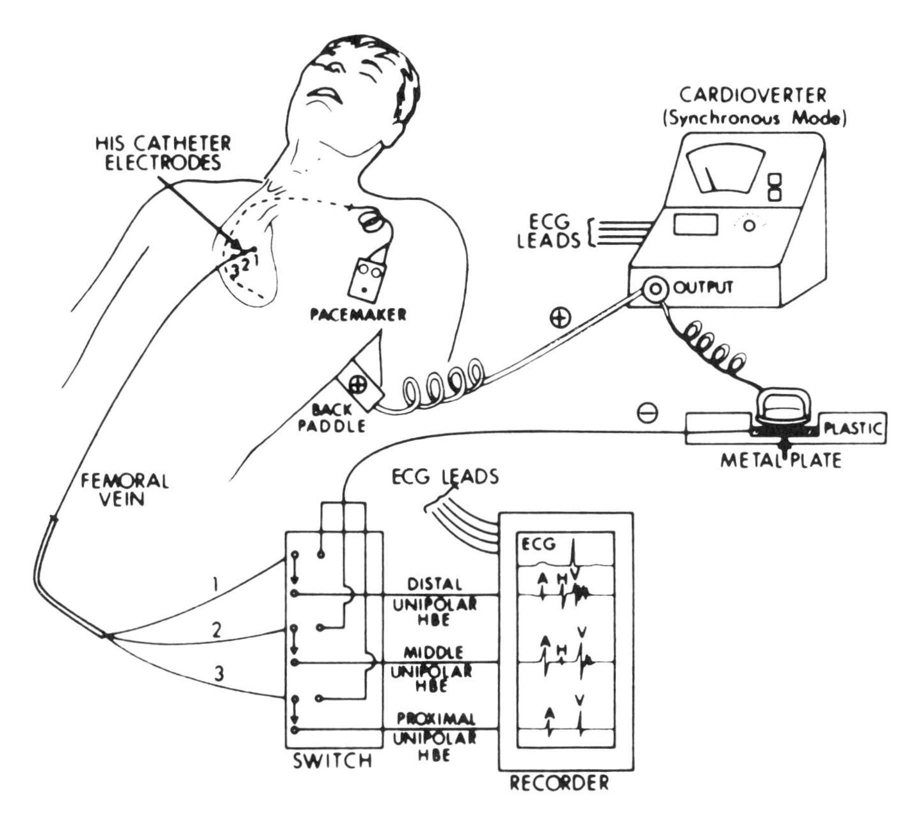

FIGURE 2.2 Catheter technique for DC catheter ablation of the AV junction. An electrode catheter is positioned via a femoral vein to record the largest His bundle deflection. Direct current is delivered to this site resulting in complete AV block. ECG = electrocardiogram, HBE = His bundle electrogram, A = atrium, H = His bundle, V = ventricle. (Reproduced from the *New England Journal of Medicine* with permission.[1])

Multicenter Percutaneous Cardiac Mapping and Ablation Registry were published. Catheter ablation of the AV junction using DC current was performed in 553 patients.[17] The mean followup time was 23 ± 18 months. Complete ablation of AV conduction was accomplished in 65% of patients with partial success achieved in an additional 20%. Early complications, including hypotension, perforation, cardiac tamponade, embolization, pericarditis, and ventricular tachyarrhythmias, occurred in 10% of the patients. During the followup period there were 46 deaths (8.3% of cases), of which 10 were sudden. In a later study, the Catheter Ablation Registry prospectively followed 136 patients who underwent DC catheter ablation of the AV junction.[18] There was a 5.1% risk of death when this technique was applied to patients with left ventricular dysfunction. The complications of this technique prompted the search for alternative energy sources for catheter ablation that, perhaps, would be safer and equally efficacious.

2.3 RADIOFREQUENCY ENERGY FOR CATHETER ABLATION

2.3.1 Biophysics of Radiofrequency Energy for Catheter Ablation

Radiofrequency (RF) current is another energy form that may be utilized for catheter ablation.[4,7,19–22] The effects of RF current (100 kHz–1.5 MHz) when passed through an electrode depend on several factors including the mode of output (unipolar vs. bipolar), the waveform (damped vs. unmodulated), the crest factor [ratio of peak voltage to root-mean-square voltage (RMS)], and the intensity of the power input. Three distinct results may be obtained: (1) electrosurgical cutting, (2) fulguration, and (3) desiccation. In electrosurgical cutting the electrode is separated from the tissue surface. Short high-energy sparks cause cells on the surface to vaporize into steam, allowing the electrode to function as an electric scalpel. For fulguration, the electrode is also separated from the tissue surfacc with long sparks to the tissue coagulating large bleeders and charring tissue. Desiccation varies from the two previous effects in that there must be adequate contact between the electrode and tissue surface. Coagulation of tissue occurs at low power without spark or significant barotrauma. The maximum voltage is low even though the current delivered is high. The delivered RMS voltage ranges from 40 to 60 V with an average impedance of 100 Ω and current of 0.2–0.6 A.[19,20] This contrasts with voltages of 2000–4000 V typically achieved with DC ablation. Therefore, unlike DC ablation, which is associated with barotrauma and spark formation, RF ablation does not stimulate neuromuscular fibers and thus general anesthesia is not required. In addition, gas bubble formation and arcing do not occur at the catheter tip, leaving the integrity of the catheter and its internal insulation intact.

When continuous, unmodulated 500–750-kHz RF energy is passed through the tip of a catheter electrode to a dispersive electrode on the patient's skin, desiccation of tissue occurs at the electrode tip as a result of localized heating, which drives water out of cells, resulting in coagulation necrosis. The mechanism for heating of the tip of the catheter is resistive or ohmic heating[23] (Fig. 2.3). Radiofrequency current passes through the low-impedance metal electrode and meets a resistive barrier at the tissue interface. Heat develops at the small contact point between the electrode and the tissue surface. At this point the current density is high and the conductivity into the tissue low. The small rim of tissue adjacent to the electrode tip is heated. Heat is then passively conducted to the surrounding tissue. Cell death results when the tissue is heated to a critical temperature (48–50 °C).[24]

The mechanism of lesion formation from RF energy is tissue heating from a point source. Several factors will influence the size of RF lesions,[20,24–27] including: delivered power, temperature at the electrode–tissue interface, tissue impedance, size of the electrode, transfer of heat into the surrounding tissue, and contact pressure of the electrode on the tissue surface. There is a linear relationship between delivered power and lesion size. However, maximum power delivery is limited by a rise in impedance that develops at higher power. A rise in

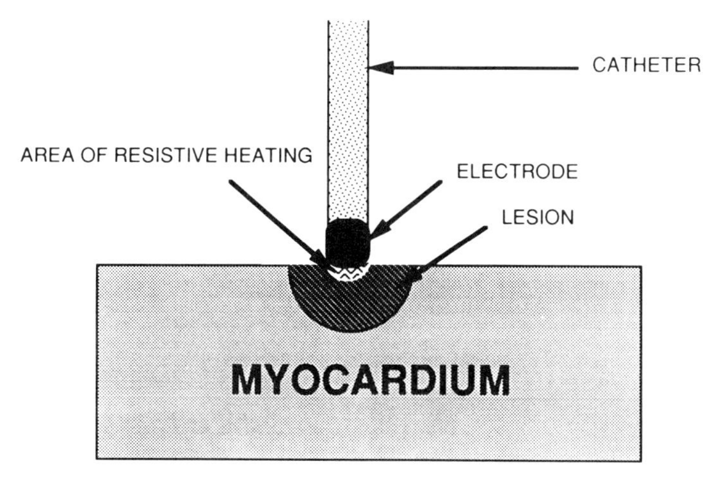

FIGURE 2.3 Mechanism of RF ablation. When RF current is delivered to the tip of a catheter electrode, resistive heating occurs along a small rim of tissue in direct contact with the electrode. A lesion is created as heat conducts passively away from this zone and the surrounding myocardium is heated to a temperature where cell death occurs (~ 50 °C). Lesion size is therefore a function of the size of the electrode and the resulting temperature at the electrode–tissue interface.

impedance is associated with coagulum formation on the catheter tip, preventing further current delivery into tissue. Experimental studies have demonstrated that whenever the tip temperature exceeds 100 °C there is boiling of plasma with resulting adherence of denatured protein to the electrode surface. This markedly diminishes delivery of RF current. Therefore, if tip temperature can be maintained at less than 100 °C, there will be no rise in impedance.

In vitro studies utilizing superperfused canine right ventricle and intact animal preparations have demonstrated that temperature at the electrode–tissue interface accurately predicts lesion volume.[25] Lesion size is directly proportional to tip temperature provided it does not exceed 100 °C. Heat at the electrode–tissue interface drops off exponentially as the distance from the tip increases. Tissue equilibration occurs rapidly with a half-time of 8–10 s for lesion growth following the delivery of RF current. Therefore, maintaining tip temperature at a minimum of 80–90 °C (but less than 100 °C) for approximately 30–60 s will produce the maximal lesion size. Since the delivery of RF current is limited by a rise in impedance, there is a theoretical maximal lesion size for any electrode geometry. Enlargement of lesion size may be accomplished by increasing the size of the electrode surface area, which will allow higher power delivery.[27] The final parameter that is critical for RF-induced lesion formation is adequate electrode–tissue contact. If there is poor contact, RF current cannot be coupled to tissue to produce the desired effect of tissue heating. RF heating of tissue occurs only within the critical rim of tissue in direct contact with the electrode. The remainder of tissue is heated conductively. Therefore, if the catheter moves even a millimeter from the tissue surface, there will be little tissue heating and lesion formation.

Histologically, RF lesions are hemispherical, measuring approximately 8–12 mm in diameter with a volume of about 400–500 mm^3.[20–22] There is homogeneous coagulation necrosis with a peripheral inflammatory response. There may be charring or pitting seen in the center of these lesions due to the higher temperatures at this point. Chronically these lesions appear as localized, whitish scars. The homogeneous nature of these lesions suggests that they are less likely to be arrhythmogenic than those created with DC energy.

2.3.2 Clinical Studies

2.3.2.1 RF Ablation of the AV Junction

Radiofrequency ablation of the atrioventricular (AV) junction has been performed in patients with drug-resistant supraventricular tachycardias. Success rates of 62–82% have been reported, which is similar to the results achieved with DC shock.[28–31] The major advantages of RF ablation are the lack of serious complication associated with the procedure and the fact that general anesthesia is not required. This technique may be safely performed in patients with poor left ventricular function. Sudden arrhythmic death has not been associated with this procedure. RF ablation of the AV junction is a safe method for controlling heart rate without the need for drug therapy in those patients who have supraventricular tachycardias with rapid AV conduction. Ablation of the AV junction and implantation of a permanent pacemaker will effectively control symptoms in these patients.

2.3.2.2 RF Ablation for Supraventricular Tachycardias

Radiofrequency ablation is well suited for the nonsurgical treatment of supraventricular tachycardia since small, well-demarcated lesions are produced with this technique. Most supraventricular tachycardias are caused by movement of electrical activation over defined anatomic circuits in the heart.[32] Such movement is often termed *reentry* because activation proceeds anterograde (forward) from atrium to ventricle over one pathway and returns to the atrium or reenters over another retrograde pathway so that the circus movement may continue. The arrhythmia is initiated whenever a premature beat blocks in one of the two pathways, conducts anterograde (forward) over one of the pathways, and returns or reenters over the other. The arrhythmia will continue endlessly over this circuit unless one of the arrhythmic pathways is interrupted. RF catheter ablation is effective in treating supraventricular tachycardias because the arrhythmic pathways participating in these arrhythmias can be precisely mapped in the catheterization laboratory. RF lesions may then be directed at these sites. The first supraventricular tachycardias targeted for RF ablation were those associated with the Wolff–Parkinson–White syndrome.[33–35] In this condition the anatomic basis for supraventricular tachycardias is an accessory connection or pathway that connects the atrium and ventricle outside the normal AV conduction pathway (Fig. 2.4). These accessory pathways cross between the atrium and the ventricle at the level of the mitral and tricuspid anulus. RF energy delivered to the mitral or tricuspid anulus either from the ventricular[33–35] or atrial[36] aspect can ablate these pathways. Success rates of over 90% have been reported using either approach.

The method for approaching an accessory pathway depends on its anatomic location. The location of accessory pathways is determined by applying activation mapping techniques in the electrophysiology laboratory. Multiple electrode catheters are percutaneously placed in the heart via either femoral or subclavian veins. In patients with an accessory pathway earliest activation of the ventricle will be recorded at the site of the pathway rather than the normal AV conduction system. These pathways may lie in either the right or left freewall or the septum. A left freewall pathway may be approached by either targeting RF to the ventricular or atrial insertion of the pathway (Fig. 2.5). The ventricular approach is performed by passing an ablation probe across the aortic valve into the left ventricle. The electrode is then directed under the mitral valve leaflet adjacent to the mitral anulus. For the atrial approach, a catheter is passed from the right atrium across the interatrial septum into the left atrium (transseptal catheterization). The ablation probe is then maneuvered to the mitral anulus. Right freewall and septal pathways are approached similarly targeting either the atrial or ventricular insertion of the pathway. When RF energy at 25–35 W is applied to a successful site, loss of accessory pathway conduction is observed within 5 s (Fig. 2.6). Using such an approach, loss of accessory pathway conduction is achieved in over 90% of patients with a recurrence rate of less than 10%.

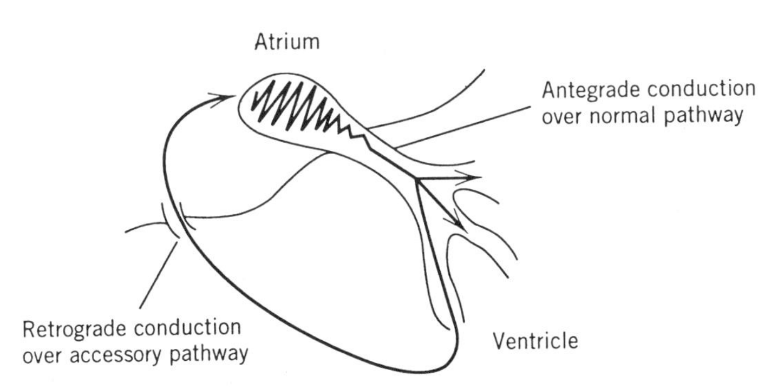

FIGURE 2.4 An arrhythmic circuit associated with the Wolff–Parkinson–White syndrome. In this syndrome, there is a connection between the atrium and ventricle outside the normal AV nodal pathway (accessory pathway). A tachycardia circuit can develop if an impulse conducts antegrade (forward) via the normal AV node pathway and is able to conduct retrograde (reverse) from the ventricle to the atrium via the accessory pathway. Catheter ablation successfully treats these arrhythmias because it interupts accessory pathway conduction without interfering with normal AV nodal conduction.

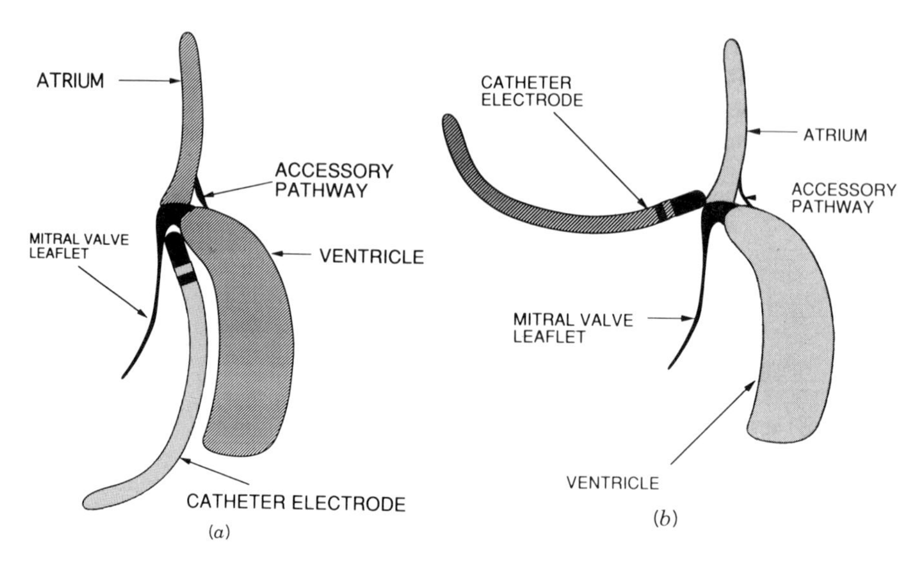

FIGURE 2.5 Diagrams of electrode positions used in RF catheter ablation of accessory pathways: (*a*) for the ventricular approach a catheter is passed retrograde across the aortic valve and positioned under the mitral leaflet; (*b*) for the atrial approach a catheter is passed across the interatrial septum (transseptal catheterization) and positioned on top of the mitral valve leaflet. Electrical mapping confirms the site of the accessory pathway prior to the delibery of RF energy.

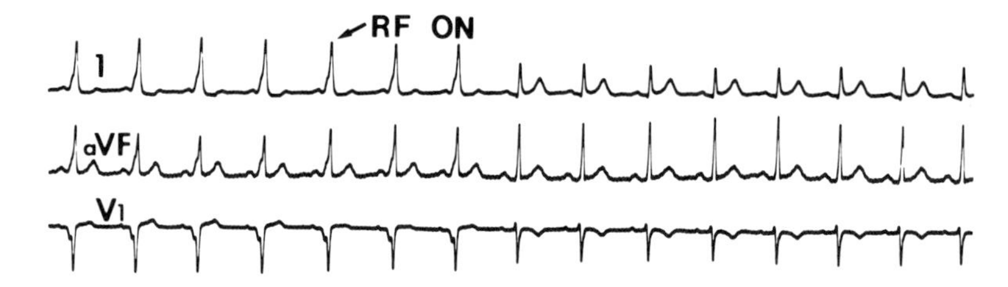

FIGURE 2.6 Electrocardiogram taken during application of RF energy to the site of an accessory pathway in a patient with the Wolff–Parkinson–White syndrome. There is loss of conduction over an accessory pathway following the delivery of radiofrequency current (RF ON). Note the change in the complexes beginning with the third beat following delivery of RF. This change in the electrocardiographic pattern represents loss of conduction over the accessory pathway with persistent conduction over the normal AV nodal pathway.

The majority of patients with supraventricular tachycardias do not have accessory pathways. Rather, these patients have reentry occurring within the AV node itself. AV node reentry (AVNRT), which is responsible for approximately 50% of supraventricular tachycardias, is due to antegrade conduction over a slowly conducting AV nodal pathway, and retrograde conduction over a rapidly conducting AV nodal pathway.[37] Anatomically, the rapid pathway resides in the region of the anterior tricuspid anulus near the bundle of His, whereas the slow pathway resides more posteriorly along the tricuspid anulus near the coronary sinus os. Initial attempts at treating AV node reentry were directed against the rapid AV nodal pathway. Success was achieved in 82–95% of patients.[19,38] However, because of the close proximity of the His bundle to the rapid pathway, heart block was seen in 8% of patients. For this reason, attempts at modifying the slow pathway were developed.

Investigators have shown that "slow" potentials may be recorded from the posterior tricuspid anulus in patients with AVNRT.[39,40] These slow potentials are felt to represent depolarization over slowly conducting AV nodal fibers or selective atrial fibers that input into the AV node. RF current at 25–35 W is delivered to the sites where these potentials are recorded. Because these potentials are recorded at sites that are posterior–inferior to the His bundle, there is little chance that lesions created in this area will lead to heart block. Therefore, this technique is safer than the anterior approach, which targets the fast AV nodal pathway.

Using the posterior approach targeting the slow AV nodal pathway, RF ablation is successful in 88–95% of cases.[39–42] Heart block is only rarely seen. Similar results have been obtained when an anatomic approach to the slow AV nodal pathway, not guided by electrical mapping for slow potentials, is used.[42,43] An anatomic approach targets RF energy to the atrial side of the tricuspid anulus at the level of the coronary sinus os, which is posterior to the compact AV node. Additional lesions are then placed more anteriorly along the tricuspid anulus if supraventricular tachycardia can still be induced using pacing techniques. After each lesion is placed, attempts are made to reinduce the tachycardia. Successive lesions are created more anteriorly until tachycardia is rendered noninducible or the region of the compact AV node is reached, which is where the His bundle electrogram is recorded.

2.3.2.3 RF Catheter Ablation for Ventricular Tachycardia

Ablation of sustained ventricular tachycardia has been much less effective[44] except in those patients with idiopathic ventricular tachycardia[45] and bundle branch reentry.[46–48] Idiopathic sustained ventricular tachycardia occurs in a patient whose heart is otherwise structurally normal. In these patients, the ventricular tachycardia arises from a distinct focus, usually the right ventricular outflow tract. RF energy delivered to this focus in the right ventricle will extinguish the tachycardia in 94% of cases. Idiopathic ventricular tachycardia arising from the left ventricle has also been effectively treated in this manner.

Bundle branch reentry ventricular tachycardia is an unusual arrhythmia that is seen in patients with cardiomyopathy, decreased left ventricular function, and conduction system disease. In these patients ventricular tachycardia may develop as a result of a macroreentrant circuit involving the left and right bundle branch. Selective RF ablation of the right bundle is effective in preventing this form of ventricular from developing.

Unfortunately, most patients who present with sustained ventricular tachycardia have coronary artery disease or decreased left ventricular function, and have had a remote transmural myocardial infarction.[49] Unlike idiopathic ventricular tachycardia and most supraventricular tachycardias where the anatomic substrate for the arrhythmia arises from a small, easily defined anatomic area, ventricular tachycardia associated with coronary artery disease is much more complex. In these patients, ventricular tachycardia results from reentry within zones of slow conduction that are present in the region of an old healed myocardial infarction.[50] Unlike in the case of supraventricular tachycardia, there may be multiple potential pathways that are operative in the same patient. In addition, these potential pathways may be located a variable distance below the endocardial surface. These two factors—a complex activation sequence during the tachycardia and a larger anatomic pathway—make RF ablation of ventricular arrhythmias more difficult. Because of the relatively small lesion size associated with RF, the success rate for ablation of sustained ventricular tachycardia associated with coronary artery disease is much lower.[19,51] Therefore, research is being directed at alternative energy forms for creating larger lesions that may be more effective in ablating ventricular tachycardia.

2.4 MICROWAVE ENERGY FOR CATHETER ABLATION

Microwave hyperthermia has been useful in radiation oncology for the treatment of various solid tumors.[52] The cardiac applications of this modality have only recently been explored. Microwave energy using either 915 or 2450 MHz has been studied in an attempt to enlarge myocardial lesions in catheter ablation.[53–56] Microwave energy is delivered down the length of a coaxial cable that terminates in an antenna capable of radiating the energy into tissue. Radiant energy will cause the water molecules in myocardial tissue to oscillate, producing tissue heating and cell death. The higher frequency of microwave energy allows for greater tissue penetration and theoretically a greater volume of heating than that possible with RF, which produces direct ohmic or resistive heating.

Whip and helical antenna configurations have been developed for microwave ablation catheters. (See Chapter 1, especially Fig. 1.4.)

Wonnell and coworkers studied the effects of microwave energy for cardiac ablation using a helical antenna mounted on a coaxial cable (2.44 mm o.d.).[55] High-frequency current at 2450 MHz was delivered via the helical antenna into a tissue-equivalent phantom model. The temperature distribution profile was measured around the antenna as well as into surrounding volume (the depth of penetration). The volume of heating for the microwave catheter system was 11 times greater than that of an RF electrode catheter at the same surface temperature. In addition, the microwave catheter penetrated an area that was twice as large as that penetrated by the RF catheter. These data suggest that microwave energy will produce larger lesions than RF because a greater volume of tissue is being heated. An additional theoretical advantage of the microwave system is that direct tissue contact is not crucial for tissue heating since heating occurs via radiation, and not via direct ohmic heating as seen with RF. Using this system, preliminary studies in six animals demonstrated that complete heart block could be achieved in all six animals by directing microwave energy (50 W at 2450 MHz for 114 ± 118 s) to the atrioventricular junction.[57]

We evaluated helical and whip antenna designs in a tissue equivalent phantom at 915 and 2450 MHz utilizing a coaxial cable (0.06 in. o.d.).[54] All catheters were studied on a network analyzer prior to placing them in the phantom model. Such analysis demonstrated the great variability in tuning of these microwave catheters. Microwave ablation catheters have suffered from imperfect tuning leading to inefficient radiation of energy. Consequently, there is generation of heat along the length of the catheter rather than radiation of energy into tissue. Little heating into tissue was observed in poorly tuned catheters. Such analysis underscores the critical importance of proper tuning of microwave catheters prior to any further studies.

A perfusion chamber containing a muscle-equivalent phantom was constructed and placed in a saline bath held at 37 °C. The muscle-equivalent phantom consisted of TX150, polyethylene powder, NaCl, and water. Ablation catheters were placed on the surface of the phantom material. Temperature measurements were performed using a 12-channel Luxtron fiberoptic thermometry system. Probes were placed beneath the surface of the phantom. Saline at a constant temperature of 37 °C was infused at a flow rate of 4 L/min across the surface of the phantom. This model simulates the heart where the phantom material has the dielectric properties of cardiac muscle and the saline properties of blood.

Temperature curves were plotted from probes placed 1, 2.5, 5, and 7.5 mm from the point of maximal heating on the microwave catheter. Thermal profiling of these catheters demonstrated volume heating. Heating was proportional to forward power duration of power application and to surface temperature. In addition to the small amount of volume heating, conductive heating was also present as a result of the increased temperature at the catheter–phantom interface. The magnitude of heating with the microwave catheters was smaller when compared to that of the RF ablation catheters.

In vivo ablation using microwaves was performed on canine left ventricular myocardium; A power of 80 W was delivered for a total of 5 min. Mean lesion size measured $435 \pm 236\ \text{mm}^3$, which was similar in size to lesions created with small-tipped RF catheters. The microwave ablation catheters, as presently designed, were not capable of producing lesions larger than those produced by RF catheters.

The theoretical advantages of microwaves as an energy source for cardiac ablation include the lack of importance of direct probe contact since tissue heating occurs by radiation, not conduction, and larger lesion size due to a volume heating effect. However, practical problems remain to be solved before microwaves become a useful clinical energy source. These problems include (1) power loss in the coaxial cable, (2) resultant heating of the coaxial cable during power delivery that has led to a breakdown in the dielectric and catheter material, (3) inefficiency of the radiating antenna, and (4) lack of a unidirectional antenna that can radiate energy into tissue and not the circulating blood pool. At the present time microwave catheter systems are poorly efficient radiators of energy into cardiac tissue. These obstacles will have to be overcome before microwaves supplant radiofrequency as the preferred energy source for cardiac ablation.

REFERENCES

1. Gallagher JJ, Svenson RH, Kassell JH, et al: Catheter technique for closed chest ablation of the atrioventricular conduction system. *New Engl J Med* 306:194, 1982.
2. Scheinman MM, Morady F, Hess DS, et al: Catheter induced ablation of the atrioventricular junction to control refractory supraventricular arrhythmias. *J Am Med Assoc* 248:851, 1982.
3. Fontaine G, Tonet JL, Frank R, et al: Clinical experience with fulguration and antiarrhythmic therapy for the treatment of ventricular tachycardia. Long term followup of 43 patients. *Chest* 95:785, 1989.
4. Haines DE: Current and future modalities of catheter ablation for the treatment of cardiac arrhythmias. *J Invas Cardiol* 4:291, 1992.
5. Boyd EGCA, Holt PM: An investigation into the electrical ablation technique and a method of electrode assessment. *PACE* 8:815, 1985.
6. Bardy GH, Coltorti F, Ivey TD, et al: Some factors affecting bubble formation with catheter-mediated defibrillator pulses. *Circulation* 73:525, 1986.
7. Bardy GH, Sawyer PL: Biophysical and anatomical considerations for safe and efficacious catheter ablation for arrhythmias. *Clin Cardiol* 13:425, 1990.
8. Ward DE, Davies M: Transvenous high-energy shock for ablating atrioventricular conduction in man: Observations on the histological effects. *Br Heart J* 51:175, 1984.
9. Kempf FC, Falcone RA, Iozzo RV, et al: Anatomic and hemodynamic effects of catheter-delivered ablation energies in the ventricle. *Am J Cardiol* 56:373, 1985.
10. Jones JL, Proskauer CC, Paull WK, et al: Ultrastructural injury to duck myocardial cells in vitro following "electric" countershock. *Circ Res* 46:387, 1980.
11. Jones JL, Lepeschkin E, Jones RE, et al: Response of cultured myocardial cells to countershock-type electrical field stimulation. *Am J Physiol* 235:H214, 1978.
12. Westveer DC, Nelson T, Stewart JR, et al: Sequelae of left ventricular electrical endocardial ablation. *J Am Coll Cardiol* 5:956, 1985.
13. Lerman BB, Weiss JL, Bulkley BH, et al: Myocardial injury and induction of arrhythmia by direct current shocks delivered via endocardial catheters in dogs. *Circulation* 69:1006, 1984.
14. Bardy GH, Sawyer PL, Johnson G, et al: The effects of voltage and charge of electrical ablation on canine myocardium. *Am J Physiol* 257:1534, 1989.
15. Cunningham D, Rowland E, Rickards AF: A new low energy power source for catheter ablation. *PACE* 9:1384, 1986.

16. Rowland E, Cunningham D, Ahsan A, et al: Transvenous ablation of atrioventricular conduction with a low energy power source. *Br Heart J* 62:361, 1989.

17. Evans GT Jr, Scheinman MM, the Executive Committee of the Registry: The Percutaneous Cardiac Mapping and Ablation Registry: Final summary of results. *PACE* 11:1621; 1988.

18. Evans GT Jr, Scheinman MM, Bardy GH, et al: Predictors of in-hospital mortality after DC catheter ablation of atrioventricular conduction. *Circulation* 84:1924, 1991.

19. Kalbfleish SJ, Langberg JJ: Catheter ablation with radiofrequency energy: Biophysical aspects and clinical applications. *J Cardiovasc Electrophysiol* 3:173, 1992.

20. Huang SKS: Use of radiofrequency energy for catheter ablation of the endomyocardium: A prospective energy source. *J Electrophysiol* 1:78, 1987.

21. Huang SKS: Advances in applications of radiofrequency current to catheter ablation therapy. *PACE* 14:28, 1991.

22. Haverkamp W, Hindricks G, Gulker H, et al: Coagulation of ventricular myocardium using radiofrequency alternating current: Biophysical aspects and experimental findings. *PACE* 12:187, 1989.

23. Haines DE, Watson DD: Tissue heating during radiofrequency catheter ablation: A thermodynamic model and observations in isolated perfused and superfused canine right ventricular free wall. *PACE* 12:962, 1989.

24. Haines DE, Watson DD, Verow AF: Electrode radius predicts lesion radius during radiofrequency heating. Validation of a proposed thermodynamic model. *Circ Res* 67:124, 1990.

25. Haines DE, Verow AF: Observations on electrode-tissue interface temperature and effect on electrical impedance during radiofrequency ablation of ventricular myocardium. *Circulation* 82:1034, 1990.

26. Hoyt RH, Huang SK, Marcus FI, et al: Factors influencing trans-catheter radiofrequency ablation of the myocardium. *J Appl Cardiol* 1:469, 1986.

27. Rosenbaum R, Greenspon AJ, Smith M, et al: Advanced radiofrequency catheter ablation in canine myocardium. *Am Heart J* (in press).

28. Jackman WM, Wang X, Friday KJ, et al: Catheter ablation of atrioventricular junction using radiofrequency current in 17 patients. Comparison of standard and large-tip electrode catheters. *Circulation* 83:1562, 1991.

29. Langberg JJ, Chin M, Schamp DJ, et al: Ablation of the atrioventricular junction with radiofrequency energy using a new electrode catheter. *Am J Cardiol* 67:142, 1991.

30. Langberg JJ, Chin M, Rosenqvist M, et al: Catheter ablation of the atrioventricular junction with radiofrequency energy. *Circulation* 80:1527, 1989.

31. Olgin JE, Scheinman MM: Comparison of high-energy direct current and radiofrequency catheter ablation of the atrioventricular junction. *J Am Coll Cardiol* 21:557, 1993.

32. Josephson ME, Kastor JA: Supraventricular tachycardia: Mechanisms and management. *Ann Intern Med* 87:346, 1977.

33. Jackman WM, Wang X, Friday KJ et al: Catheter ablation of accessory atrioventricular pathways (Wolff-Parkinson-White syndrome) by radiofrequency current. *New Engl J Med* 324:1605, 1991.

34. Calkins H, Sousa J, El-Atassi R, et al: Diagnosis and cure of the Wolff-Parkinson-White syndrome or paroxysmal supraventricular tachycardia during a single electrophysiologic test. *New Engl J Med* 324:1612, 1991.

35. Schluter M, Geiger M, Siebels J, et al: Catheter ablation using radiofrequency current to cure symptomatic patients with tachyarrhythmias related to an accessory atrioventricular pathway. *Circulation* 84:1644, 1991.

36. Swartz JF, Tracy CM, Fletcher R: Radiofrequency endocardial catheter ablation of accessory atrioventricular pathway atrial insertion sites. *Circulation* 87:487, 1993.

37. Josephson ME: Supraventricular tachycardias. In *Clinical Cardiac Electrophysiology Techniques and Interpretations*, 2nd ed., Lea & Febiger, Philadelphia, 1993, p. 181.

38. Lee MA, Morady F, Kadish A, et al: Catheter modification of the atrioventricular junction using radiofrequency energy for control of atrioventricular nodal reentry tachycardia. *Circulation* 83:827, 1991.

39. Jackman WM, Beckman KJ, McCelland JH, et al: Treatment of supraventricular tachycardia due to atrioventricular nodal reentry, by radiofrequency catheter ablation of slow-pathway conduction. *New Engl J Med* 327:313, 1992.

40. Haissaguerre M, Gaita F, Fischer B, et al: Elimination of atrioventricular nodal reentrant tachycardia using discrete slow potentials to guide radiofrequency energy. *Circulation* 85:2162, 1992.

41. Kay GN, Epstein AE, Dailey SM, et al: Selective radiofrequency ablation of the slow pathway for the treatment of atrioventricular nodal re-entrant tachycardia. *Circulation* 85:1675, 1992.

42. Jazayeri MR, Hemple JL, Sra JS et al: Selective trans-catheter ablation of the fast and slow pathways using radiofrequency energy in patients with atrioventricular nodal reentrant tachycardia. *Circulation* 85:1318, 1992.

43. Wathen M, Natale A, Wolfe K, et al: An anatomically guided approach to atrioventricular node slow pathway ablation. *Am J Cardiol* 70:886, 1992.

44. Morady F, Scheinman MM, DiCarlo LA Jr, et al: Catheter ablation of ventricular tachycardia with intra cardiac shocks: results in 33 patients. *Circulation* 75:1037, 1987.

45. Klein LS, Shih HT, Hackett K, et al: Radiofrequency catheter ablation of ventricular tachycardia in patients without structural heart disease. *Circulation* 85:1666, 1992.

46. Tchou P, Jazayeri M, Denker S, et al: Transcatheter electrical ablation of right bundle branch. A method of treating macroreentrant ventricular tachycardia attributed to bundle branch reentry. *Circulation* 78:246, 1988.

47. Caceres J, Jazayeri M, McKinnie J, et al: Sustained bundle branch reentry as a mechanism of clinical tachycardia. *Circulation* 79:256, 1989.

48. Langberg JJ, Desai J, Dullet N, et al: Treatment of macroreentrant ventricular tachycardia with radiofrequency ablation of the right bundle branch. *Am J Cardiol* 63:1010, 1989.

49. Josephson ME, Gottleib CD: Ventricular tachycardia associated with coronary artery disease. In *Cardiac Electrophysiology: From Cell to Bedside*, Zipes DP, Jalife J (eds.): Saunders, Philadelphia, 1990, p. 571.

50. Stevenson WG, Weiss JN, Wiener I, et al: Slow conduction in the infarct scar: Relevance to the occurrence, detection, and ablation of ventricular reentry circuits resulting from myocardial infarction. *Am Heart J* 117:452, 1989.

51. Morady F, Harvey M, Kalbfleish SJ, et al: Radiofrequency catheter ablation of ventricular tachycardia in patients with coronary artery disease. *Circulation* 87:363, 1993.

52. Satoh T, Stauffer PR: Implantable helical coil microwave antenna for interstitial hyperthermia. *Internatl J Hyperthermia* 4:497, 1988.

53. Walinsky P, Rosen A, Greenspon AJ: Method and apparatus for high frequency catheter ablation. U.S. Pat. 4,641,649.

54. Rosenbaum RM, Greenspon AJ, Hsu S, et al: RF ablation for the treatment of ventricular tachycardia. *IEEE-MTT-S International Microwave Symposium Digest* Vol. 2, 1993, p. 1155.

55. Wonnell TL, Stauffer PR, Langberg JJ: Evaluation of microwave and RF catheter ablation in a myocardial equivalent phantom model. *IEEE Trans Biomed Eng* 39:1086, 1992.

56. Whayne JG, Haines DE: Comparison of thermal profiles produced by new antenna designs for microwave catheter ablation (abstract). *PACE* 15:580, 1992.

57. Langberg JJ, Wonnell TL, Chin M, et al: Catheter ablation of the atrioventricular junction using a helical microwave antenna: A novel means of coupling energy to the endocardium. *PACE* 14:2105, 1991.

CHAPTER THREE

The Efficacy of Transurethral Thermal Ablation in the Management of Benign Prostatic Hyperplasia

ARYE ROSEN, Ph.D., *Department of Electrical and Computer Engineering, Drexel University, Philadelphia, PA*
HAREL D. ROSEN, M.D., *University of Medicine and Dentistry of New Jersey, New Brunswick, NJ*

3.1 INTRODUCTION

Benign prostatic hypertrophy or hyperplasia (BPH) is one of the most common medical problems experienced by men over 50 years old. In fact, urinary tract obstruction due to prostatic hyperplasia has been recognized since the earliest days of medicine. Hyperplastic enlargement of the prostate gland often leads to compression of the urethra, resulting in obstruction of the urinary tract and the subsequent development of symptoms including frequent urination, decrease in urinary flow, nocturia, pain, discomfort, and dribbling. The association of BPH with aging has been shown by the incidence of BPH in 50% of men over 50 years of age, increasing to over 75% in men over 80 years of age. Symptoms of

New Frontiers in Medical Device Technology, Edited by Arye Rosen and Harel Rosen
ISBN 0-471-59189-0

urinary obstruction occur most frequently between the ages of 65 and 70, when approximately 65% of men have prostatic enlargement.

Currently, the surgical procedures available for treating BPH are not totally satisfactory. Patients suffering from the obstructive symptoms of this condition are provided with few options: continue to cope with the symptoms (i.e., conservative management), submit to drug therapy at early stages, or submit to surgical intervention. More than 400,000 patients per year in the United States undergo surgery for removal of prostatic tissue. However these patients represent only a small percentage of men exhibiting clinically significant symptoms.

Those suffering from BPH are often elderly men, many with additional health problems that increase the risk of surgical procedures. Surgical procedures for the removal of prostatic tissue are associated with a number of hazards, including anesthesia-related morbidity, hemorrhage, coagulopathies, pulmonary emboli, and electrolyte imbalances. The procedures performed currently can also lead to cardiac complications, bladder perforation, incontinence, infection, urethral or bladder-neck stricture, retention of prostatic chips, retrograde ejaculation, and infertility. Because of the extensive invasive nature of the current treatment options for obstructive uropathy, the majority of patients delay definitive treatment of their condition. Such delay can lead to serious damage to other structures secondary to the obstructive lesion in the prostate (i.e., bladder hypertrophy, hydronephrosis, dilation of the kidney pelves, chronic infection, dilation of the ureters, etc.). Therefore, a significant number of patients with symptoms that are sufficiently severe to warrant surgical intervention are concurrently poor operative risks, and thus poor candidates for prostatectomy. In addition, younger men suffering from BPH who do not desire to risk complications such as infertility are often forced to avoid surgical intervention. Thus the need, importance, and value of improved surgical and nonsurgical methods for treating BPH are unquestionable.

Excluding drug therapy, some of the minimally invasive alternative approaches for the treatment of BPH are listed below. These procedures, predicted to capture medical interest in the 1990s as replacements for such "gold standards" as prostatectomies and transurethral resection of the prostate (TURP) include:

1. Laser procedures, such as visual laser ablation of the prostate (VLAP), transurethral ultrasound-guided laser-induced prostatectomy (TULIP), and semiconductor laser ablation (SCLA)—a future technology for the treatment of BPH discussed later
2. Balloon dilatation and microwave balloon procedures (see Chapter 2)
3. The use of stents, which are tubes of flexible wire mesh and/or tubes made of biodegradable polymers that are placed in the urethra either after balloon dilatation, or directly to preserve the opening of the urethral channel
4. Microwave and RF thermotherapy, such as transurethral microwave therapy (TUMT)

5. Transurethral needle ablation (TUNA), an interstitial ablating method currently utilizing RF and that can be extended to utilize microwave and semiconductor lasers
6. Acoustic ablation

This chapter focuses on the TUMT and TUNA techniques, in addition to the possible future use of semiconductor laser technology in the treatment of BPH.

3.2 TRANSURETHRAL NEEDLE ABLATION (TUNA)

3.2.1 Equipment and Method[1–4]

A new concept of thermotherapy, using an interstitial ablation approach, is being pursued by VidaMed, Inc. This technology uses RF (460 kHz) with excellent control of the RF thermal energy applied to the tissue. The TUNA catheter used is 24.1 cm long and 21 French (Fig. 3.1). Through the tip of the catheter, two needles (electrodes), oriented 40 ° apart, can be deployed. The electrode–needles are shaped to facilitate passage through tissue. They are thin, and thus can be directed from the catheter through intervening tissue with a minimum of trauma to normal tissue. Each electrode–needle is enclosed within a longitudinally adjustable sleeve acting as a shield to prevent exposure of the tissue adjacent to the sleeve

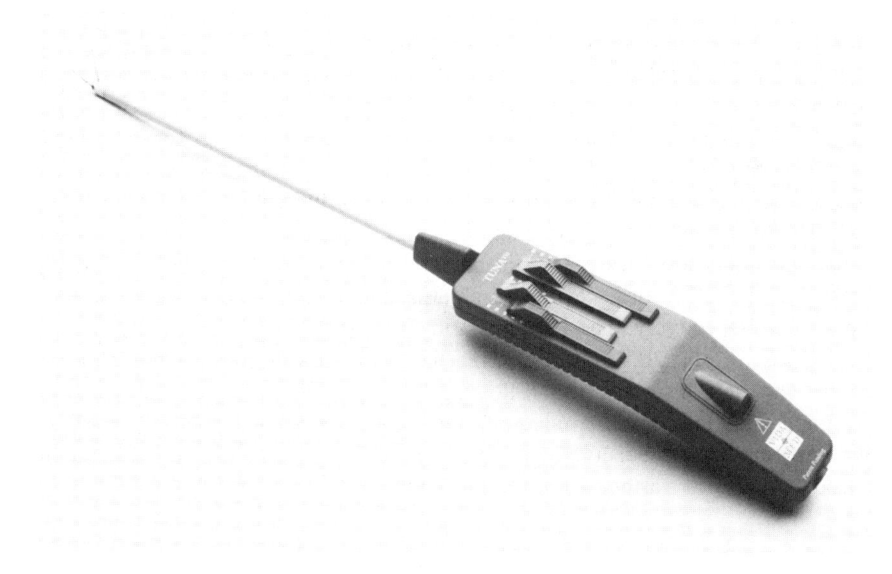

FIGURE 3.1 The TUNA catheter's handle containing controls for the needles and shields, and connections to the RF unit. A knob located between the handle and catheter can rotate the bullet head into the desired position. (With permission of VidaMed, Inc.).

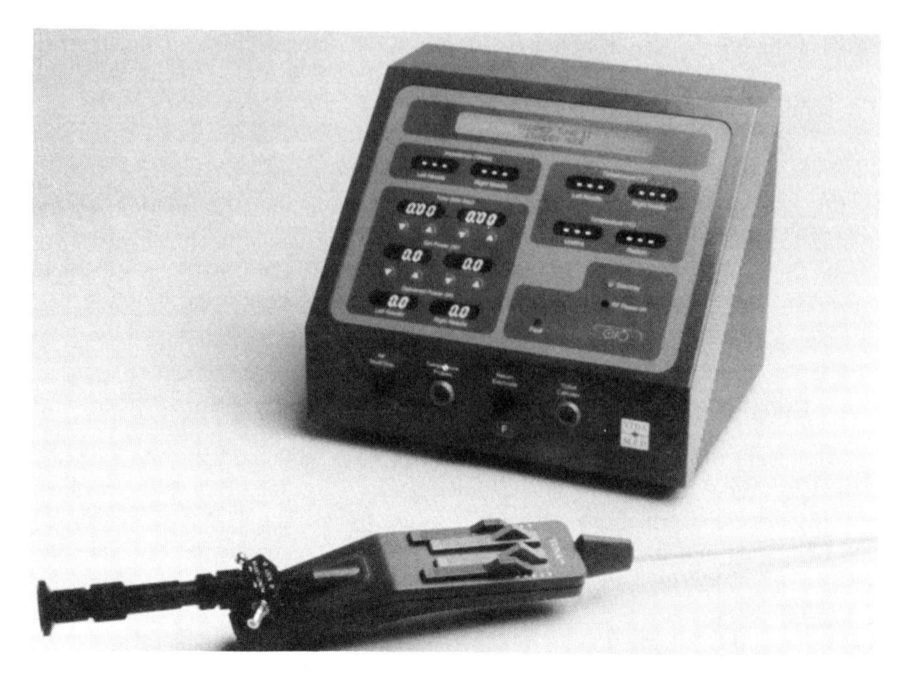

FIGURE 3.2 TUNA RF generator unit (with permission of VidaMed, Inc.).

to the RF current, thus preserving the urethra by reducing the possibility of a rise in its temperature. The sleeve is also used to control the tissue interface, and therefore the ablation volume. Both the electrode–needle and the sleeve are advanced or retracted by controls on the catheter handle (Fig. 3.1) and can be locked into position. The TUNA catheter needle acts as the thermal electrode, and a grounding pad that is placed in the back of the subject under treatment closes the RF circuit to the power supply (Fig. 3.2).

Thermocouples are located at the shield tip below each needle, and at the catheter bullet head in order to record ablation temperature and prostatic urethral temperature, respectively. The RF unit (VidaMed, Inc.) includes an RF generator that delivers up to 20 W at 490 kHz, RF power-level control and time-set control. The unit also has the following readouts: RF power level, ablation time, impedance, and six thermocouple readouts (Fig. 3.2). The TUNA catheter (Fig. 3.3*a*) includes direct fiberoptic vision, as well as provisions for introducing electrode–needles at various angles (Fig. 3.3*b*).

3.2.2 Generation of Tissue Ablation by RF Energy

The RF generator is a source of RF voltage between two electrodes. When the generator is connected to the tissue to be ablated, current will flow through the tissue between the active and dispersive electrodes. The active electrode

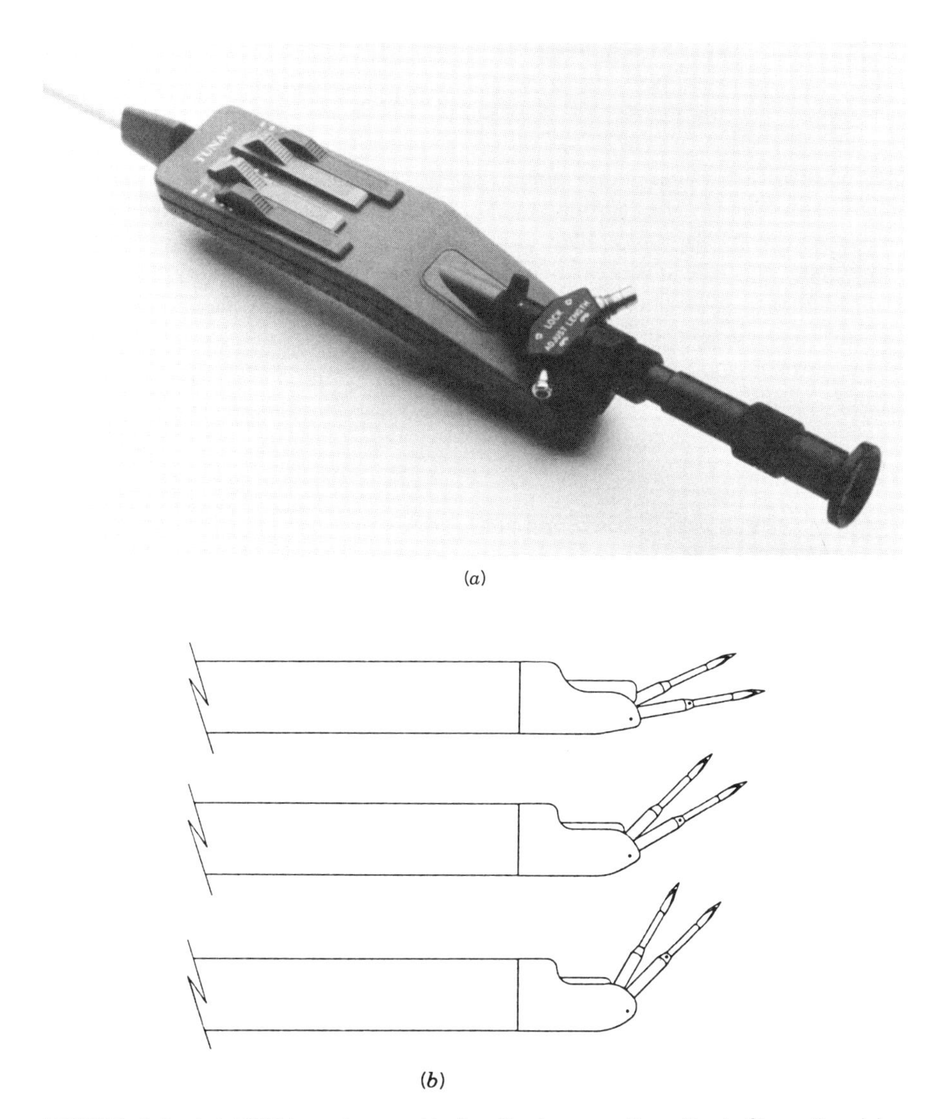

(*a*)

(*b*)

FIGURE 3.3 (*a*) TUNA catheter with handle incorporating direct fiberoptic vision (with permission of VidaMed, Inc.). (*b*) Electrodes and needles at various angles (with permission of VidaMed, Inc.).

is connected to the tissue volume where the ablation is to be made, and the dispersive electrode is a large-area electrode forcing a reduction in current density in order to prevent tissue heating. The total RF current, I_{RF} is a function of the applied voltage between the electrodes connected to the tissue and the tissue conductance. The heating distribution is a function of the current density. The greatest heating takes place in regions of the highest current density, J. The mechanism for tissue heating in the RF range of hundreds of kilohertz is primarily ionic. The electrical field produces a driving force on the ions in the tissue electrolytes, causing the ions to vibrate at the frequency of operation. The current density is $J = \sigma E$, where σ is the tissue conductivity. The ionic motion and friction heats the tissue, with a heating power per unit volume equal to J^2/σ. The equilibrium temperature distribution, as a function of distance from the electrode tip, is related to the power deposition, the thermal conductivity of the target tissue, and the heat sink, which is a function of blood circulation. The lesion size is, in turn, a function of the volume temperature. Many theoretical models to determine tissue ablation volume as a function of tissue type are available, but none is as good as actual data. Chapter 2, which deals with RF ablation of cardiac tissue, relates catheter parameters to lesion size.

3.2.3 Results[1–4]

Twenty patients were treated utilizing TUNA prior to scheduled retropubic prostatectomy. Three procedures were performed under general or spinal anesthesia, and the other 17 patients were treated without anesthesia. Schulman et al.[4] reported that the patients treated without anesthesia tolerated the procedure well, which is an important advantage in the treatment of BPH. Table 3.1 depicts the average rectal, urethral, and proximal lesion temperatures at different power settings obtained from 31 lesions in the 20 patients treated. The maximum temperatures reached were 39.1 °C in the rectum, 52.6 °C in the urethra, and 75.6 °C proximal to the lesion, indicating local heating at the site of treatment.

3.2.4 Pathology and Imaging Studies[4]

The surgical prostatic specimens were received at various times after TUNA, ranging from 1 h to 29 days. The specimens were step-sectioned and examined

TABLE 3.1 Average Temperatures with TUNA Treatment Related to Power Delivered

	Average Temperature (°C)			
Power (W)	Rectal	Urethral	Proximal Lesion	Central Lesion (estimated)[a]
4	35	37	42	82
5	35	39	44	84
6	35	42	50	90
7	35	41	49	89
8	35	41	46	86
10	37	42	53	93

[a]Estimate by infrared thermal imaging system in an ex vivo tissue model.
Source: Reproduced with permission form Dr. Schulman and Vidamed, Inc.

macroscopically and microscopically. Macroscopic examination showed localized lesions with sharply defined margins and a sharp transition between pathologic changes and unchanged prostatic tissue (Figs. 3.4, 3.5). Table 3.2 shows the average macroscopic lesion size as a function of power level. Power levels of 5–10 W are sufficient to produce large lesions.

Microscopic examination of the specimens showed extensive coagulative necrosis extending far beyond the macroscopic lesion. Magnetic resonance

FIGURE 3.4 Macroscopic appearance of necrotic lesions (*arrows*) in a step-sectioned prostate specimen 15 days after TUNA. Note the necrotic lesions close to the uretha.

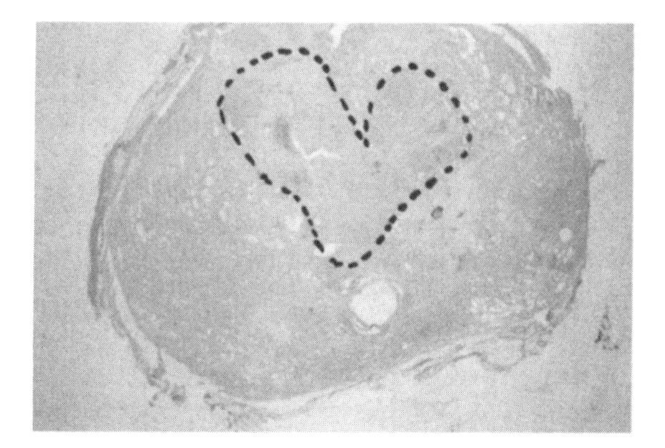

FIGURE 3.5 Macrosection of a prostate 15 days after TUNA. Note the extensive coagulative necrosis (marked) with preservation of the urethra and capsule.

TABLE 3.2. Macroscopic Lesion Size Produced with TUNA at Different RF Power Levels

Average Power (W)	Average Macroscopic[a] Lesion, mm (length × width)
4	8×5
5	12×7
6	12×6
7	15×8
8	10×7
9	13×9
10	15×8

[a]Microscopic lesion size was approximately twice that measured macroscopically.

Source: Reproduced with permission from Dr. Schulman and Vidamed, Inc.

imaging (MRI) studies (see also Chapter 8) of TUNA-treated prostates in vivo agreed well with the pathology studies (the specimens were removed) 18 days after the TUNA treatment (Fig. 3.6).

3.3 TRANSURETHRAL MICROWAVE THERMOTHERAPY

A detailed account of the use of microwave energy utilizing a small coaxial applicator (antenna) has been discussed in Chapters 1 and 2. (See also Refs. 5–11 in this chapter and the additional list of Prostatron publications in the Appendix to this chapter.) At this time, there are several types of microwave delivery systems on the market, of which share the goal of delivering volume heating to the prostate gland in order to cause coagulation necrosis followed by shrinkage of the enlarged prostate, thus reducing BPH symptoms. The only device of this type currently evaluated by the FDA (U.S. Food and Drug Administration) in the United States, and already used routinely (since 1991) in many countries, is the Prostatron (Fig. 3.7), by Technomed Medical Systems, with over 20,000 patients treated worldwide. The objective of the thermotherapy treatment utilizing a transurethral catheter is to selectively destroy a tissue target area without endangering the structures surrounding the treated area, and to be deliverable without anesthesia. The temperature of the prostatic tissue located beyond the few-millimeter-thick urethral mucosa is raised to well above 45 °C for about one hour without significantly raising the temperature of either the rectal wall or the urethral tissue (currently, temperatures as high as 60–70 °C are routinely achieved during a TUMT session).

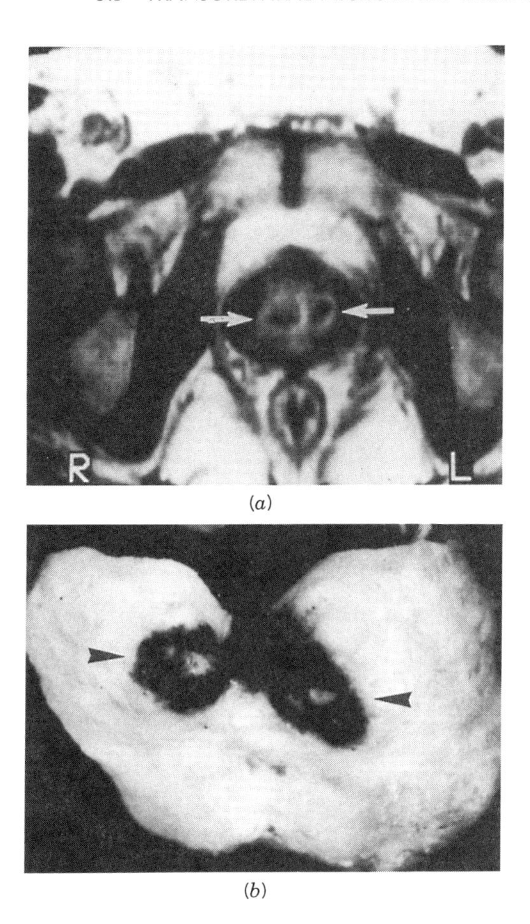

(a)

(b)

FIGURE 3.6 (*a*) MRI studies of TUNA treated in vivo; (*b*) pathologic studies (With permission of VidaMed, Inc.).

The Prostatron system consists of five subsystems: a power delivery system, a microwave generator, a cooling system, a fiberoptic system for temperature measurements, and a monitoring system (Fig. 3.8). The building blocks of the power delivery system are the urethral catheter, the rectal probe, and a connecting box (Fig. 3.9). The urethral catheter (Fig. 3.10) consists of (1) a coaxial cable terminated by a microwave antenna located below a Foley-type balloon, which is inflated prior to the procedure to secure the catheter into position during the procedure; (2) a temperature sensor located at the surface of the catheter in the radiation zone; (3) a cooling system to ensure a good thermocoagulation; and (4) a rectal probe supporting three thermosensors to monitor the temperature during the therapeutic phase. The connecting box allows for the integration of the subassemblies, such as the fiberoptic thermometry system, the microwave delivery system, the cooling system, and the rectal probe.

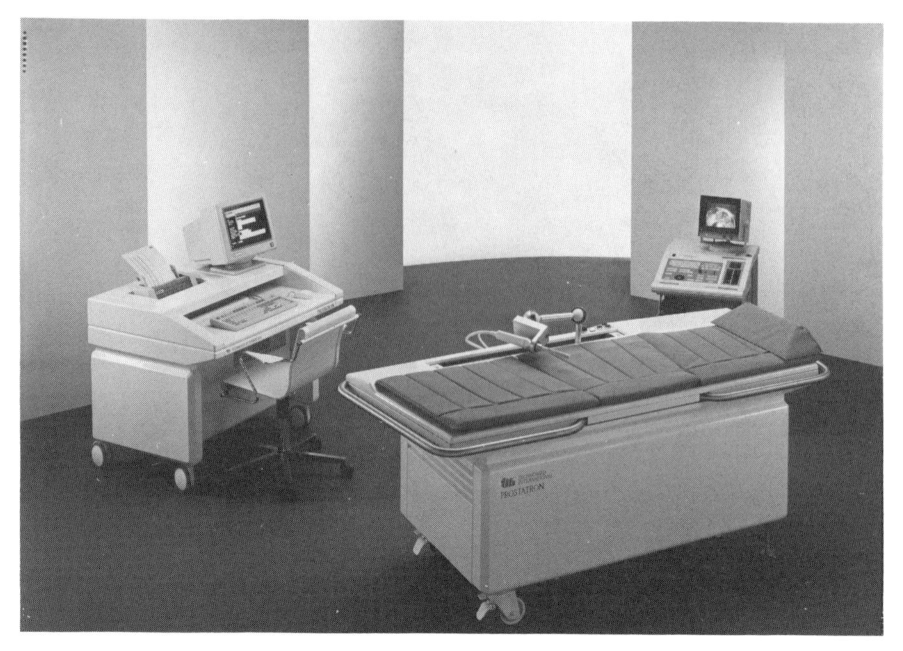

FIGURE 3.7 The Prostatron (with permission of Technomed Medical Systems).

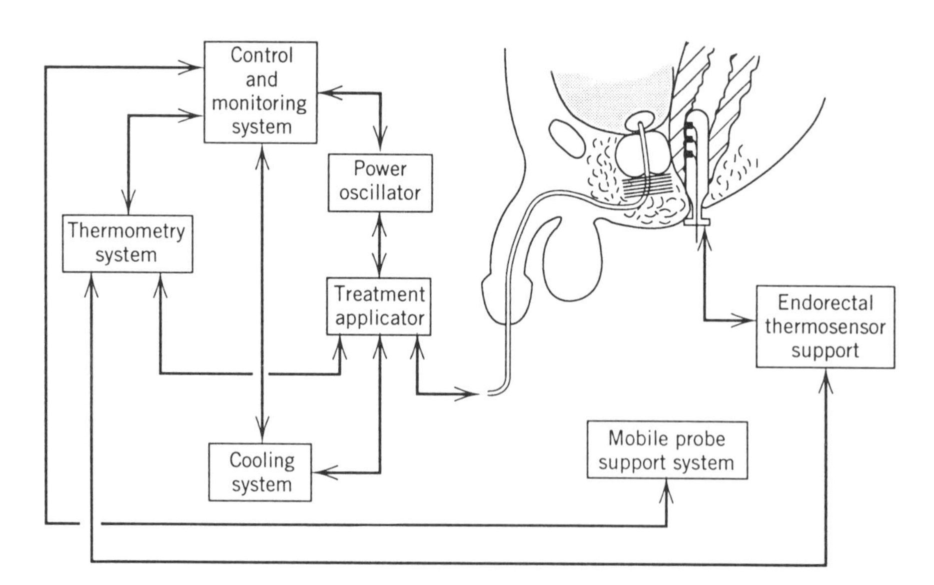

FIGURE 3.8 Prostatron treatment functional diagram (with permission of Technomed Medical Systems).

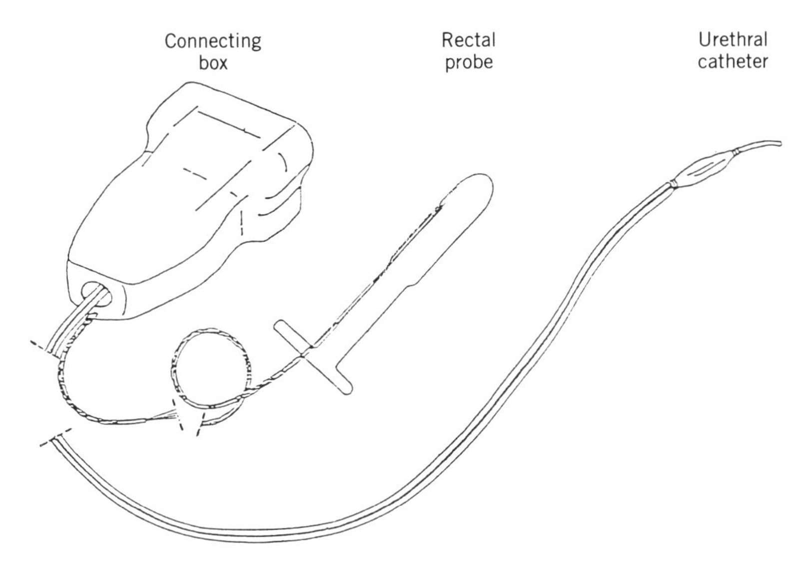

FIGURE 3.9 Schematic representation of the treatment catheter (with permission of Technomed Medical Systems).

A fiberoptic thermometry system was chosen originally for safety reasons, because of its specific feature of noninterferance with microwaves. This aspect is of key consideration, especially when working at high temperatures. Using the Prostatron system, scientists at Charing Cross Hospital, London, calculated and later verified a temperature profile (Figs. 3.11, 3.12). Figures 3.13 and 3.14 depict prostatic heating with cooling (prostate lateral view) at two frequencies: (1) 915 MHz and (2) 1296 MHz, respectively. These profiles underline the usefulness of the coolant circulating inside the catheter. The 915-MHz frequency was abandoned at an early stage because of the finding of a significant "heating tail," especially when working with high-temperature protocols, using this frequency, thus overheating the external sphincter zone. Cases of swelling of the scrotum, incontinence, or destruction of the cavernous body of the penis have been sporadically reported following treatment with different equipment working at 915 MHz. The 1296-MHz frequency was subsequently proposed in order to confine and limit the heating pattern to the very transition zone of the gland, at the origin of BPH obstruction, and was eventually chosen for clinical evaluation. Clinical trials performed in various centers in Europe and Japan,[5–10] led to a unanimous recommendation of 1296 MHz for the Prostatron system. This frequency allows the use of power at such levels that prostatic tissue could be selectively destroyed without risking unwanted damage by overheating the periprostatic area.

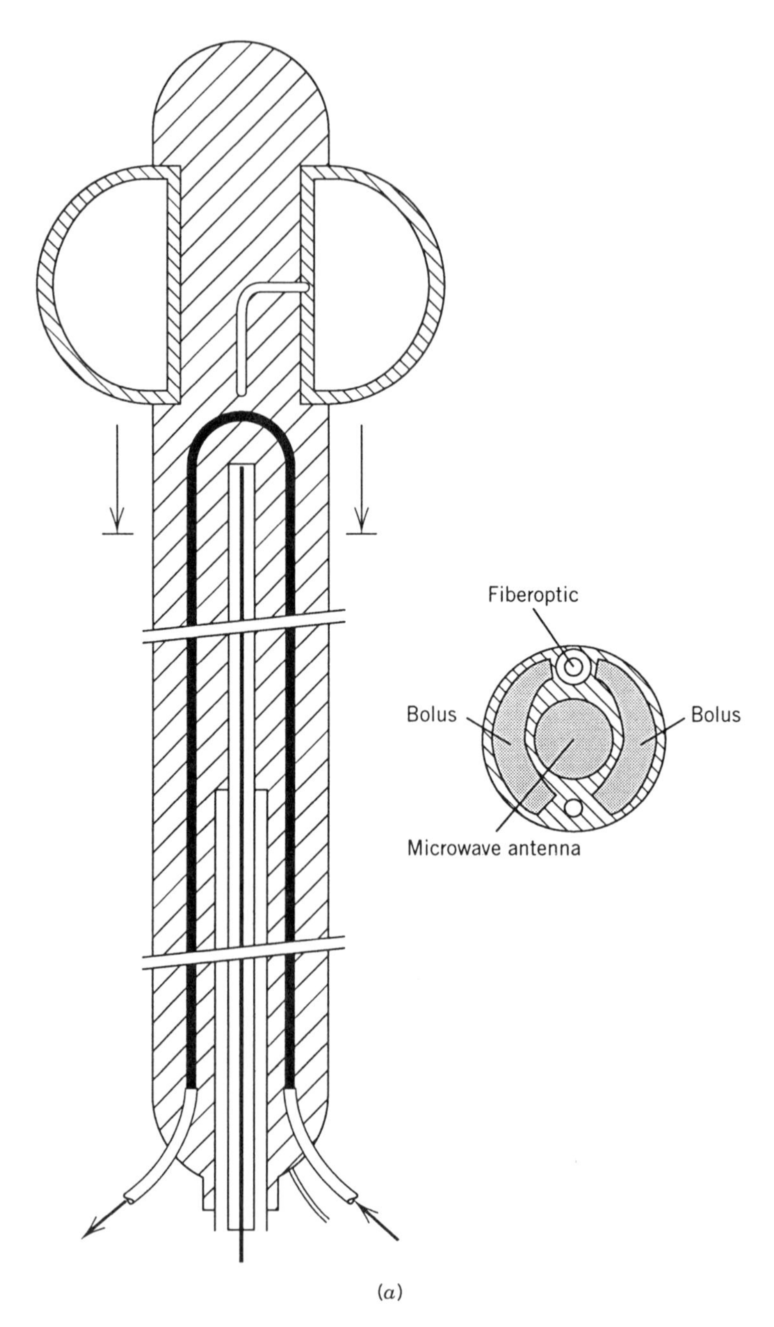

FIGURE 3.10 (*a*) The catheter; (*b*) treatment applicator tip (with permission of Technomed Medical Systems).

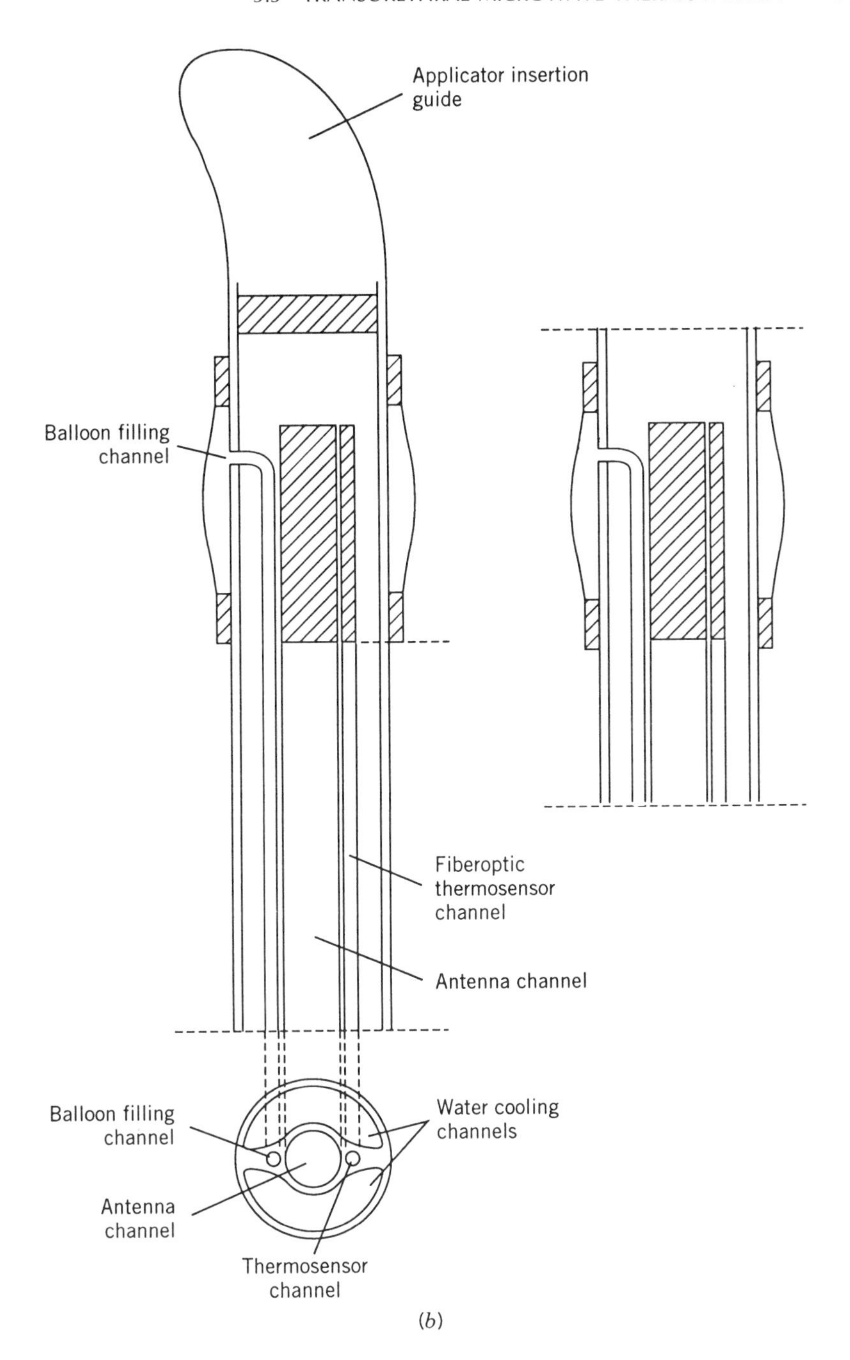

(*b*)

FIGURE 3.10 (*Continued*).

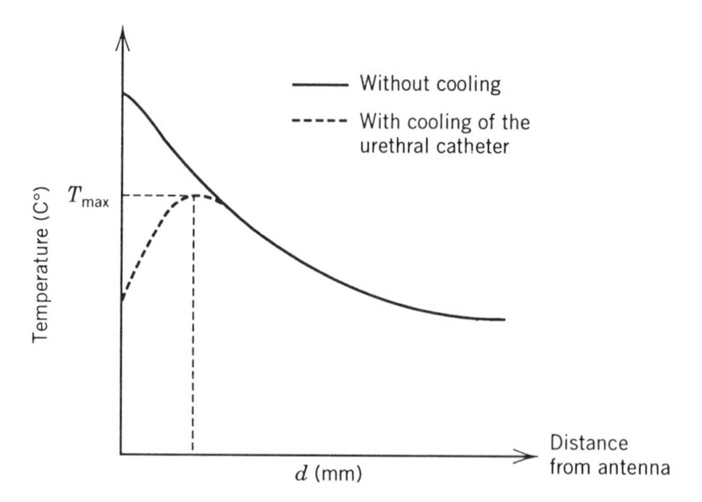

FIGURE 3.11 Temperature profiles obtained during the propagation of microwaves in the prostate (theoretical curves) (with permission of Technomed Medical Systems).

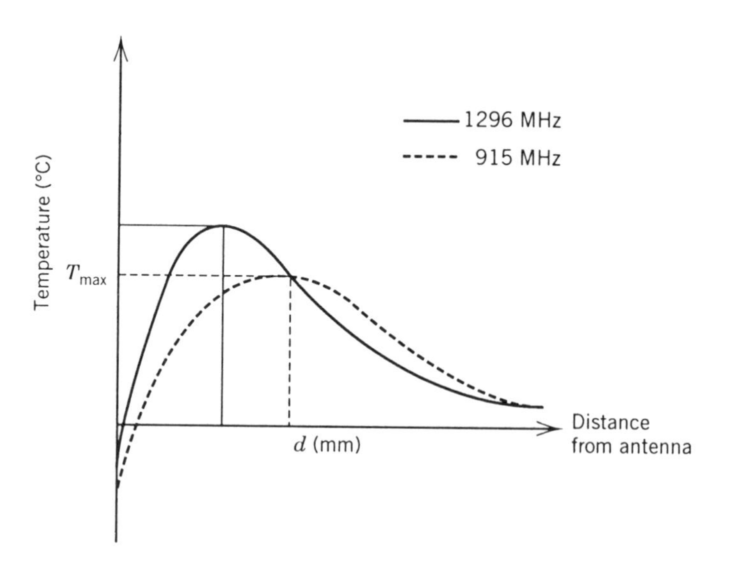

FIGURE 3.12 Temperature profiles obtained during the propagation of microwaves in the prostate using the Prostatron catheter at 915 and 1296 MHz (theoretical curves) (with permission of Technomed Medical Systems).

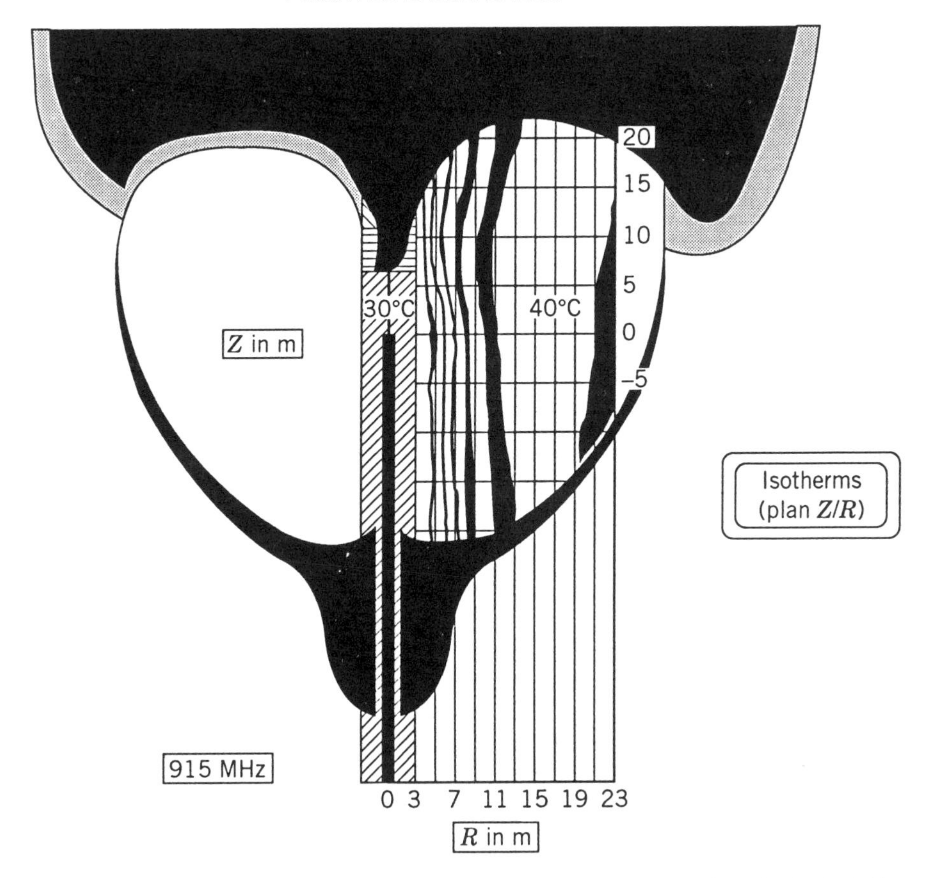

FIGURE 3.13 Urethral heating with cooling at 915 MHz (prostate lateral view) (with permission of Technomed Medical Systems).

3.4 FUTURE TECHNOLOGY: TISSUE ABLATION BY MEANS OF SEMICONDUCTOR LASER ENERGY (TUSLA)

3.4.1 Semiconductor Laser Background[12,13]

The high efficiency of semiconductor lasers is a result of improvements over the decade in material quality, processing, and epitaxial design. Low threshold current densities and high conversion of electrical to optical power in AlGaAs semiconductor lasers are the result of epitaxial structures with graded-index (GRIN) waveguide layers to provide optical confinement with a single-quantum-well (SQW) active layer for electrical confinement [separate confinement

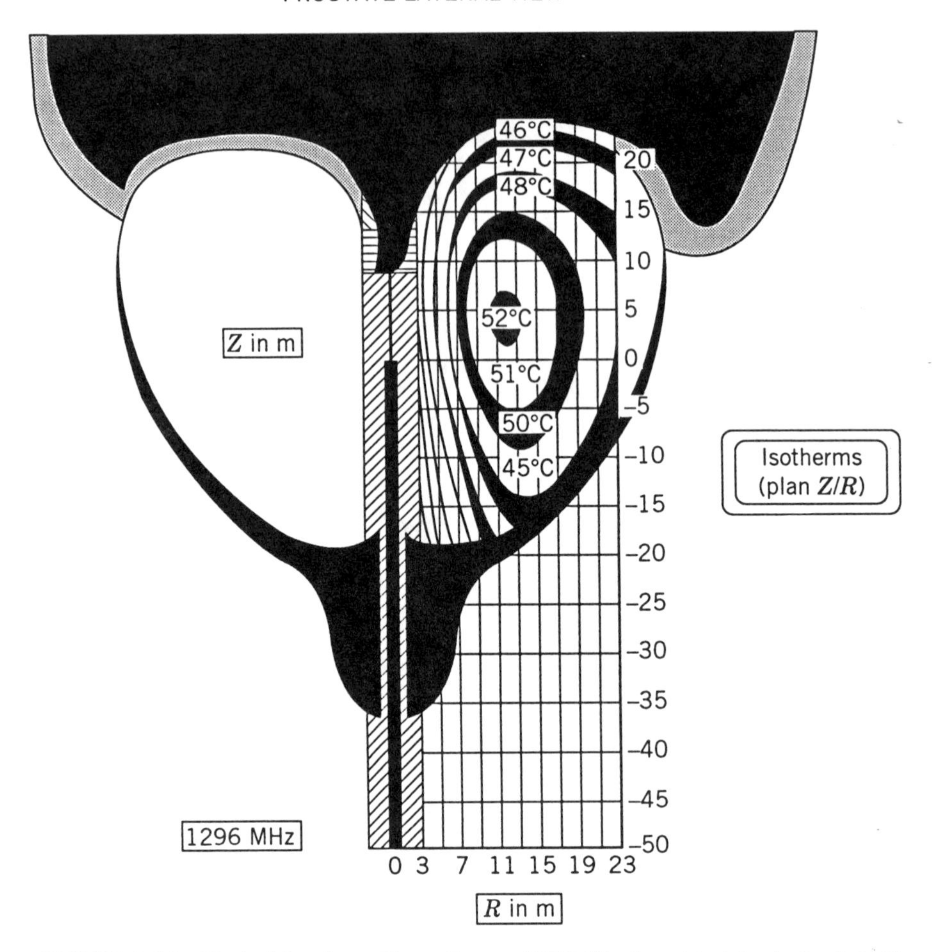

FIGURE 3.14 Urethral heating with cooling at 1296 MHz (prostate lateral view) (with permission of Technomed Medical Systems).

heterostructure (SCH)]. This structure is often referred to as a "GRINSCH–SQW." The epitaxial layers in the laser structure require enough impurity doping for a low overall series resistance while minimizing free-carrier losses in the optical waveguide region.

Similarly, the emission wavelength can be increased from 0.88 μm to about 1.1 μm by increasing the amount of indium in an InGaAs quantum well embedded in the same (or similar) GRINSCH waveguide mentioned above. The addition of indium to the quantum well changes the lattice constant of the semiconductor, thus introducing strain in the quantum-well epitaxial layer. The slight, controlled

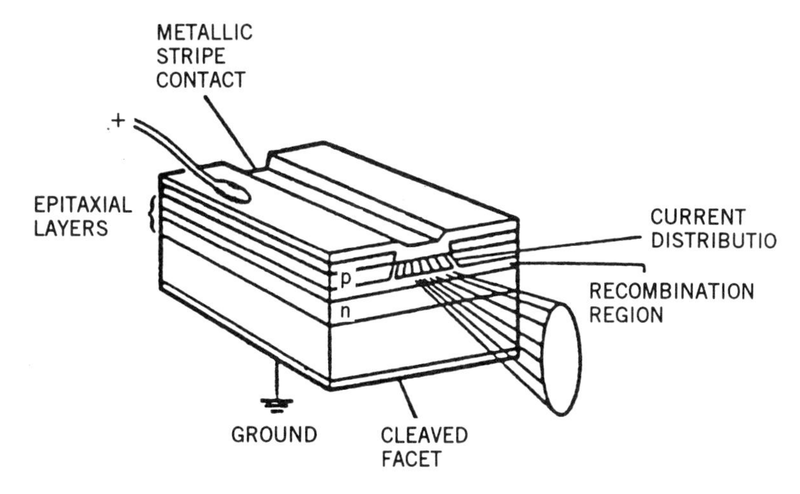

FIGURE 3.15 Typical AlGaAs laser diode features.

strain has positive effects on device performance, improving efficiency, lowering threshold, and improving device lifetime.

Because of the improved performance due to the strained quantum well in the 0.9–1.0-μm-wavelength lasers, the use of strained InAlGaAs quantum wells for emission in the 0.78–0.88-μm region is presently being pursued.

A typical laser diode structure is shown in Figure 3.15. The optimum stripe width is determined by a tradeoff between the optimum use of material and the resulting high output density, efficiency, and thermal dissipation. This has been done both experimentally, by testing the linear arrays with different spacings and stripe widths; and theoretically, by employing a detailed thermal analysis.

3.4.2 Laser Fabrication

The mesa structure shown in Figure 3.16*a*, *b* is fabricated by means of photolithography, chemical etching, and silicon nitride deposition. The slots formed provide optical, as well as electrical, isolation between adjacent stripes.[13] After another photolithography step and the removal of the silicon nitride from the contact region, a zinc diffusion step is carried out to minimize contact resistance. After zinc diffusion, photolithography is again utilized followed by electron-beam (e-beam) deposition of Ti–Pt–Au and a metal liftoff. The wafer is then thinned to 75 μm and the *n* contact (Ge–Au–W–Au) is evaporated onto the wafer using e-beam deposition. Just prior to facet coating, the wafer is cleaved into bars that are then placed in a special fixture in an e-beam evaporator. The rear facet of the bars is coated with a six-layer stack composed of alternating layers of Al_2O_3 and silicon. The output facet is coated with a single layer of silicon, the thickness of which is adjusted to give a reflectivity of about 15%.

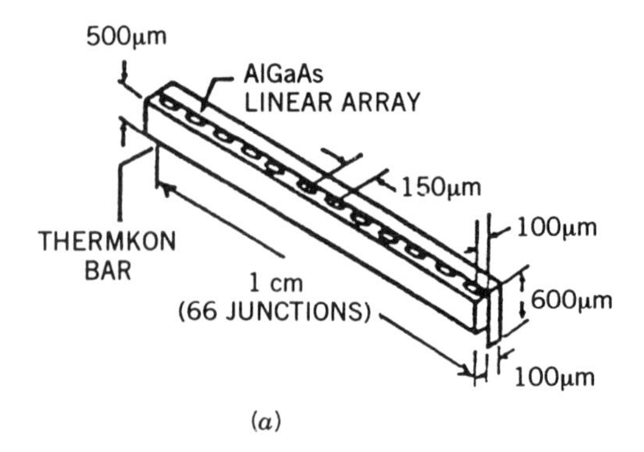

(*a*)

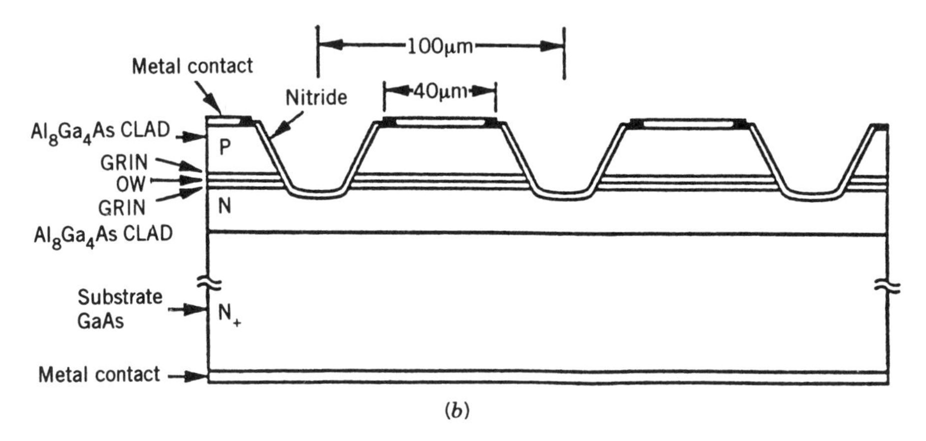

(*b*)

FIGURE 3.16 (*a*) Laser bar; (*b*) Mesa structure.

The bar of laser junctions can then be further cleaved to meet the power requirements of the application. A minimum of 1 W CW per junction has been measured.

The laser-based technique used in the treatment of BPH has recently received a great deal of attention. This treatment involves the introduction into the urethra of a fiberoptic cable, the proximal end of which is connected to a laser energy source. The distal end of the fiberoptic cable is directed toward the prostate. The laser is pulsed (or CW), and the resulting high-energy light traverses the fiberoptic cable and exits, from the distal end, to ablate a portion of the prostate growth. Problems with this system include the difficulty in matching the characteristics of available lasers with the characteristics of fiberoptic cables, and with the absorption characteristics of the tissue to be ablated. In particular, the wavelengths at which fiberoptic cables—having properties suitable for use in catheters—carry light energy with low losses do not necessarily correspond

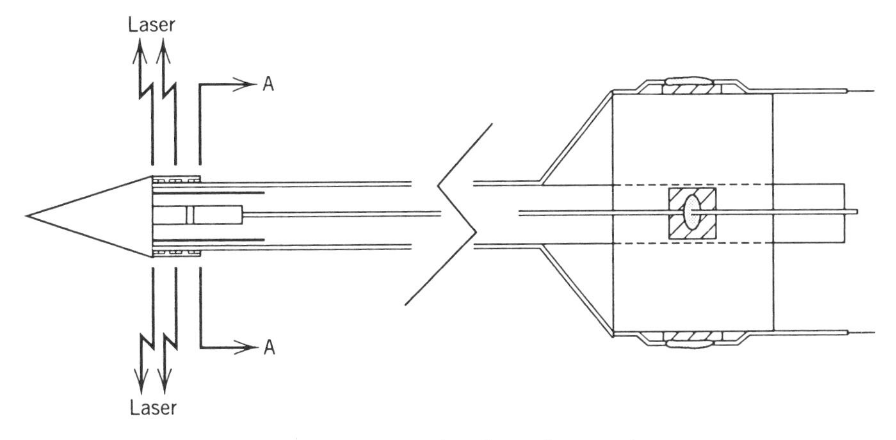

FIGURE 3.17 Semiconductor laser probe.

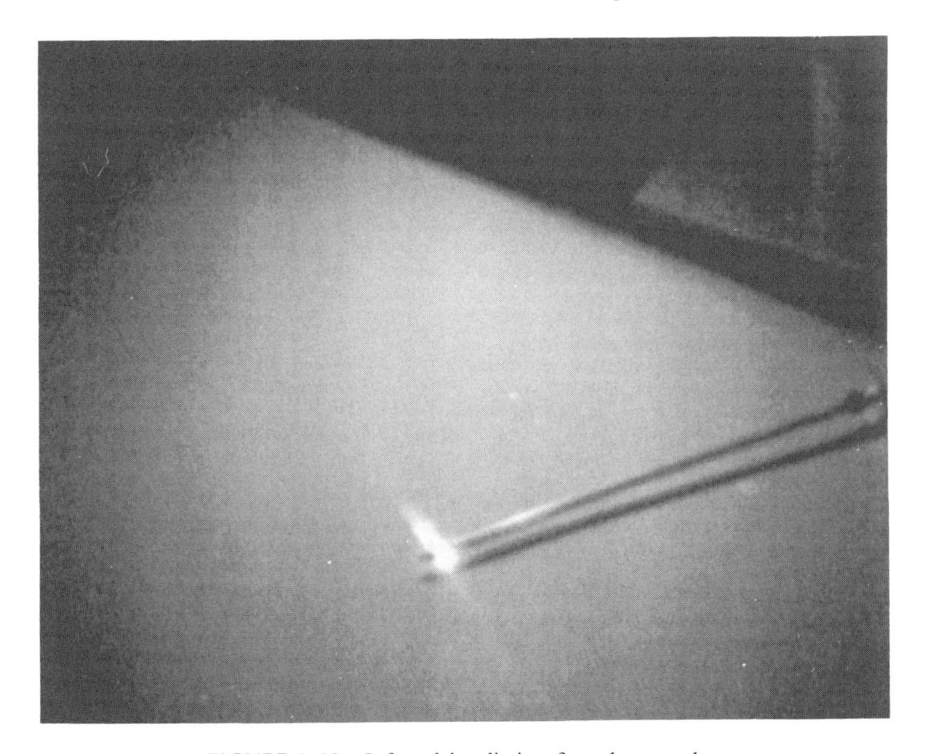

FIGURE 3.18 Infrared irradiation from laser probe.

to the wavelengths at which lasers radiate the maximum energy. Furthermore, the connections (couplings) by which light is coupled from a laser source to the proximal end of the fiberoptic cable of the catheter may introduce losses. In addition, the system's cost reduces the availability of lasers for the treatment of BPH.

In a particular embodiment of U.S. Patent Number 4,998,932, "Catheter with distally located integrated circuit radiation generator,"[12] and of a patent disclosure entitled "Medical probe apparatus with laser and/or microwave monolithic integrated circuit" (Vidamed, Inc.) one or more lasers are positioned on one electric conductor and spaced apart from the others around the periphery of the probe as shown in Figures 3.17 and 3.18. The second conductor is attached to the top of the laser. If desired, the lasers could emit light of different wavelengths to match the target tissue absorption characteristics more closely.

Acknowledgments

Much of the information offered in this chapter has been provided by Mr. Stuart Edwards, President, and Mr. Hugh Sharkey, Vice President, of VidaMed, Inc., Menlo Park, California. The chapter also includes material provided by Mr. Andre Cohen, past President, and Mr. Eric Poincelet, Director of Marketing, of Technomed Medical Systems, Vaux en Velin, France.

The authors gratefully acknowledge the contributions of both VidaMed and Technomed and wish to thank Mr. Sharkey and Mr. Poincelet for reviewing this chapter before publication.

An expression of thanks is also included for Daniella Rosen's support in typing and revising the manuscript.

REFERENCES

1. Goldwasser B, Ramon J, Engelberg S, et al: Transurethral needle ablation (TUNA) of the prostate using low-level radiofrequency energy: An animal experimental study. *Eur Urol* 24:400, 1993.
2. Ramon J, Goldwasser B, Shenfeld O, et al: Needle ablation using radiofrequency current as a treatment for benign prostatic hyperplasia: Experimental results in ex vivo human prostate. *Eur Urol* 24:406, 1993.
3. Rasor JS, Zlotta AR, Edwards SD, et al: Transurethral needle ablation (TUNA): Thermal gradient mapping and comparison of lesion size in a tissue model and in patients with benign prostatic hyperplasia. *Eur Urol* 24:411, 1993.
4. Schulman CC, Zlotta AR, Rasor JS, et al: Transurethral needle ablation (TUNA): Safety, feasibility, and tolerance of a new office procedure for treatment of benign prostatic hyperplasia. *Eur Urol* 24:415, 1993.
5. Devonec M, Ogden C, Perrin P, et al: Clinical response to transurethral microwave thermotherapy is thermal dose dependent. *Eur Urol* 23:267, 1993.
6. Laduc R, Bloem FAG, Debruyne FMJ: Transurethral microwave thermotherapy in symptomatic benign prostatic hyperplasia. *Eur Urol* 23:275, 1993.
7. Van Cauwelaert RR, Castillo OC, Aquirre CA, et al: Transurethral microwave thermotherapy for treatment of benign prostatic hyperplasia: Preliminary experience. *Eur Urol* 23:282, 1993.

8. Tubaro A, Paradiso Galatioto G, Trucchi A, et al: Transurethral microwave thermotherapy in the treatment of symptomatic benign prostatic hyperplasia. *Eur Urol* 23:285, 1993.
9. Dahlstrand C, Geirsson G, Fall M, et al: Transurethral microwave thermotherapy versus transurethral resection for benign prostatic hyperplasia: Preliminary results of a randomized study. *Eur Urol* 23:292, 1993.
10. Perachino M, Bozzo W, Puppo P, et al: *Eur Urol* 23:299, 1993.
11. Cathaud M, Poincelet E: Transurethral microwave thermotherapy, 1296 MHz versus 915 MHz. Technomed Medical Systems, company communication, Feb. 1992.
12. Rosen A, Rosen H: Catheter with distally located integrated circuit radiation generator. U.S. Pat. 4,998,932, March 12, 1991.
13. Rosen A, Zutavern F (eds): *High Power Optically Activated Solid State Switches.* Artech House, Boston, 1993.

APPENDIX: PROSTATRON PUBLICATIONS*

1994

Blute M, Patterson D, Segura J, et al: Transurethral microwave thermotherapy versus sham: A prospective double-blind randomized study. *J Urol* 151:752, 1994.

Carter S, Ogden C: Intraprostatic temperature v. clinical outcome in T.U.M.T. Is the response heat-dose dependent? *J Urol* 151:756, 1994.

Choi N, Soh SH, Yoon TH, et al: Clinical experience with transurethral microwave thermotherapy for chronic nonbacterial prostatitis and prostatodynia. *J Endourol* 8:61, 1994.

Cockett AT, Aso Y, Denis L, et al: Recommendations of the International Consensus Committee. 2nd International Consultation on BPH by World Health Organisation, Paris, June 27–30, 1993. *Proceedings*, Ed. SCI, Paris, 1994, pp.553–564.

Dahlstrand C, Geirsson G, Walden M, et al: Two-year follow-up of transurethral microwave thermotherapy versus transurethral resection for benign prostatic obstruction. *J Urol* 151:753, 1994.

Dahlstrand C, Walden M, Pettersson S, et al: Pressure flow studies in BPH patients: A useful tool for treatment selection. *J Urol* 151:1123, 1994.

Devonec M, Houdelette P, Colombeau P, et al: A multicenter study of sham versus thermotherapy in benign prostatic hypertrophy. *J Urol* 151:751, 1994.

Höfner K, Kramer G, Kuczyk M, et al: The changes of outflow obstruction and bladder power utilisation after transurethral microwave thermotherapy. *J Urol* 151:757 1994.

Jonas U, Ogden C, De la Rosette J, et al: Symptom score and flow rate: Independent parameters in clinical response to microwave thermotherapy. *J Urol* 151:394, 1994.

*Courtesy of Technomed Medical Systems.

Nickel C, Sorensen R: Randomized double blinded placebo controlled group to evaluate the effect of transurethral microwave thermotherapy in patients with complaints of prostatitis. *J Urol* 151:708, 1994.

Robinette M, Mahoney J, Buckley R, et al: Results of transurethral microwave thermotherapy (TUMT) in patients with symptomatic BPH and urinary retention. *J Urol* 151:754, 1994.

Tubaro A, Ogden C, De la Rosette J, et al: The prediction of clinical outcome from thermotherapy by pressure-flow study. Results of a European multicenter study. *J Urol* 151:758, 1994.

Tazaki H, Deguchi N, Baba S, et al: Objective evaluation of microwave thermotherapy, laser ablation and high-intensity focused ultrasound for benign prostatic hyperplasia (BPH). *J Urol* 151:759, 1994.

1993

Berg C, Choi N, Colombeau P, et al: Responders versus non-responders to thermotherapy in BPH: A multicenter retrospective analysis of patient and treatment profiles. *J Urol* 150:149, 1993.

Blute M: Studies show that transurethral microwave thermotherapy for BPH is safe. *AUA Today* 6:1, 1993.

Blute M, Tomera K, Hellerstein D, et al: Transurethral microwave thermotherapy: An alternative of benign prostatic hypertrophy. One year clinical results of the Prostatron Study Group. *J Urol* 150:143, 1993.

Blute M, Tomera K, Hellerstein D, et al: Transurethral microwave thermotherapy for management of benign prostatic hyperplasia: Results of the United States Prostatron Cooperative Study. *J Urol* 150:1591, 1993.

Carter S, Patel A, Reddy P, et al: *Single-session Transurethral Microwave Thermotherapy for the Treatment of Benign Prostatic Obstruction. Therapeutic Alternatives in the Management of Benign Prostatic Hyperplasia.* Thieme Medical Publishers, New York, 1993, pp. 127–133.

Dahlstrand C, Fall M, Geirsson G, et al: Transurethral microwave thermotherapy versus transurethral resection for benign prostatic hyperplasia: Results of a randomized study. *J Urol* 150:148, 1993.

Dahlstrand C, Geirsson G, Fall M, et al: Transurethral microwave thermotherapy versus transurethral resection for benign prostatic hyperplasia: Preliminary results of a randomized study. *E Urol* 23:292, 1993.

Debruyne F, Bloem F, De la Rosette J, et al: Transurethral microwave thermotherapy (TUMT) in benign prostatic hyperplasia: Placebo vs TUMT. *J Urol* 150:146, 1993.

Devonec M, Fendler JP, Nasser M, et al: The clinical response to transurethral microwave thermotherapy is dose-dependent: From thermo-coagulation to thermo-ablation. *J Urol* 150:142, 1993.

Devonec M, Ogden C, Perrin P, et al: Clinical response to transurethral microwave thermotherapy is thermal dose dependent. *Eur Urol* 23:267, 1993.

Devonec M, Tomera K, Perrin P: Review: Transurethral microwave thermotherapy in benign hyperplasia. *J Endourol* 7:255, 1993.

Harzmann R, Weckermann D: Benign prostatic hyperplasia: Newer approaches. *Curr Opinion Urol* 3:10, 1993.

Hoefner K, Gruenwald Y, Truss M, et al: Influence of transurethral microwave thermotherapy on outflow obstruction assessed by computerized pressure flow analysis. *J Urol* 150:1008, 1993.

Homma Y, Aso Y: Transurethral microwave thermotherapy in benign hyperplasia: A 2-year follow-up study. *J Endourol* 7:261, 1993.

Laduc R, Bloem FAG, Debruyne F: Transurethral microwave thermotherapy in symptomatic benign hyperplasia. *Eur Urol* 23:275, 1993.

Ogden C, Reddy P, Johnson H, et al: Sham vs TUMT. A randomized study with cross over. *J Urol* 150:147, 1993.

Ogden C, Reddy P, Johnson H, et al: A prospective study of sham versus transurethral microwave thermotherapy in symptomatic prostatic bladder outflow obstruction. *Lancet*, 341:14, 1993.

Perrachino M, Bozzo W, Puppo P, et al: Does transurethral thermotherapy induce a long-term alpha blockade? An immunohistochemical study. *Eur Urol* 23:299, 1993.

Sorensen R, McGarragle M, Chen R, et al: Transurethral microwave thermotherapy (TUMT): A non-surgical alternative for the treatment of benign prostatic hyperplasia (BPH). *J Urol* 150:144, 1993.

Sorensen R, McGarragle P, Grignon D, et al: Transurethral microwave thermotherapy (TUMT) using the prostatron: A histopathological evaluation of the thermal effects on carcinoma of the prostate. *J Urol* 150:73, 1993.

Steven A, Kaplan S, Olsson CA: State of the art: Microwave therapy in the management of men with benign prostatic hyperplasia: Current status. *J Urol* 150:1597, 1993.

Tazaki H, Baba S, Deguchi N, et al: Transurethral microwave thermotherapy in BPH; long term follow-up results. *J Urol* 150:145, 1993.

Tomera K, Selmy G, Corcos J, et al: Effects of transurethral microwave thermotherapy on erectile performance. *J Urol* 150:254, 1993.

Trucchi A, Begani A, Trucchi E, et al: Transurethral microwave thermotherapy and benign prostatic obstruction: The urodynamic standpoint. *J Urol* 150:576, 1993.

Tubaro A, Paradiso Galatioto G, Trucchi A, et al: Transurethral microwave thermotherapy in the treatment of symptomatic benign prostatic hyperplasia. *Eur Urol* 23:285, 1993.

Van Cauwelaert R, Castillo OC, Aquirre CA, et al: Transurethral microwave thermotherapy for the treatment of benign prostatic hyperplasia: Preliminary experience. *Eur Urol* 23:282, 1993.

1992

Beaven AJ, Ogden C, Reddy P, et al: The treatment of urinary retention with transurethral microwave thermotherapy. *Urology* (Monduzzi Editore), 1992.

Blute M: Transurethral microwave thermotherapy for benign prostatic hypertrophy. *Mediguide to Urol* 1992.

Blute M, Tomera K, Hellerstein D, et al: Early results of TUMT for benign prostatic obstruction: Mayo foundation experience. *Mayo Clin Prog* 67:417, 1992.

Carter S, Ogden C, Patel A, et al: Long-term results of transurethral microwave thermotherapy for benign prostatic obstruction. *Urology* (Monduzzi Editore), 1992.

Carter S, Patel A, Beaven T, et al: Experience of transurethral microwave thermotherapy for the treatment of benign prostatic obstruction. In Fitzpatrick JM (ed): *Non Surgical Treatment of BPH.* Churchill Livingstone, 1992, pp. 207–224.

Devonec M, Berger N, Cathaud M, et al: Historical development of transurethral microwave thermotherapy (TUMT): Short and long-term results in benign prostatic hyperplasia. In Fitzpatrick JM (ed): *Non Surgical Treatment of BPH.* Churchill Livingstone, 1992, pp. 187–206.

Devonec M, Berger N, Fendler JP, et al: Role of natural and artificial thermoregulation on thermotherapy lesion in benign prostatic hypertrophy. *J Endourol* 6(suppl):104, 1992.

Devonec M, Berger N, Fendler JP, et al: Transurethral thermotherapy of benign prostatic hypertrophy: Clinical results. *J Endourol* 6(suppl): 104, 1992.

Devonec M, Cathaud M, Berger N, et al: Transurethral thermotherapy of benign prostatic hypertrophy: Clinical results. *J Urol* 149:147, 1992, 305A.

Devonec M, Cathaud M, Bringeon G, et al: Transurethral microwave thermotherapy of the prostate: Principles and early clinical results. In Jackse G et al (eds): *Benign Prostatic Hyperplasia.* Springer Verlag, 1992, pp. 143–151.

Devonec M, Cathaud M, Dutrieux-Berger N, et al: Histology of thermal injury induced by transurethral microwave thermotherapy of benign prostatic hyperplasia. In Bichler, Strohmaier, Wilbert, (eds): *Hyperthermia of the Prostate—State of the Art.* Verlagsgruppe, 1992, pp. 98–103.

Devonec M, Tomera K, Perrin P: Transurethral microwave thermotherapy (TUMT) in benign prostatic hyperplasia. Synopsis of 10th World Congress on Endourology and ESWL, Singapore, Sept. 1992.

Devonec M, Tomera K, Perrin P: Transurethral microwave thermotherapy (TUMT) in benign prostatic hyperplasia. *Urology* (Monduzzi Editore), 239, 1992.

Devonec M, Tomera K, Perrin P: Transurethral microwave thermotherapy (TUMT). In Stamey TA (ed): *Monographs in Urology.* Vol. 13, pp. 77–95, 1992.

Ersev D, Ilker Y, Simsek F, et al: Preliminary results of transurethral microwave thermotherapy in the treatment of benign prostatic hyperplasia. *Eur Urol* 21:187, 1992.

Harzmann R, Weckermann D: New technical alternatives for the treatment of symptomatic BPH. *Urologe A* 31:150, 1992.

Homma Y, Aso Y: Transurethral microwave thermotherapy for benign prostatic hyperplasia. *Jpn J Endourol ESWL*, 5:5, 1992.

Ogden C, Johnson H, Patel A, et al: A single blind prospective study of SHAM versus TUMT in symptomatic prostatic bladder outflow obstruction. Initial results. *Urology* (Monduzzi Editore), 1992.

SIU Symposium Report Nov. 2, 1991: Second International Symposium on Prostatron Thermotherapy. *Eur Urol Today*, June 1992.

Smith P, Chaussy C, Conort P, et al: New technology in management of benign prostatic hyperplasia. *Report of the committee on Other Non Medical Treatments. The International Consultation on BPH by World Health Organization, Paris, June 26–27, 1991.* Cockett ATK, et al (eds). Ed. SCI, Paris, 1992, pp. 223–257.

Tomera K, Hellerstein D: Transurethral microwave thermotherapy (TUMT). *Contemp Urol* 4: 1992.

1991

AUA Symposium Report June 2, 1991: Prostatron Thermotherapy. *Eur Urol Today*, Sept. 1991.

Berger N, Devonec M, Bringeon G, et al: Histologie des lesions induites par thermothérapie micro-ondes sur l'hypertrophie benigne de la prostate: Aspects à court et long terme. *Progrès Urologie* 1:B11, 1991.

Blute M, Lewis R: Local microwave thermotherapy as a treatment alternative for benign prostatic hyperplasia. *J Androl*, 1991.

Carter S, Patel A, Reddy P, et al: Single session transurethral microwave thermotherapy for treatment of benign prostatic obstruction. *J Endourol* 5:137, 1991.

Devonec M: Long term results with TUMT in patients with benign prostatic hypertrophy. *J Urol* 148:145, 1991.

Devonec M: Short and long term histological effects of TUMT on benign prostatic hypertrophy. *J Urol* 148:145, 1991.

Devonec M: Long term results with TUMT in patients with benign prostatic hypertrophy. *J Endourol* 5(Suppl):100, 1991.

Devonec M: Short and long tern histological effects of TUMT on benign prostatic hypertrophy. *J Endourol* 5(Suppl):101, 1991.

Devonec M, Berger N, Perrin P: Transurethral microwave heating of the prostate—or from hyperthermia to thermotherapy. *J Endourol* 5:129, 1991.

Devonec M, Bringeon G, Dujardin T, et al: Traitement de l'hypertrophie bénigne de la prostate (HBP) par radiations non ionisantes: Bases physiques et biologiques. Comparison des caractéristiques techniques des machines. *Progrés Urologie* 1:B10, 1991.

Devonec M, Bringeon G, Dujardin T, et al: Résultats à court et long terme de la thermothérapie micro-ondes chez les patients porteurs d'une hypertrophie Bénigne prostatique (HBP). *Progrés Urologie* 1:B12, 1991.

Harzmann R, Weckermann D: Local microwave hyperthermia in benign hyperplasia of the prostate gland. Progress or Placebo? *Deutches Ärtzblatt* 11:859, 1991.

Perez-Castro E, Carbonero M, Mancebo JM, et al: Treatment of benign prostatic hypertrophy (BPH) by transurethral thermotherapy. *Archivos Espanoles Urologia* 5:637, 1991.

1990

Carter S, Patel A, Ramsay J, et al: Objective clinical results of transurethral microwave thermotherapy for benign prostatic hypertrophy. *J Endourol* 4(Suppl):134, 1990.

Carter S, Patel A, Ramsay J, et al: The physics of transurethral microwave thermotherapy. *J Endourol* 4(Suppl):133, 1990.

Devonec M, Cathaud M, Carter S, et al: Transurethral microwave therapy (TUMT) in patients with benign prostatic hypertrophy. *J Endourol* 4(Suppl):135, 1990.

Devonec M, Cathaud M, Carter S, et al: Transurethral microwave application: Temperature sensation and thermokinetics of the human prostate. *J Urol* 143:414A, 1990.

CHAPTER FOUR

Localized Heating of Deep-Seated Tissues Using Microwave Balloon Catheters

FRED STERZER, Ph.D., *MMTC, Inc., Princeton, NJ*

4.1 INTRODUCTION

Microwave balloon catheters are catheters designed for the simultaneous ballooning and heating of deep-seated tissues. Such catheters can heat larger volumes of tissues to more uniform temperature distributions than can conventional microwave catheters. Furthermore, by simultaneously heating and applying pressure to tissues, microwave balloon catheters can produce biological stents in vessels, and can liquify fat in fatty plaques and squeeze this fat from the plaque. For these reasons, microwave balloon catheters are expected to find applications in angioplasty, hyperthermia treatment of cancer, and the treatment of benign prostatic hypertrophy. Figure 4.1 shows a schematic drawing of a basic microwave balloon catheter. The catheter consists a thin plastic tubing that is terminated by an inflatable balloon. The tubing has two lumens: one for inflating the balloon, and a second one for inserting a miniature coaxial microwave cable that is terminated in a microwave antenna. Additional lumens for cooling liquids, thermocouples, and suctioning of plaque are often added.

The tissues surrounding the balloon are heated with microwave power that is fed into the coaxial cable and is broadcast into the tissues by the antenna. The microwave frequencies most commonly used are 915 and 2450 MHz. Both of

New Frontiers in Medical Device Technology, Edited by Arye Rosen and Harel Rosen
ISBN 0-471-59189-0

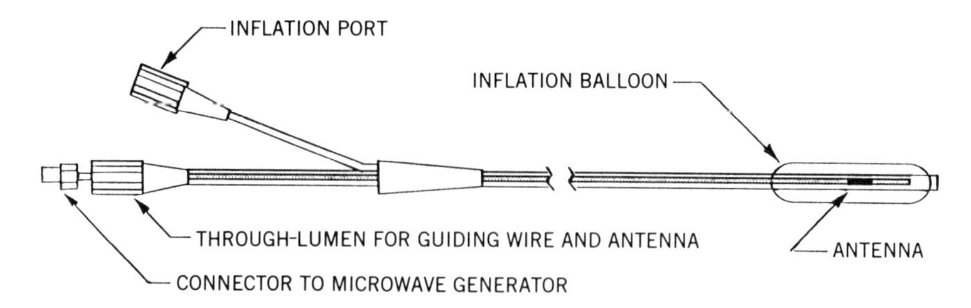

FIGURE 4.1 Schematic diagram of a basic balloon catheter.

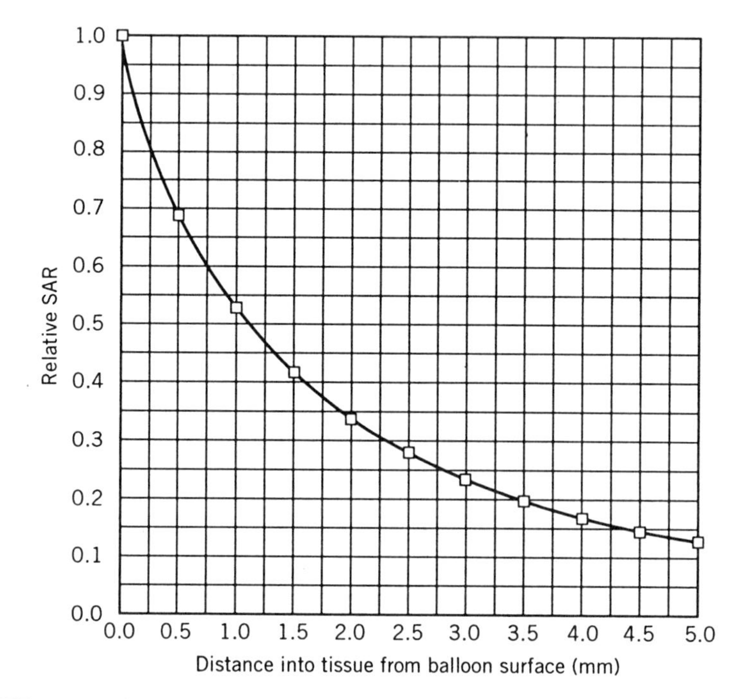

FIGURE 4.2 Calculated relative specific absorption rate (SAR) versus radial distance from the surface of an inflated microwave balloon catheter. The calculations assume that the inflated balloon has a diameter of 3 mm, that the balloon is surrounded by tissues with a water content similar to that of muscle, that a microwave frequency of 2450 MHz is used to heat the tissues, and that the antenna is a half-wave dipole giving rise to a modified-cosine distribution of the radiated field.

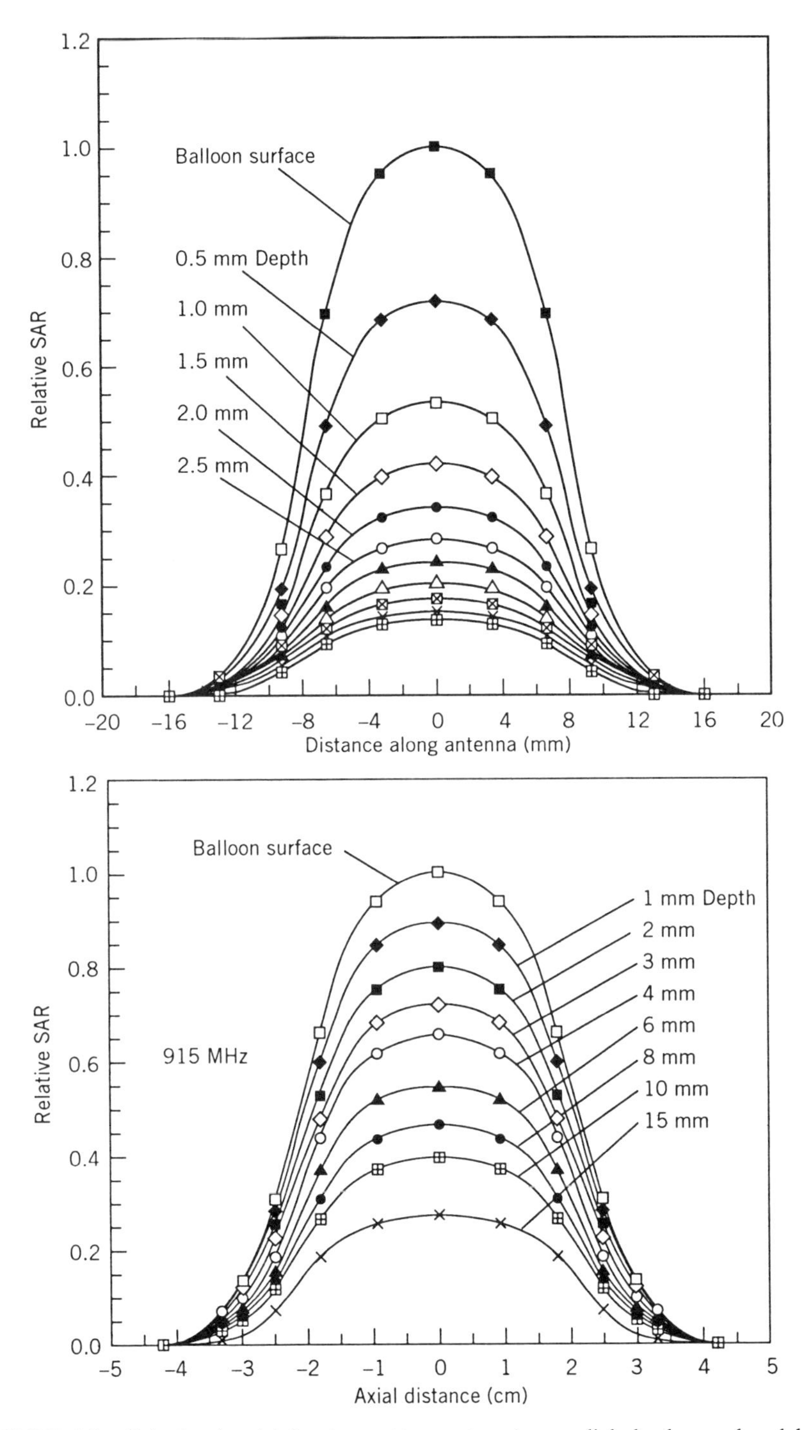

FIGURE 4.3 Calculated axial heating patterns at various radial depths produced by a microwave balloon catheter. The assumptions stated in Figure 4.2 apply, except that Figure 4.3*b* was calculated for an inflated balloon diameter of 1.2cm and a frequency of 915MHz.

these frequencies are approved by the Federal Communications Commission for use in medical applications. The antenna may be a whip, gap, helical or other configuration depending mostly on the desired location of the maximum power deposition relative to the location of the antenna.[1] For example, helical antennas tend to have end-fire patterns that heat preferentially at the tip of the antenna, whereas gap antennas tend to maximize heating in the vicinity of the gap.

The amount of heat that is generated in the tissues by the microwave power decreases exponentially with the distance from the surface of the balloon. This is illustrated in Figure 4.2 which is a plot of calculated specific absorption rate, that is, the calculated relative rate at which microwave power is absorbed by the tissues, as a function of radial distance from the surface of the balloon. Microwaves are attenuated as they traverse and heat the tissues and spread with radial distance from the balloon. The larger the diameter of the balloon, the less the effect of spreading of the wavefronts, and the smaller the decrease in heating with distance from the surface of the balloon. Figure 4.3 shows the calculated axial heating patterns for half-wave dipole antennas. Figure 4.4 shows the calculated isothermal heating contours for one of these antennas. Figures 4.2–4.4 were calculated by D.D Mawhinney of MMTC, Inc. using simplifying assumptions about the radiating fields produced by the antenna. For rigorous analysis of the fields produced by antennas suitable for microwave balloon catheters, see, for example, Refs. 2 and 3.

Although calculations of the microwave power that is absorbed by the tissues surrounding a microwave balloon catheter are very useful, they cannot be used to accurately predict actual increases in tissue temperature. This is because tissues are rarely homogeneous, and the increase in tissue temperature depends not only on microwave heating but also on the amount of cooling provided by blood flow, by thermal conduction to adjacent tissues, and by cooling liquids in the balloon. In practice it is therefore almost always necessary to measure tissue temperatures. This is usually done with either thermocouples, thermistors, or fiberoptic thermometers that are incorporated into the microwave balloon catheters or are invasively placed into tissues that are being heated. An alternate approach is to use microwave radiometers that timeshare the antenna with the microwave power source.[4,5]

4.2 MICROWAVE BALLOON ANGIOPLASTY

Arteriosclerosis, or “hardening of the arteries,” is a disease that causes the walls of arteries to thicken and loose their elasticity. Arteriosclerotic vessels cannot carry as much blood as similar healthy vessels. In advanced disease, the channel (or lumen) of the artheriosclerotic vessel can become completely blocked, so that no blood can flow through the vessel.

Balloon catheters are widely used to increase the blood flow in partially occluded atheriosclerotic vessels. A catheter with a deflated balloon at its tip is inserted into the diseased vessel, and the deflated balloon is pushed into the

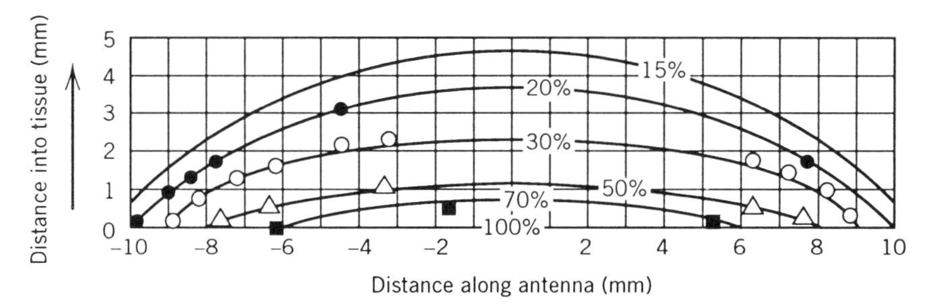

FIGURE 4.4 Calculated isothermal heating contours produced by a microwave balloon catheter. The assumptions stated in Figure 4.2 apply.

narrowed part of the vessel. The balloon is then briefly inflated to expand the narrowed walls.[6]

An important application of balloon catheters is in the treatment of coronary heart disease, where these catheters are used to open partially occluded coronary arteries: arteries that supply blood to the heart muscle. The technical term for this application of balloon catheters is *percutaneous transluminal coronary angioplasty* [PTCA].[7]

PTCA produces less trauma, carries less risk, and is less expensive than open-heart surgery. Because of these advantages, PCTA has become a widely used alternative to open-heart surgery. There are, however, problems with PCTA. In 2–4% of patients undergoing PCTA there are acute problems such as abrupt closure of the treated vessel by tissue flaps, spasms, or blood clots that require emergency bypass surgery. Furthermore, anywhere from 25 to 50% of patients who undergo PCTA need another procedure because the lumen of the treated vessel shrinks again (restenosis) as a result of elastic recoil of the ballooned walls or new plaque formation at the traumatized ballooned sites.

Several new revascularization techniques are being developed to overcome the limitations of conventional PCTA. These new techniques include laser angioplasty (combination of balloon and lasers),[8] directional coronary atherectomy (combination of balloon and miniature rotating cutting device), coronary stents (metallic wire tubes that are implanted at the site of the narrowing), and microwave balloon angioplasty.[9,10]

In microwave balloon angioplasty (MBA) the balloon and the microwave antenna of a microwave balloon catheter of the type shown in Figure 4.1 are positioned in the narrowed part of the coronary vessel to be treated. The narrowed walls are then heated with microwaves and the lumen expanded by inflating the balloon. Microwave heating can be administered prior to and during the inflation of the balloon.

4.2.1 In Vitro Experiments

Figure 4.5 illustrates the effect of simultaneous microwave heating and ballooning on animal vessels. Figure 4.5*a* shows a vessel before heating and ballooning. The vessel is flaccid, and its lumen is partially collapsed. No observable change results if the vessel is heated without ballooning, or ballooned without heating. However, after microwave heating to 45 °C and simultaneous ballooning treated vessels become stiff with wide-open lumens, becoming, in effect, "biological stents" (Fig. 4.5*b, c*).

Figure 4.6 shows the effect of simultaneous heating and pressure on a piece of human fatty arterial plaque. Figure 4.6*a* is a photograph of the plaque when it was heated to 35 °C by conduction from a hot plate while pressure was applied to it by means of a glass microscope slide. While maintaining the pressure on

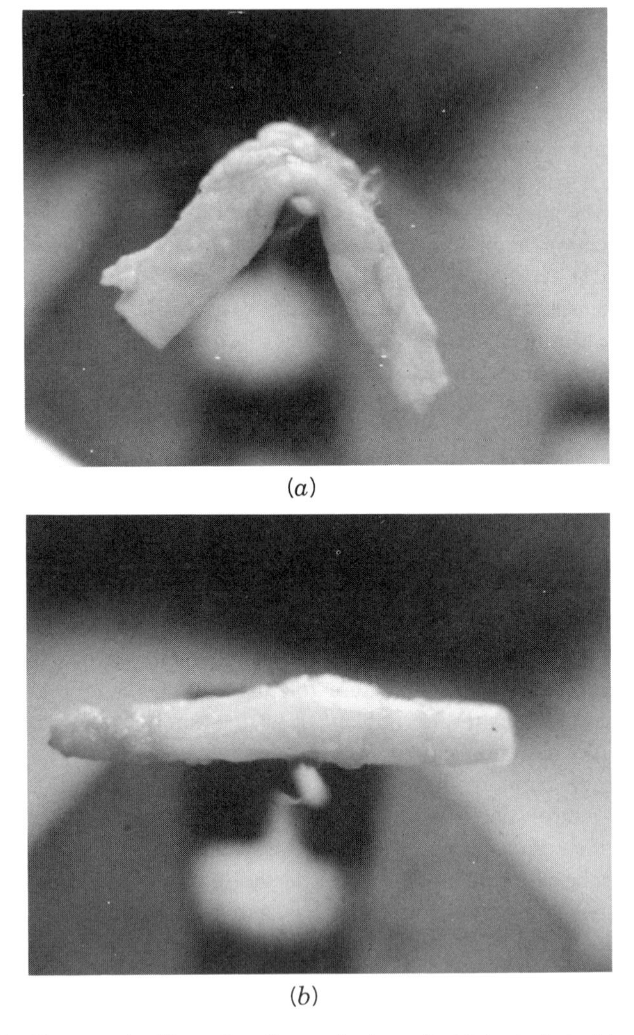

(*a*)

(*b*)

FIGURE 4.5 Photographs illustrating the production of a biological stent by the simultaneous application of ballooning and microwave heating: (*a*) vessel before ballooning and microwave heating; (*b, c*) vessel after ballooning and microwave heating.

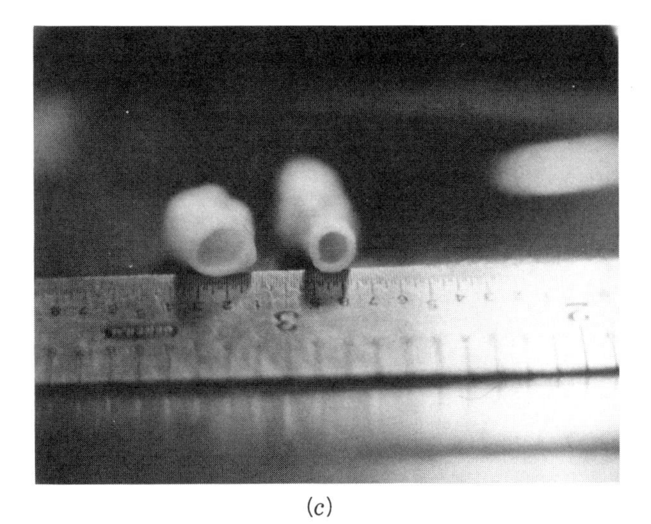

(c)

FIGURE 4.5 (*Continued*).

the plaque, its temperature was first raised to 42 °C (Fig. 4.6*b*) and then to 44 °C (Fig. 4.6*c*). Note that the size of the plaque increased with increasing temperature, in other words, the higher the temperature, the more the piece of plaque was flattened.

When arterial fatty plaque is heated in vitro to 44 °C or higher while pressure is applied to it, not only does the plaque flatten out as shown in Figure 4.6, but in addition liquid fat droplets begin to appear on the surface of the plaque.

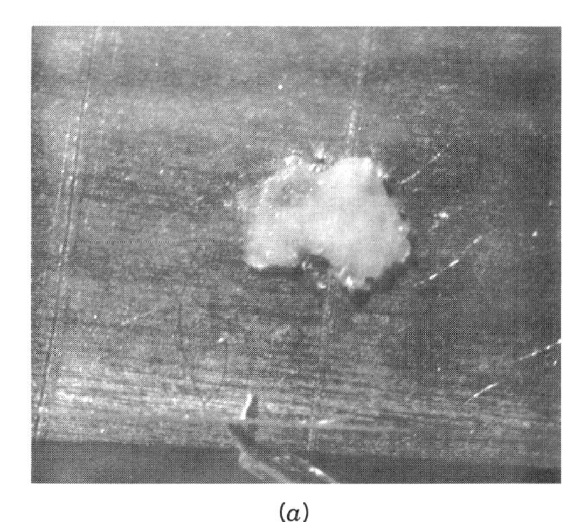

(a)

FIGURE 4.6 Photographs showing the effect of simultaneously applying pressure and heat to fatty plaque from an aorta: (*a*) 35 °C; (*b*) 42 °C; (*c*) 44 °C.

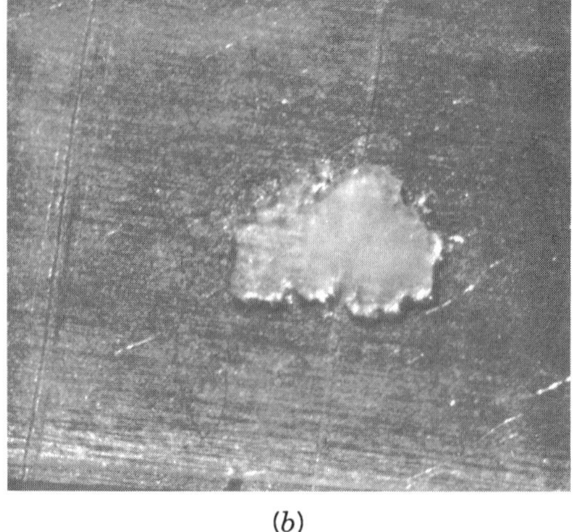

(*b*)

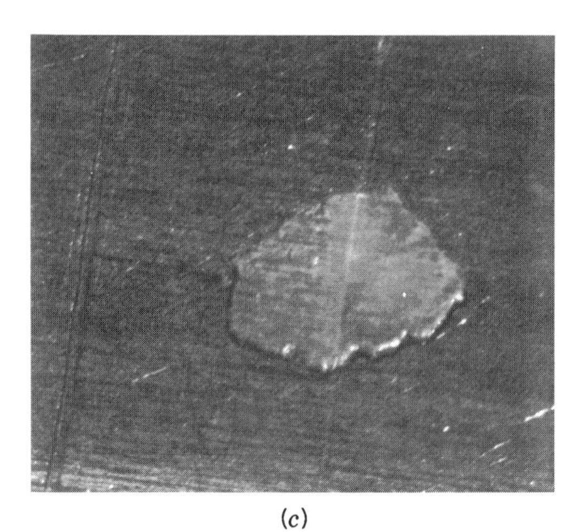

(*c*)

FIGURE 4.6 (*Continued*).

These fat droplets can be suctioned into a specially designed microwave balloon catheter, or can be removed by a sleeve on the catheter that is made from a material that absorbs fat.[11] Calcified plaque can be broken into small pieces by means of an ultrasonic vibrator in the catheter and can then be suctioned out.

4.2.2 In Vivo Experiments

In vivo studies of MBA have been carried out in New Zealand white rabbits and in dogs.[12,13] In the rabbit studies the animals were administered a high-cholesterol diet. After 2 weeks on this diet, endothelial denudation of the external iliac arteries was performed in order to induce the formation of artherosclerotic lesions. Then 4–6 weeks after the denudation MBA was performed in one of the atherosclerotic external iliac artery and conventional balloon angioplasty (CBA) in the contralateral artery. Quantitative angiography immediately after the angioplasty and 4 weeks afterward showed that the lumens of the arteries treated with MBA (microwave heating at 2450 MHz for 30 s to a balloon temperature of 85 °C) were consistently larger than those treated with conventional angioplasty. Histologic analysis showed that in the arteries treated with MBA there was a decrease in lipid-laden cells and replacement of these cells with a hypocellular fibrotic matrix.

In the dog studies, the efficacy of MBA in resolving thrombus was investigated. (A thrombus may be present before angioplasty is started or may develop during the procedure. A thrombus can result in complete occlusion of the treated vessel despite an adequate dilation of the vessel.) A thrombus was artificially induced in the left anterior descending coronary artery of the dogs, and the arteries were then treated with either CBA or MBA. In the MBA microwave heating was administered at a frequency of 2450 MHz for 30 s to a balloon temperature of 85 °C. Five out of six animals treated with conventional angioplasty showed evidence of thrombus after the procedure, while only two out of six animals treated with microwave angioplasty showed similar evidence. Histologic examination of a particular artery following microwave angioplasty revealed a coagulated thrombus with peripheral lamination. A central channel through the thrombus was formed with an adequate lumen for blood flow.

4.3 LOCALIZED HYPERTHERMIA TREATMENT OF CANCER WITH MICROWAVE BALLOON CATHETERS

Localized hyperthermia is used in a number of institutions around the world to treat solid malignant tumors.[14–16] The treatment consists of heating the tumors until they reach temperatures several degrees Celsius above core temperature (typically 42.5 °C), and maintaining the tumors at the elevated temperature for typically 30–45 min. The procedure is usually repeated several times at intervals of 2 or more days. Localized hyperthermia is usually administered in combination with ionizing radiation (thermoradiotherapy) or chemotherapy (thermochemotherapy), or surgery.

Studies of the response of spontaneous animal tumors to thermoradiotherapy have shown that the response rate of these tumors depends strongly on the minimum temperature reached in any part of the tumor during the therapy: the higher the minimum temperature, the better the response rate. High minimum tumor temperatures in humans can be achieved in practice only if the tumor temperatures during the hyperthermia treatment are fairly uniform. If the tumor temperatures are highly nonuniform, it becomes impossible to reach high minimum tumor temperatures without causing unacceptable burning of the hotter part of the tumor or of the healthy tissues surrounding it. Whole-body hyperthermia, and hyperthermia of limbs using heated blood can produce nearly uniform tumor temperatures, but with these techniques the temperatures of healthy tissues is raised to the same value as that of the malignant tissues, which limits the maximum permissible treatment temperature. With noninvasive localized hyperthermia using microwaves, radiofrequencies, or ultrasound it is possible to preferentially heat tumors and thus minimize heating of healthy tissues, but tumor temperatures are often highly nonuniform, particularly when the tumors are deep-seated. This is because currently available apparatus for noninvasively producing localized hyperthermia in deep-seated tumors cannot adequately compensate for local variations in power absorption and heat transfer in the tumors being treated.

With hyperthermia apparatus using arrays of invasive (interstitial) applicators it is possible to heat deep-seated tumors much more uniformly than with apparatus using noninvasive applicators.[17,18] Interstitial applicators can guide microwave energy directly into the interior of tumors. The temperature produced in the small tissue volume heated by each applicator can be kept constant by monitoring tissue temperatures with a thermistor or other small temperature sensor, and using feedback to control the microwave power being fed into the applicator (see Fig. 4.7).

Most of the energy broadcast by the antennas of interstitial applicators is absorbed in tumor tissues, and only a small fraction of the energy reaches the healthy tissues surrounding the tumors. The malignant tissues can therefore be preferentially heated to higher temperatures than healthy tissues. Thus it becomes possible to damage or destroy malignant tissues by heating them to high

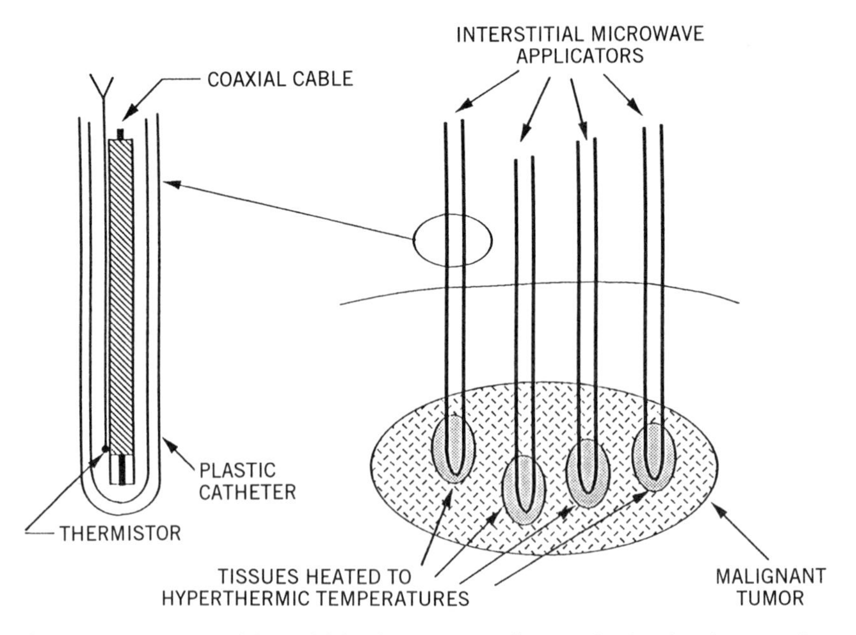

FIGURE 4.7 Arrays of interstitial microwave applicators for heating large malignant tumors to hyperthermic temperatures.

hyperthermic temperatures while sparing healthy tissues. This preferential damage or destruction of malignant cells by microwave heating is often amplified because malignant cells are often more sensitive to insult by heat than are healthy cells. This is because malignant cells, particularly those in the interiors of tumors, are often in a state of nutritional deprivation, low pH, and chronic hypoxia. Furthermore, the microvasculature of tumors, being not as well developed and resistant to insult as healthy microvasculature, is more easily damaged than healthy microvasculature. Cells fed by microvasculatures that have been damaged by heat become, in turn, weakened and more sensitive to heat.

Interstitial hyperthermia is usually combined with radiation therapy using radioactive seeds that are inserted into the tumor via the same tubing that is used for the hyperthermia[19] (Fig. 4.8). A typical treatment sequence is brachytherapy (irradiation of the tumor with the seeds) followed by hyperthermia, followed again by brachytherapy. The interstitial hyperthermia enhances the efficacy of brachytherapy because (1) hyperthermia interferes with the repair of cells that have been sublethally damaged by the ionizing radiation, (2) cells in the S phase of the cell cycle and hypoxic tumor cells tend to be resistant to ionizing radiation but sensitive to heat, and (3) hyperthermia can be effective in oxygenating radiation-resistant hypoxic cells.

Interstitial arrays using conventional applicators are useful only for treating small tumors, because each interstitial applicator can heat and irradiate only a small volume of tissues, and the number of applicators that can be inserted into

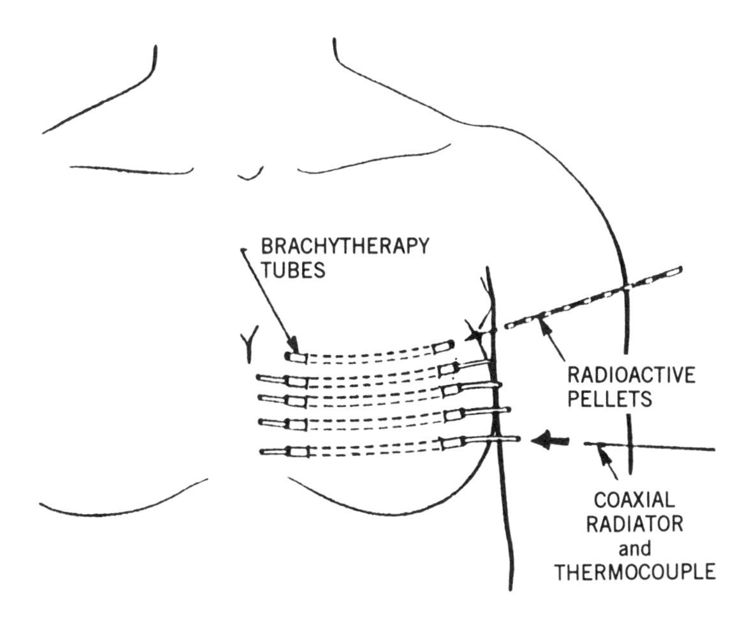

FIGURE 4.8 Interstitial hyperthermia combined with brachytherapy for treating breast cancer.

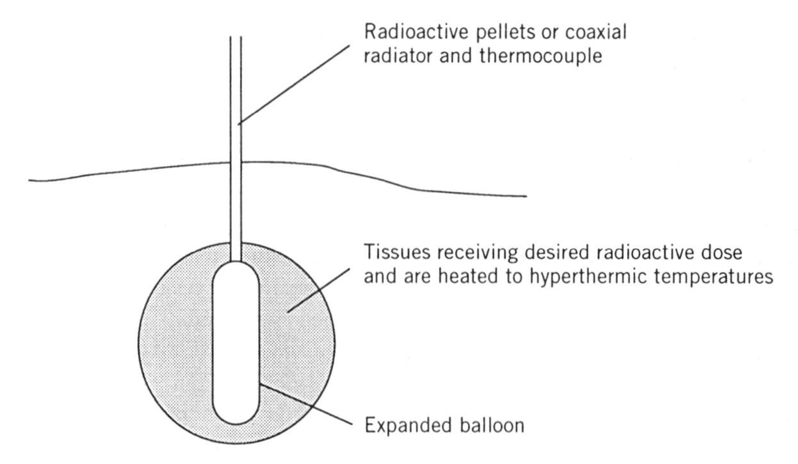

FIGURE 4.9 Interstitial microwave balloon catheter for combined localized hyperthermia and brachytherapy treatments of cancer.

a tumor is limited because of their invasive nature. Interstitial applicators using balloons, on the other hand, can heat much larger volumes of tissues than can conventional interstitial applicators, making it possible to treat larger tumors. A catheter with a deflated balloon at its tip is inserted into the tumor volume to be heated, or where applicable, into a natural opening of the body such as the urethra, rectum, or vagina, the balloon is inflated, and either radioactive seeds or a microwave antenna are inserted through the center lumen of the balloon catheter (see Fig. 4.9).

4.4 TREATMENT OF BENIGN PROSTATIC HYPERTROPHY WITH MICROWAVE BALLOON CATHETERS

The prostate is a chestnut-shaped gland that surrounds the urethra of males immediately below the bladder. The gland usually grows from birth until the end of the third decade of life, and then remains fairly constant in volume until the middle of the fifth decade. At this time it may progressively increase in volume as a result of benign hypertrophy, a condition called *benign prostatic hypertrophy* (BPH). BPH is characteristically a condition of older men, but is occasionally seen in men younger than 40. During an 80-year lifespan, there is a 10% probability that a man will need treatment of BPH.

The clinical manifestations of BPH include a decrease in the caliber and force of the urinary stream, difficulty in initiating and stopping urination, overflow incontinence (dribbling after voiding), and frequent voiding. Often there is a sensation of incomplete emptying of the bladder; with complete obstruction the patient is unable to urinate.[20]

The most common definitive treatment of BPH is surgical removal of the obstructing tissue (adenoma) by transurethral resection, but there are risks associated with this and other surgical procedures, particularly in elderly patients. For this reason, other options for treating BPH are currently being investigated. These include laser surgery, drugs, balloon dilation,[21] localized microwave of RF hyperthermia, and simultaneous microwave hyperthermia and balloon dilation.[22]

Localized microwave hyperthermia has been used for more than a decade to treat cancer of the prostate[23–25] and since 1985 to treat BPH.[26] The initial hyperthermia treatments used microwave applicators that heated the prostate via the rectum, but today transurethal applicators are favored.[27,28] Transurethal applicators are usually placed inside liquid-cooled catheters; the temperatures produced inside the treated prostate can be non-invasively measured with a microwave radiometer.

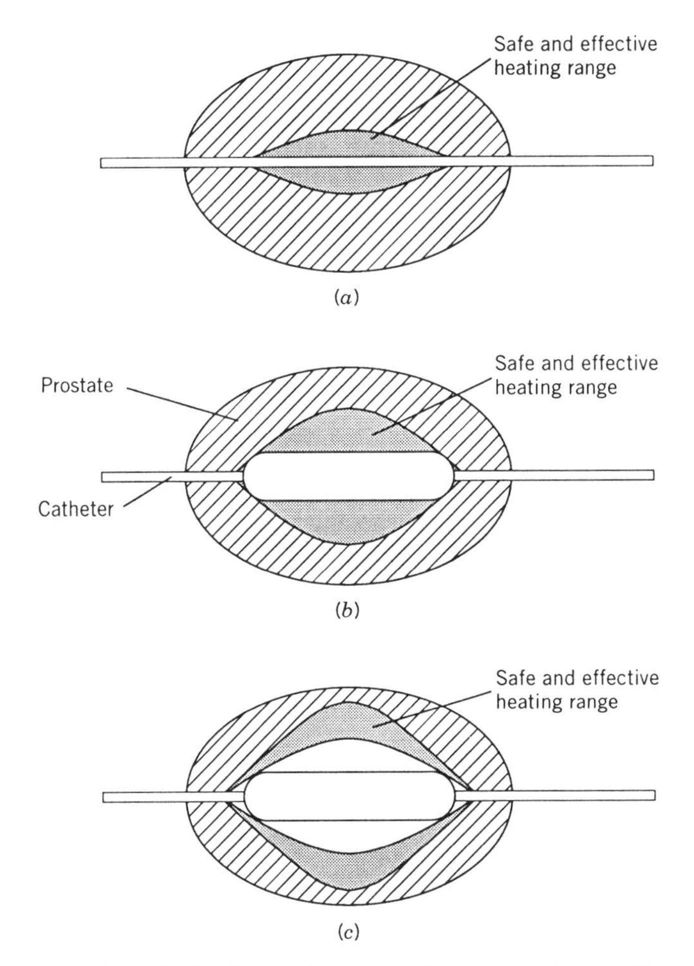

FIGURE 4.10 Safe and effective heating ranges in prostate glands with regular microwave catheters and with microwave balloon catheters. Microwave heating patterns: (*a*) with regular catheter; (*b*) with balloon catheter; (*c*) with balloon catheter and water cooling.

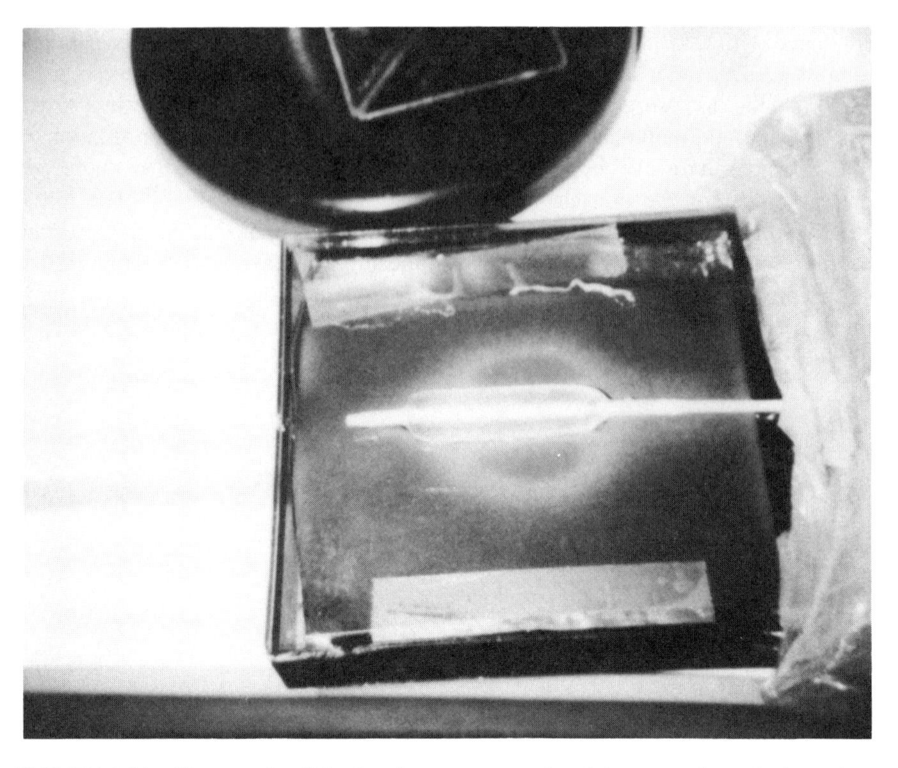

FIGURE 4.11 Photograph of the heating pattern produced in a muscle-equivalent phantom material by a transurethral microwave balloon catheter.

With balloon catheters it is possible to produce high therapeutic temperatures throughout the prostate gland without causing burning of tissues and to produce biological stents in the urethra in a single treatment session. Compared to conventional microwave catheters, the distances microwaves have to travel through the prostate to reach the outer surface of the gland are reduced by the use of balloon catheters, as is the radial spreading of the microwave energy (see Fig. 4.10). Furthermore, compression of the gland tissues reduces blood flow in the gland and the local cooling caused by it. Also, since balloons make excellent contact with the urethra, much better than do conventional catheters, the urethra is well cooled by the cooling liquid and therefore well protected from thermal damage. Figure 4.11 is a photograph of the heating pattern produced in a muscle-equivalent phantom material by a transurethal microwave balloon catheter.

4.5 CONCLUSIONS

Microwave balloon catheters are promising new medical devices that can simultaneously heat and apply pressure to deep-seated tissues. Potential applications include the production of biological stents in partially obstructed vessels and in the urethra, liquifying of the fat in fatty plaques and squeezing this fat out of the plaques, and heating deep-seated tumors to hyperthermic temperatures.

Acknowledgments

The author wishes to thank Dr. Arye Rosen and Dr. Paul Walinsky of Thomas Jefferson University Hospital, Philadelphia, PA, and Mr. Daniel D. Mawhinney and Mr. Adolph Presser of MMTC, Inc. Princeton, NJ for their many invaluable contributions to the concepts and techniques described in this chapter.

REFERENCES

1. Tumeh AM, Iskander MF: Performance comparison of available interstitial antennas for microwave hyperthermia. *IEEE Trans MTT* 37: 1989.
2. King RW, et al: The electromagnetic field of an insulated antenna in a conducting or dielectric medium. *IEEE Trans MTT* 31: 1983.
3. Casy JP, Bansal R: Near field of an insulated dipole in a dissipative dielectric medium. *IEEE Trans MTT* 34: 1986.
4. Sterzer F: Apparatus and method for hyperthermia treatment. U.S. Pat. 4, 190,053, Feb. 26: 1980.
5. Sterzer F: Microwave radiometers for non-invasive measurements of subsurface tissue temperatures. *Automedica* 8: 1987.
6. Dotter CT, Judkins MP: Transluminal treatment of arteriosclerotic obstruction: Description of a new technique and a preliminary report of its application. *Cirulation* 30:654, 1964.
7. Vliestra RE, Holmes DR, Jr: *Percutaneous Transluminal Coronary Angioplasty*. FA Davis, Philadelphia, 1987.
8. Spears JR, et al: Percutaneous coronary laser balloon angioplasty: Initial results of a multicenter experience. *J Am Coll Cardiol* 2:293, 1990.
9. Rosen A, Walinksy P: Percutaneous transluminal microwave catheter angioplasty. U.S. Pat. 4,643,186, Feb. 17, 1987.
10. Rosen A, Walinsky P, Smith D, et al: Percutaneous transluminal microwave balloon angioplasty. *IEEE Trans MTT* 38:1, 1990.
11. Sterzer F: Angioplastic method for removing plaque from a vas. U.S. Pat. 4,924,863, May 15, 1990.
12. Smith DL, et al: Microwave thermal balloon angioplasty in the normal rabbit. *American Heart Journal* 123:1516–1521, June 1992.
13. Rosen A, Walinsky P, Nardone D, et al: Treatment of intracoronary thrombus with microwave thermal balloon angioplasty. *IEEE MTT-S Digest,* 1992.
14. Sterzer F: Localized hyperthermia treatment of cancer. *RCA Rev* 42:727, 1981.
15. Hahn GM: *Hyperthermia and Cancer*. Plenum Press, New York, 1982.
16. Storm FK: *Hyperthermia in Cancer Therapy*. GK Hall, Boston, 1983.

17. Taylor LS: Implantable radiators for cancer therapy by microwave hyperthermia. *Proc IEEE* 68: 1980.
18. Winter A, et al: Microwave hyperthermia for brain tumors. *Neurosurgery* 17:387, 1985.
19. Borok TL: Microwave hyperthermia radiosensitized iridum-192 for recurrent brain malignancy. *Med Dosimetry* 13:29, 1988.
20. Paulson D: *Diseases of the Prostate.* Clinical Symposia, Ciba-Geigy Corp., 1989.
21. Burhenne HJ, et al: Prostatic hyperplasia: Radiological intervention. *Radiology* 152: 1984.
22. Sterzer F: Catheters for treating prostate disease. U.S. Pat. 5,007,437, April 16, 1991.
23. Petrowicz O, et al: Experimental studies on the use of microwaves for the localized heat treatment of the prostate. *J Microwave Power* 14: 1979.
24. Mendecki J, et al: Microwave applicators for localized hyperthermia treatment of cancer of the prostate. *Internatl J Radiat Oncol Biol Phys* 6: 1980.
25. Sterzer F, Paglione RW: Nonsymetrical bulb applicator for hyperthermia treatment of the body. U.S. Pat. 4,311,154, Jan. 19, 1982.
26. Yerushalmi A, et al: Localized deep microwave hyperthermia in the treatment of poor operative risk patients with benign prostatic hyperplasia. *J Urol* 133: 1985.
27. Astrahan MA, et al: Microwave applicator for transurethal hyperthermia of benign prostatic hyperplasia. *Internatl J Hyperthermia* 5: 1989b.
28. Roehrborn CG, et al: Temperature mapping in the canine prostate during transurethally-applied local microwave hyperthermia. *Prostate* 20: 1992.

CHAPTER FIVE

Microwave Surgery

LEONARD S. TAYLOR, Ph.D., *Departments of Electrical Engineering and Radiation Oncology, University of Maryland, College Park, MD*

5.1 INTRODUCTION

The use of intense heating to effect coagulation and stem the flow of blood was known to primitive peoples its use in ancient times is documented in Egyptian papyri. Because heating by conduction is simple to employ, these ancient methods involved thermal conduction using heated solids applied to the patient. The application of electromagnetic fields to achieve hemostasis began in this century following the discoveries of the nineteenth century pioneers in electromagnetism. Those discoveries excited the popular imagination and led to many experiments and claims for various therapeutic applications of electromagnetic fields. It was, however, the codiscovery of tissue heating by electromagnetic fields by Tesla and d'Arsonval in 1891, following the discovery of electromagnetic radiation in 1866 by Herz, that led through a series of later workers to the first patented electrosurgical apparatus by DeForest in 1907. Many experiments performed during the early period of research in electromagnetism involved the eradication of cells rather than the coagulation of blood; the use of the contrivances was directed at the removal of unwanted tissue (cancers, hemorrhoids, tonsils, etc.) or at the destruction of tissue to effect cutting, rather than coagulation alone. The early history of electromedicine is laden with claims and counterclaims of discoveries and cures. The review articles cited [Licht[1] and Geddes[2]] provide an extended history of this era.

New Frontiers in Medical Device Technology, Edited by Arye Rosen and Harel Rosen
ISBN 0-471-59189-0

As far as electrosurgical devices were concerned, DeForest's apparatus was ignored by surgeons in the United States, although it was used successfully in Europe. As radiofrequency (RF) techniques and generators improved and higher powers and shorter wavelengths became available, a succession of better devices appeared. The first RF electrosurgical apparatus to achieve widespread use, however, was developed by the physicist W. T. Bovie, who worked with the renowned neurosurgeon, Harvey Cushing. Together in 1926–1928 they developed the system that is still in common use and is generally called *the bovie*. The bovie is a dual electrosurgical unit in which one RF current waveform is used for cutting and the other is used for coagulation. The decades following the introduction of the bovie saw a succession of improvements in both the design and clinical techniques used with this instrument, but it was not until 1975 that new coagulating devices were introduced into surgery with the arrival of both laser and microwave techniques. Because the operation of microwave coagulating devices is advantageously discussed by comparison with the bovie and laser surgical tools, the operation of these devices will be discussed at some greater length in succeeding sections.

5.2 RADIOFREQUENCY COAGULATORS

Radiofrequency electrosurgery depends on the flow of high-frequency currents through small localized volumes of tissue to produce high temperatures and tissue destruction in the volume. The type of tissue destruction depends on the RF current parameters: The result may be cutting by the explosion of cells in the volume, or it may be desiccation and coagulation. Coagulation is produced by using damped sinusoidal waveforms. The RF current is usually applied through a spark between a small probe and the patient, and the current then very rapidly spreads out through the patient to ground. Poor ground contact can lead to burns at the sites at which the current exits to ground, so that a large metal "butt plate" on which the patient rests is usually employed. To prevent bleeding from severed vessels, the technique that is often employed involves grasping the vessel with a pair of (conducting) forceps and touching the forceps with the probe. The frequencies employed by the RF devices are in the low megahertz range (0.5–4 MHz) and, because of the waveforms employed, fall over a broad band. Even when pure sinusoids are provided by the generator, low-frequency spectral content will result from system nonlinearities, such as rectification at the probe tip. Bovie RF generator power levels are about 5–10 W. A foot-pedal control for the bovie is most usual, but the devices are available with a switch on the handheld probe.

Certain basic restrictions on the operation of the RF bovie coagulator follow from the nature of the spark. First, because the spark current spreads out rapidly once the spark has entered the tissue, the depth of coagulation is limited to a fraction of a millimeter; usually this is sufficient for hemostasis and is a useful feature of the device. Second, the intense spark current density cannot be maintained if the tissue is covered by blood or if the probe tip is immersed in blood. To extend the operation of the bovie to situations in which active bleeding is covering the tissue with blood, the *argon-jet* bovie has been recently introduced. In this system, a jet of inert argon gas blows away the blood and presents a dry surface to the RF spark.

The leads from the generator to the bovie probe are usually unshielded. Anyhow, radiation from the RF spark itself cannot be prevented, so that other electronic instruments in the operating room, in particular the anesthesiologist's patient monitors, are inoperable during the intervals in which the bovie is in use. The possibility of neuromuscular stimulation by the RF current also exists,[3,4] and there have been occasional observations of violent muscular contractions when the bovie current is initiated. These contractions can be dangerous to the patient during certain surgical procedures, such as transurethral resection bladder surgery using an endoscopic device inserted through the urethra. Additionally, although the danger of explosions initiated by the RF spark has been virtually eliminated since the introduction of nonexplosive gases into anesthesiology, some remote dangers do remain: for example, the ignition of the patient's own colonic gas, can create an internal explosion.

5.3 LASER SURGICAL DEVICES

The invention of the laser by Townes in 1975 began a new chapter in medical technology. A variety of laser-tissue interactions have since found their use in medicine,[5] and the application of laser irradiation for the photothermal production of heat, producing tissue destruction and cauterization, has found its use in nearly all surgical specialities.

The basis of laser action in surgical procedures is the application of a relatively small amount of radiated power to a small surface area. Thus, the effective power density is quite high and the effect is instantaneous. The capability for effecting this concentration of power results from the fact that the atomic sources of the laser energy are radiating coherently and the energy is emitted in the form of a beam wave that will maintain its small cross section for long distances. Typical parameters for laser surgical devices include radiated powers varying widely between one and ninety watts, depending on the laser type, and irradiances of hundreds of joules per centimeter squared.

The critical parameter for laser surgery is the wavelength of the source, and, most important, its relation to the effective depth of penetration of the laser beam into the tissue. The penetration depth δ of a wave into an absorbing medium is defined by assuming a Beer law relation for the radiation energy density $W(z)$ at depth z in the medium:

$$W(z) = W(0)\exp\left(\frac{-z}{\delta}\right)$$

Thus, approximately 63% of the beam energy is absorbed with a distance δ of the surface.

At wavelengths below 0.8 μm, hemoglobin is highly absorbing and the argon laser wavelengths, $\sim$ 0.5 μm, are effectively absorbed within a fraction of a millimeter. Thus, this wavelength is highly effective in producing a thin coagulated layer of blood to seal small vessels. At wavelengths above $\sim$ 1.5 μm, water absorption of laser energy effectively limits penetration to a fraction of a millimeter, so that the CO_2 infrared laser frequencies are highly effective in cutting nonvascular tissue. The cutting is effected as the laser beam energy is absorbed, vaporizing cellular water and producing shock waves that mechanically explode the cells.[6] At wavelengths between 0.8 and 1.5 μm, absorption is only a few percent per millimeter, and the Nd:YAG (neodymium YAG) laser at 1.06 μm penetrates to relatively larger depths, producing tissue damage rapidly to depths several times that of the other lasers mentioned. Because of the very large penetration depth into nonvascular tissue, the application of this device requires great care. The relative effectiveness of this laser wavelength for applications of interest here can be judged from the following data:[7]

Laser	Wavelength (μm)	$\delta(H_2O)$	δ (hepatic Parenchyma)
Argon	0.5	66.7 m	0.35 mm
Nd:YAG	1.06	2.7 cm	0.8 mm
CO_2	10.6	0.01 mm	0.05 mm

(Hepatic parenchyma is highly pigmented and vascularized.) The variety of penetration depths available for different laser wavelengths provides specific tools for different surgical applications.

5.4 MICROWAVE SURGICAL DEVICES

The application of microwave fields in surgery was initiated in the 1970s in Japan, where experimenters used a series of coagulated punctures produced by coaxial needle antennas for cutting. The thick eschars produced by the microwave fields furnished layers that effectively stopped hemorrhage. It was realized that this type of action would be useful in liver and spleen surgery. Ordinarily, surgeons try to control bleeding through sutures. Repairing either the liver or spleen by stitching alone can be a difficult procedure, and it does not always succeed. This is because the stitches used to repair the organ may tear and damage it further, producing even more bleeding. In addition, not every small blood vessel in the liver or spleen that is cut during surgery can be sutured. Moreover, the coagulating action of the RF and laser surgical devices do not provide more than a thin eschar that will not usually control bleeding from these large, highly vascularized, parenchymal organs. If ligation does not succeed, the surgeon may have to remove the spleen or part of the liver. In the case of the spleen, because of the immunologic functions performed in the body by the spleen, this procedure may increase a patient's chances of sustaining a severe or even fatal infection later in life (postsplenectomy sepsis). Consequently, every effort is usually made to save a bleeding spleen. Nevertheless, more than 30,000 splenectomies are performed annually in the United States, and there are relatively few splenic repairs performed because of the difficulties in controlling bleeding from this organ. It is estimated that about 700 deaths per year can be ascribed to infections that occur in those who lack spleens. In the case of the liver, the human liver is a vital organ that cannot be removed in its entirety. However, the liver is the most common site for the appearances of metastases of colon cancer. As a result, a large number of liver resections occur annually. These are difficult procedures with a high mortality rate, due in part to the danger of uncontrollable bleeding. Thus, microwave surgical devices find their application in important problem areas of surgery that have not been effectively handled in other ways. Certain other areas, for example, pelvic surgery, also involve the danger of uncontrolled hemorrhage and are possible candidates for the application of microwaves in surgery.

The microwave frequency employed for surgical systems is 2.45 GHz. This choice is based on the fortunate circumstance that this is a frequency designated for medical applications, that the thickness of the char layer produced using this frequency is appropriate to the applications, and that ordinary microwave-oven magnetrons can readily be converted to serve as the microwave power generator tubes. The free-space wavelength at 2.45 GHz is 12.24 cm. However, the wavelength in tissue is less than 2 cm because of the high relative dielectric constant. This later dimension is also consistent with the design of small devices for this application.

Related considerations for any surgical system are the questions of cost, testing, and maintenance. For the most part, the microwave generators employed for surgical applications use standard microwave-oven power circuits and parts. These components are inexpensive, especially when compared to the costs of the components of high-power laser surgical equipment. The microwave systems are simple to test and can be maintained by ordinary electronics technicians. The operation of a system can easily be checked by medical personnel themselves by measuring the rate at which the device raises the temperature of a styrofoam cupful of normal saline solution. This easy procedure is in contrast to more elaborate procedures required for laser and RF coagulating devices.

Discussions of the technical aspects of microwave–tissue interactions are based on the complex dielectric permittivity of tissue and the effective depth of absorption of the energy. Animal tissues are generally classified as either *low-water-content* (fat, bone, lungs, etc.) or *high-water-content* (muscle, organs, blood). It is the latter type that is significant with respect to microwave surgical devices. Microwave absorption in tissue is due to a broad absorption band caused by water molecule resonance that covers the entire microwave region. High-water-content tissue is significantly more absorbing than the low-water types. It should be noted that the microwave heating process is effected not only through free water molecules in the tissue but also by the water molecules bound to the structure of large biomolecules.

The electromagnetic properties of absorbing media are determined by the dielectric permittivity ε and conductivity σ. The *relative complex permittivity* is defined by the expression $\varepsilon_r - j\sigma/2\pi f \varepsilon_0$, where f is the frequency, $\varepsilon_r = \varepsilon \div \varepsilon_0$, and $\varepsilon_0 = 8.854 \cdot 10^{-12}$ [farads per meter (F/m)] is the *permittivity of free space*. For high-water-content tissue, the complex relative dielectric permittivity is approximately $48 - j18$ at 2.45 GHz, the frequency used for microwave surgical devices. It follows from Maxwell's equations that a plane wave entering such a medium would be absorbed and be attenuated (by a formula similar to Beer's law) as $\exp(-z/\delta)$, where z is depth into the tissue and δ is the electromagnetic skin depth for energy absorption and δ is the distance into the medium at which the power density has been reduced by absorption to 37% of the power density entering the medium and is given by the expression

$$\frac{c}{2\pi f}\left[\frac{2}{(\varepsilon' - \varepsilon'')^{1/2} - \varepsilon'}\right]^{1/2}$$

where ε' and ε'' are respectively the real and imaginary parts of the complex dielectric permittivity. Although some variation of complex permittivity with individual samples and tissue types exists, the figure $48 - j18$ quoted above is a useful approximation and leads to $\delta \approx 7.6$ mm. The fact that this parameter is approximately the same for many tissues, including blood, is a significant advantage of microwave surgical devices. It should be noted that because tissue dielectric and conductive properties vary slowly over the microwave band, the skin depth is not simply inversely proportional to frequency. For example, at 0.915 GHz, the skin depth is only about 11.8 mm. The skin depth decreases as the frequency increases.

The skin-depth parameter is a convenient and simple numerical figure for descriptions of the effectiveness of microwave energy in penetrating tissue. However, it should be recognized that its application in this context (as well as in the laser-beam context) is not scientifically justified, because the coagulating fields are not plane waves, but are instead antenna near fields. Such fields include inhomogeneous wave components, and it is known that their depth of absorption may differ considerably from those of plane waves at the same frequency. At microwave frequencies, the radiated energy is absorbed in the near field, with no appreciable radiation into the far zone (despite the compression of the near zone because of the shortened wavelength in the high dielectric). Only in the far zone can the field be represented as a plane wave. The properties of an antenna's near field are generally quite different from those of its far field. For example, even in free space the angular radiation patterns differ significantly for the few simple radiator configurations that have been analyzed. For simple dipole and loop radiators in absorbing media, the theoretical expressions are available, but difficult to evaluate. These theoretical expressions have, in fact, received little attention in the design of surgical devices. The reason is that such expressions are based on models that bear little relation to the dynamic physical situation that results from the application of the high powers required for coagulation. The microwave energy progressively heats and denatures the tissue in the vicinity of the radiator, forming a carbonized layer with unknown dielectric properties. Thus the radiation is absorbed in a dynamic, inhomogeneous medium that bears little relation to the theoretical models.

The design of microwave surgical devices has proceeded by experimental trial and observation of the heating and coagulation patterns produced. Many considerations that dominate the design of microwave radiators for communications, surveillance, remote sensing, and even microwave diathermy do not apply to the coagulating tools. In part this is due to the necessity of using flexible coaxial transmission lines because other types of transmission lines would be too large for this application. Furthermore, it is impractical to achieve the impedance match (eliminating reflection of power back to the generator) usually sought for microwave devices in other applications because of the rapid change of dielectric properties of the tissue and the formation of the layer of char on the radiator during use. The high powers employed and the practical limitations on the size of the coagulating devices also have prohibited the use of matching techniques. The practical answer has been to use crudely matched devices (VSWR $\sim$ 2–5) with power reflection losses of 5–50% and to use enough generator power to meet the requirement for rapid coagulation despite reflection and cable losses and the cooling effect of blood flow.

It is sometimes asked whether it would not be advantageous to employ a much higher frequency than 2.45 GHz to obtain a thinner char layer. On the theoretical level, this appears feasible. However, there are some definite reasons, besides the higher cost of an appropriate generator, that make this appear impractical. In particular, the energy absorption loss in the flexible coaxial cables required to transmit the power from generator to patient is inversely proportional to frequency. Using a higher frequency would require higher input power and would lead to problems of excessive cable heating as well as possible arc-overs at the connections.

Because the microwave devices operate based on different physical mechanisms than the familiar RF electroscalpel, there are differences in both the effects and the surgical techniques involved. Among the differences are the facts that the microwave devices do not require grounding the patient using a butt plate, that no spark is involved in the normal operation of the microwave devices, and that the microwave generator does not interfere with the operation of other electronic equipment in the operating room. The last consideration is particularly significant because of the susceptibility of anesthesiologic monitoring equipment to RF interference.

Patient safety requires the maintenance of a sterile field over the operating table. Thus, any equipment that the surgeon contacts must be sterilized. This can be accomplished by sterilizing the microwave tools and connecting cable and providing them to the surgical team in sterile bags. The bags are opened, appropriate connections are made, and the generator end of the cable passed out (without direct contact) to the technician, who connects the cable to the generator, places the foot pedal in position for the surgeon, and turns on the generator.

Autoclaving is the preferred method for sterilizing surgical equipment. Autoclave temperatures are high enough to melt soft plastics; Teflon parts and cables are preferable. Little difficulty has been experienced in providing microwave devices and generator cables designed to be fully autoclavable. It is also appropriate to mention that although anecdotal accounts of a microwave antibacterial effect have sometimes appeared in the technical literature, the author's own experience has led to the conclusion that bacterial spores are unaffected by even very high microwave field levels. A series of experiments was carried out in which spore strips were exposed to intense microwave fields for periods of several minutes. These experiments were carried out by placing the slide bearing the spore samples at the center of an S-band waveguide through which 300 W of 2.45-GHz power was transmitted to a matched load. No sterilizing effect of the microwave fields on bacterial spores was observed, and it was concluded that any microwave antibacterial effects were due to prolonged heating in hydrated bacteria.

5.5 BRAIN TISSUE COAGULATION USING MICROWAVES

The first in vivo medical experiments aimed at the application of microwave fields in surgery were carried out in Japan by Kosugi.[8] Experiments using dogs were carried out with the aim of developing a technique for using the heating power of microwave fields to coagulate and destroy brain tumors. These experiments were intended to establish a technique with the dual purpose of destroying the cancerous tissue through coagulation and inhibiting the spread of cancer cells "liberated" by the surgery. In these experiments, the microwave antenna was a "needle" dipole above a ground plane, obtained by extending the inner conductor of a coaxial line a short distance (0.5–3.0 cm) above a small ground plane attached to the other conductor (see Fig. 5.1). Antennas with bare and partially dielectric-covered needles were tested. (An antenna of this type is, of course, useful only when the volume to be irradiated and heated is available just

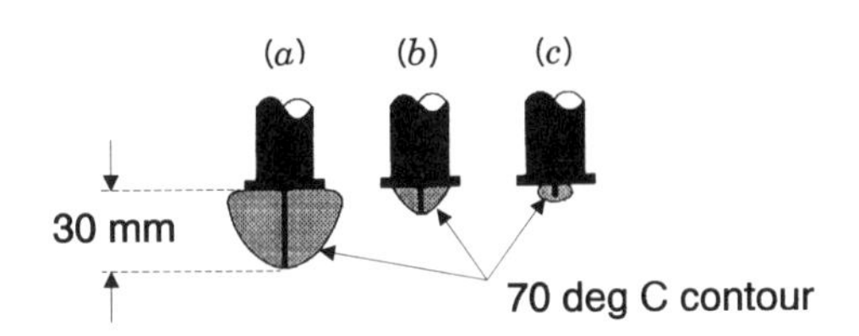

FIGURE 5.1 Thermal distributions in egg white created by needle antennas:[8] (a) $L =$ 30 mm, 105 W, 142 s; (b) $L = 10$ mm, 120 W, 12 s; (c) $L = 5$ mm, 60 W, 22 s.

below the surface of the organ.) The microwave power source was a 2.45-GHz generator; transmitted and reflected powers were measured using a pair of coaxial directional couplers and power meters. The heating pattern of the needle antennas of various lengths was determined by the following method: It was observed that microwave power would coddle egg white when the temperature of the albumin reached 70 °C. Contours of constant power deposition were obtained by applying power (at a level of 110 W) for intervals of few seconds to several minutes and observing the contour of coddled egg white. The contours were seen to begin with a roughly cylindrical shape close to the needle and to rapidly spread out to a hemispherical form. The contours agreed reasonably well with a theoretical calculation based on a simple model for the current distribution on the needle (see Fig. 5.2). These in vitro observations were confirmed in animal experiments carried out in vivo using canine brains. Similar heating patterns, as evidenced by the charred tissue volumes, were obtained. The histological evidence was encouraging; a clear line of demarcation between the charred tissue and normal tissue was found.

Despite the successful attainment of the program objectives and the demonstration of the effectiveness of microwave coagulation, there were no immediate medical applications of the technique. In retrospect it appears that the target choice was not apt; there was little medical incentive to apply the technique to the coagulation of cancerous brain tissue lying close to the surface, where it was available to the surgeon's scalpel and where the bovie RF electroscalpel provided adequate coagulation. Penetration to inaccessible deeper regions of the brain was not possible with the type of radiator used by Kosugi; the ground plane limited the insertion depth of the coaxial inner conductor. A coaxial radiator without a ground plane that could be inserted to arbitrary depths into the solid organ tissue was apparently not considered a possibility. This type of device (*microwave syringe*) was later introduced into use in microwave hyperthermia,[9] where relatively low powers are employed. The high microwave power levels required for tissue coagulation in the presence of blood flow, however, could not be transmitted through a miniature coaxial waveguide suitable for insertion into tissue.

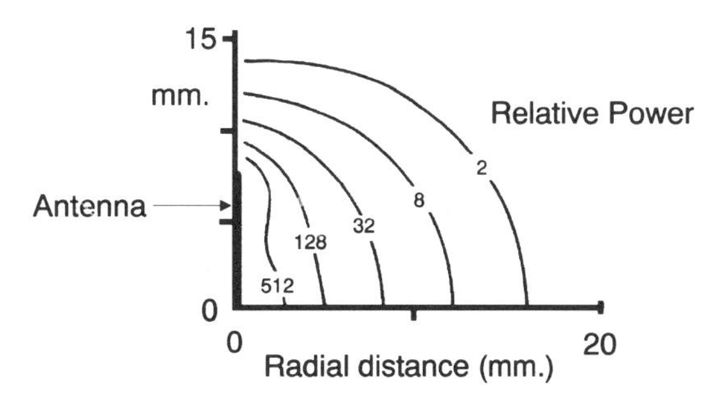

FIGURE 5.2 Theoretical distribution of relative dissipated 3-GHz power in tissue near a needle antenna.[8]

5.6 MICROWAVE SURGERY USING NEEDLE RADIATORS

The medical pioneer in the surgical applications of microwave radiation is K. Tabuse. In his seminal paper[10] Tabuse reported on the first animal experiments using microwave fields for coagulation in liver surgery. This was the first recognition that the relatively deeper penetration depth of microwave radiation into tissue could produce a thick char layer and prevent hemorrhage during hepatic surgery. A variety of surgical techniques had been tried without complete success in controlling hemorrhage or the oozing of blood and bile following liver surgery.

5.6.1 Animal Experiments

The technique introduced by Tabuse employed the same type of needle antenna used by Kosugi. The initial animal trials consisted of the transection of the livers of rabbits, performed by creating a continuous charred belt of tissue across the liver. The belt was created by repeated punctures and the formation of charred regions about 1 cm in diameter about each puncture using microwave power (see Fig. 5.3). The needle antennas employed a coaxial line with 1 cm outer diameter (o.d.) and needle lengths varying between 4 and 31.5 mm; the power source was a 2.45-GHz CW generator designed for medical diathermy. Microwave power incident at the antenna was 25–65 W and was applied for 20 s at each puncture. The antennas were not matched to the tissue. After the charred belt was completed, the liver was transected through the belt using an ordinary scalpel. About 10 min was required to transect a liver lobe by this process. Two lobes were transected in each rabbit, removing about 30% of the liver in each case. The coagulated stumps were not sutured.

From the viewpoint of microwave engineering, a surprising feature of the needle antenna applicators is that they operate quite similarly, independent of length. The author's own measurements confirm this feature. Tests of a needle

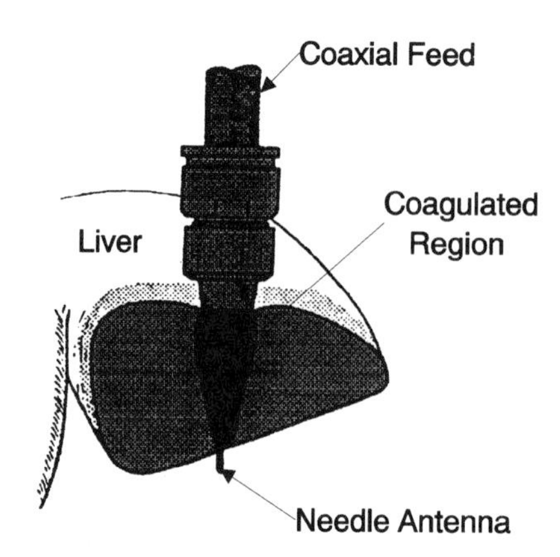

FIGURE 5.3 Microwave needle antenna applied to rabbit liver.[10]

antenna constructed in 8-in diameter semirigid coaxial waveguide with a 0.5-in radius ground plane showed that the 2.45-GHz power transmission factor into tissue varied only over the range of 0.5 ± 0.1 for a wide variation of needle lengths. The current in the needle and the effective radiation field diminish with distance down the length of the needle as power is absorbed into the tissue. A uniform current distribution is not possible, since the current must go to zero at the needle end, excluding conduction currents in the tissue itself. The current density and, consequently, the charred region is thickest near the ground plane. (We remark here that the needle antenna configuration is sometimes referred to as a *microwave scalpel*. In this chapter, to avoid confusion, we have reserved this term for a configuration described later in which a loop radiator is contained in a cutting blade.)

The animal tests proved the feasibility of the microwave technique. Twenty-six rabbits were used. After each resection, the stump surfaces were dry without blood or bile leakage. Blood vessels up to 3 mm in diameter were successfully coagulated. Autopsies of rabbits sacrificed at the ends of periods of one day and up to several weeks afer the hepatic lobectomies displayed a process in which the initial congestion of the liver subsided, the limited region of necrotic tissue was encapsulated, and the necrotic tissue was then absorbed. Histologic studies showed a gradual return to complete functionality in the stump, and no abscess formation was observed. The presence of the coagulated tissue in the body over the long period required for its absorption did not induce any antigenic effects or cause systemic damage.

5.6.2 Clinical Experience

To date, Tabuse has performed more than 300 hepatectomies on human patients, and a similar number of pancreatectomies, using the microwave technique he introduced.[11] Many similar procedures have been carried out by other surgeons in Japan. The clinical trials have employed variations of the basic needle applicator and 2.45-GHz (150-W) generators produced by Heiwa Electronic Industries, Ltd. (Osaka, Japan), under the "Microtaze" tradename (see Figs. 5.4, 5.5). The surgical techniques employed have been greatly refined and extended over the course of this experience, although the basic radiator shape and microwave power parameters have remained relatively constant. The Microtaze generator includes one noteworthy special feature, a dissociation apparatus used to ease withdrawal of the needle radiators from the coagulated tissue.[12, 13] Tissue that has been coagulated produces a char that adheres to the radiator; for the needle radiator form, this char is most pronounced and thickest around the base of the needle where the radiation intensity is highest. It presents the danger of causing additional tissue tears and bleeding as the radiator is removed. To prevent this, a DC current of 15–20 mA from a second electrode is applied near the base of the needle radiator and induces hydrolysis, facilitating removal of the needle radiator. The needle radiators are not matched to the tissue and the length of

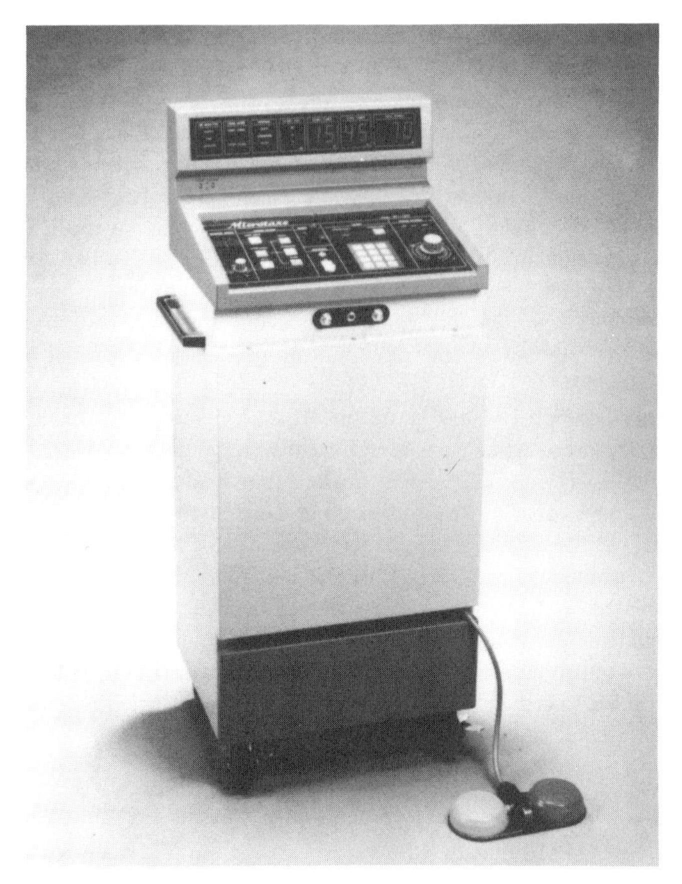

FIGURE 5.4 "Microtaze" Model OT 110M microwave surgical generator manufactured by Helwa Ind., Ltd. (Courtesy of K. Tabuse.)

time required to coagulate around each puncture is determined by the surgeon, based on observation of the surface near the puncture.

The surgical techniques described by Tabuse[12–14] include the use of ultrasonagrams to map the tumor mass and the positions of hepatic vessels to avoid injury to major vessels. Direction and depth of the coagulated puncture lines are then chosen according to the tissue volume to be resected and the positions of the major vessels. The needle radiators are chosen according to the depth elected and are successively inserted and power applied at each predetermined point. The level of power employed has varied from 60 to 80 W (CW), with the period of each needle puncture coagulation varying from 30 to 60 s. The coagulated volume appears on the surface of the tissue as a blanched (yellow-white) circular area with an approximate 1-cm diameter around the base of the radiator. Successive punctures are made to create a continuous boundary surrounding the tumor (see Fig. 5.6) and to produce hemostasis in the tumor volume before the tumor is

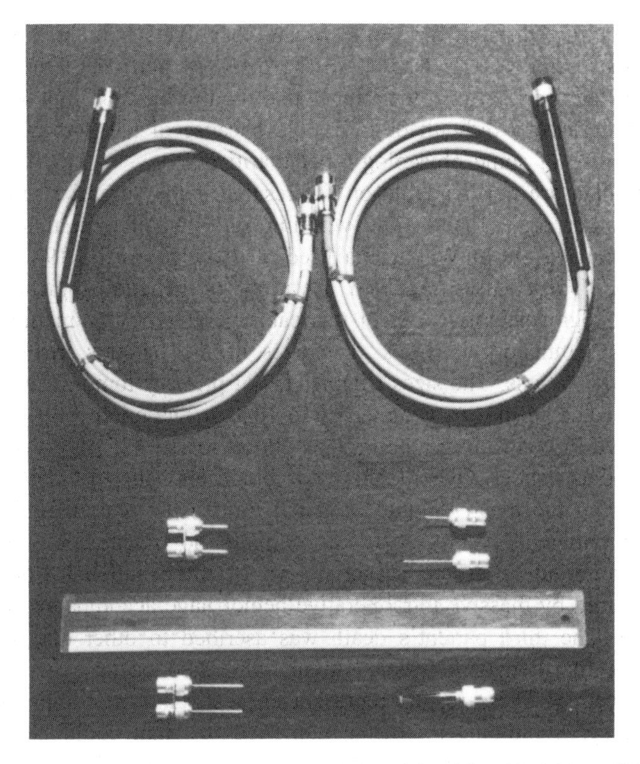

FIGURE 5.5 Needle radiators, handsets, and coaxial cables (Courtesy of K. Tabuse.)

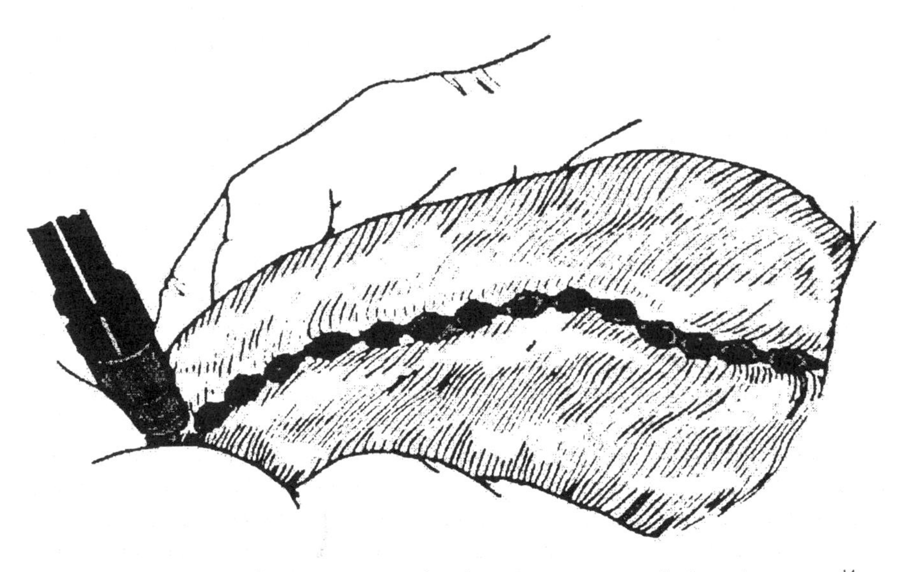

FIGURE 5.6 Needle antenna coagulated puncture sequence in hepatic surgery.[14]

removed by cutting through the center of the coagulated belt. In this procedure, ligation of the major vessels leading to the diseased portion of the liver is not required. Conventional surgical techniques that require ligation impose a condition that is often extremely difficult or infeasible because of the presence of a large tumor or because of some physiologic peculiarities.

Complete data are not available on the results of the surgeries undertaken by Tabuse and other practitioners now working in Japan. The results of the first 60 surgeries carried out at Wakayama Medical College during 1980–1983 were published.[15–17] These cases involved a variety of hepatic surgical procedures (lobectomies, segmentectomies, etc.) to treat several types of malignancies and benign and inflammatory diseases. The ages of the patients treated ran from infants to octogenarians. The sizes (diameters) of primary tumors varied from 2.5 to 11 cm. Resected specimens weighed up to 1200 g. The microwave devices were highly effective in controlling hemorrhage. Blood loss (1425 mL) was highest in cases of bile duct cancer. Overall, the mean blood loss during the 60 procedures was 950 mL and the mean volume of blood transfusion was 485 mL, less than one unit per patient. Seventeen patients required no transfusion. Four patients died within a month after the surgery. Two of these suffered from cirrhosis; death was caused by massive intraoperative hemorrhage accompanied by multiple organ failure. The third patient's death was ascribed to other diseases, not the operation; the fourth patient was an infant who died because of renal failure and hyperglycemia. These results represent an acceptable mortality rate for this type of surgery. Symptoms indicating complications that could be ascribed to the microwave technique were postoperative pyrexia (fever) lasting more than 1 week (13 patients) and bile leakage and abscess formation (eight patients). These symptoms were alleviated within 2 weeks using standard medical procedures. The most important technical problem in liver surgery, control of bleeding without seriously reducing postoperative liver function, was solved by using the microwave surgical technique. No instance of postoperative bleeding from the resected stump occurred.

The principal difficulty recognized by Tabuse was the length of time required using the technique to complete the belt of coagulated punctures. Although the published papers do not record the actual times required, we may estimate that, considering the length of the coagulated zones and the time required for each coagulation, 30–40 min is required to carefully complete the process. Moreover, when the surface area of the resected stump is large, a second and even third series of coagulated punctures deeper into the liver is required,[14] multiplying the time required accordingly.

Uda has reported on experiments using a simple double-needle configuration to reduce the time required for producing the series of coagulated punctures.[18] The configuration consists of two needle radiators placed in parallel using a BNC power splitter in the cable connection to the generator. The maximum needle spacing allowing the production of a continuous char between the radiators was 1.5 cm, corresponding to a char section of about 2.5 cm along the line of the radiators. Successful performance depended on raising the output power level to

100 W, and it was suggested that it would be necessary to cool the liver during surgery to prevent heat damage.

An unknown number of hepatic surgeries using the needle antenna technique have also been performed in the Peoples' Republic of China (PRC). A microwave generator and needle antenna surgical tool set, quite similar to the Heiwa Microtaze equipment, is being manufactured and marketed by the Advanced Science and Technology Development Company (Joint Ventures) in Qianshan, PRC.

5.6.3 Endoscopic and Laparoscopic Applications

Many human trials of endoscopic applications of microwave coagulation have been performed in Japan by a number of medical investigators at several institutions. Unfortunately, the data on these procedures are not readily available; the articles appear in Japanese journals or workshop proceedings that are not translated, or even available in the United States. These experiments have been aimed at the development of microwave surgical techniques both for coagulation and to destroy malignant tissues.

Microwave-induced endoscopic or laparoscopic hemostasis/tissue necrosis is available using instruments that have a 3-mm-diameter channel available for a miniature flexible coaxial cable. The restriction to instruments with this large channel is due to the limited power-carrying capabilities of small microwave coaxial cables. At 2.45 GHz, a 3-ft section of Teflon-filled coaxial cable such as RG-179U, with 2.6 mm outer diameter and 125 W power rating, typically has somewhat more than 1 dB of attenuation and, considering that the power reflection coefficient using a needle antenna is approximately 0.5, can be used to deposit up to nearly 50 W into the tissue. In contrast, RGU-178-U, a similar, slightly smaller cable with 1.8 mm o.d. (0.7-dB/ft attenuation and 45-W rating) could only be used to deposit about 20 W.

Endoscopic treatment of hemorrhagic ulcers using the microwave needles[12] began after initial tests in dogs. In the canine tests it was found that by applying a 2–3-mm long needle to the exposed blood vessel and by irradiating with 30 W applied to the needle, coagulation could be achieved in 15–30 s without tissue charring. Histologic studies showed tissue regeneration in the coagulative lesion over the weeks following the procedure.

The initial paper describes the clinical trials and the results of endoscopic microwave coagulative therapy applied to 19 patients with bleeding in the upper digestive tracts due to gastric ulcers, polyps, or cancer. Emergency hemostasis was always achieved and was maintained for more than 24 h in 16 of the 19 cases. Hemostasis was maintained for more than 72 h in 15 of the 19 cases. In one case in which hemostasis was not maintained, the patient's condition was complicated by severe liver cirrhosis; in another case, the site of bleeding was obscured by a sarcoma. The principal advantages claimed for microwave, as compared to laser and RF, endoscopic coagulation is the possibility to control bleeding from larger veins and arteries, up to 3 mm and 2 mm in diameter, respectively, plus the freedom from the danger of inadvertent perforation that does exist when the RF or laser devices are used. Photographs of the devices employed are shown in Figures 5.7 and 5.8.

Recently, Tabuse has introduced a monopole radiator with a cutting edge designed for tissue dissection with coagulation.[11] The radiator may have either a sickle or blade shape. These devices were employed in 72 laparoscopic cholecystectomies during February 1971–March 1993. The instruments were employed through a laparoscope flexible near the object end. In these trials, the microwave energy was applied more gradually, providing a slow, easily controllable coagulation process on the peritoneal reflection over the cystic duct and artery to facilitate dissection of these vessels and to avoid damage to important structures such as the common bile duct. The use of the microwave devices bypasses the application of RF or laser devices in this application. Tabuse

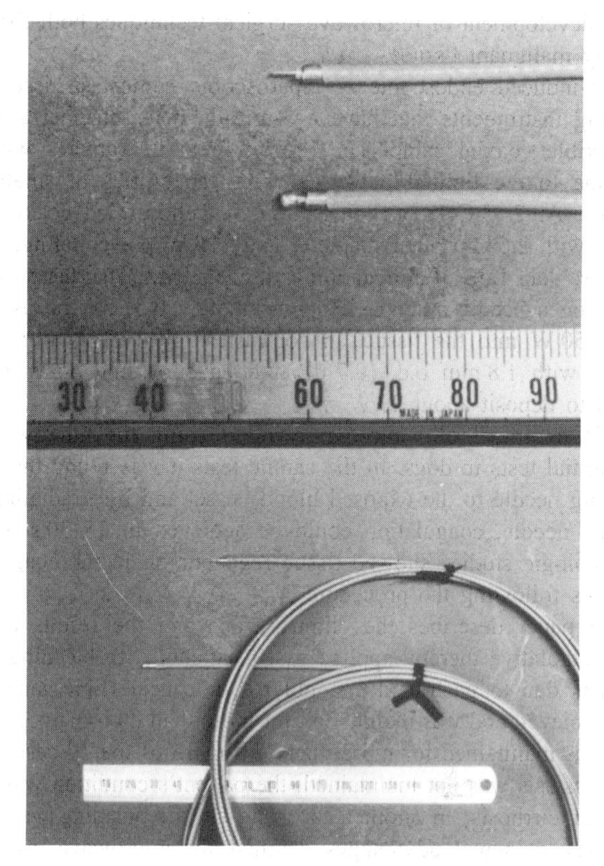

FIGURE 5.7 Needle radiators used in upper gastroenterologic endoscopic surgery. (Courtesy of K. Tabuse.)

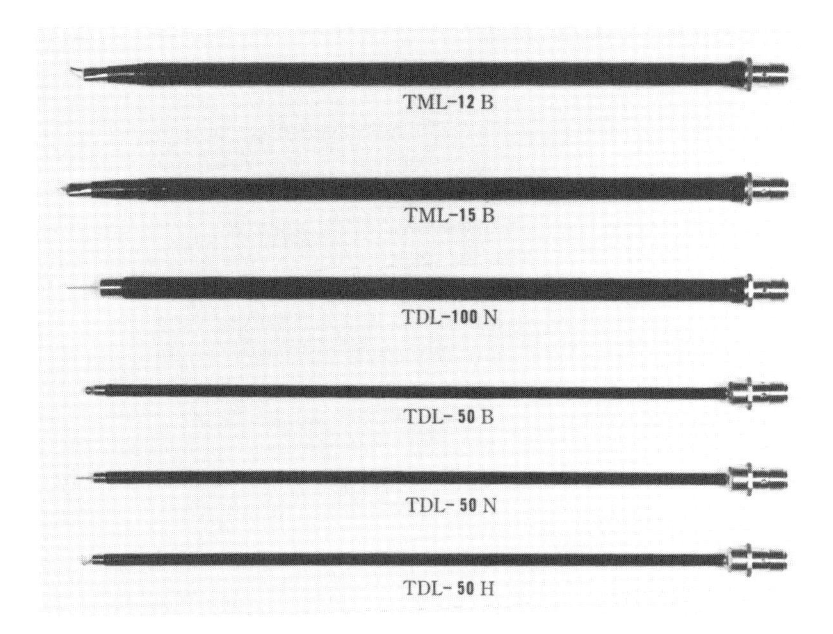

FIGURE 5.8 Needle radiators used in laparoscopic surgery (Courtesy of K. Tabuse).

reports that seven of the cases required conversion to open cholecystectomy. These included three due to common bile duct injuries not associated with the microwave device, two due to uncontrollable bleeding from the cystic artery, one severe adhesion, and one stone incarceration in the cystic duct. There were no fatalities.

In addition, the microwave needle technique has been employed through endoscopes and laparoscopes to cause the necrosis of malignant tissue. The microwave field is not used for hyperthermia, that is, a gentle elevation of tissue temperature by a few degrees centigrade to sensitize the malignancy to ionizing radiation or chemotherapy, but is instead used at a power level sufficient to destroy the tumor by direct heating to coagulate the malignant mass. The clinical trials undertaken in Japan by Tabuse and other medical investigators have been sufficient to encourage a continuing evaluation of this form of cancer therapy. The process of evaluation of a surgical technique such as this is lengthy, because of the number of different cases that must be treated, the variation of individual cases, the long time required to determine the efficacy of the treatment, and the need to compare the results with those obtained by several other therapeutic methods. The microwave coagulative technique has shown itself to be useful in reducing inoperable gastric, rectal and esophageal tumors. Other applications reported by Tabuse include coagulation therapy for benign prostatic hypertrophy and prostatic cancer, gallstone lithotripsy, and partial nephrectomy.

5.7 THE MICROWAVE SCALPEL

The *microwave scalpel* is a device that integrates the effect of a localized microwave field with a scalpel blade or dull-edged dissecting tool. The concept of linking a microwave radiator to a scalpel[19,20] was developed with the idea of streamlining Tabuse's needle instrument and providing a device capable of producing continuous charred transections. Microwave energy is fed from the generator through a flexible cable and then through a coaxial line in the handle. It is then radiated by a small loop antenna in the blade (Fig. 5.9) into the tissue, creating a thick eschar that is effective in stopping leakage and can also serve to anchor ligatures. The thickness of the eschar layer is determined primarily by the power level and by varying the rate of cutting and the time of contact of the tissue with the radiating loop. An inherent safety feature of the device is that the microwave power is not radiated when the loop is removed from tissue but is instead reflected back to the generator. Also, because the microwave field radiated by the loop has an energy density that is too low to create the cellular explosions effected by high-energy density devices such as the laser scalpel, the microwave device does not cut effectively; it acts to coagulate. Thus, to pierce the capsule of the spleen or liver, the blade of the scalpel must be used as a (mechanical) blade by the surgeon, or an ordinary blade or bovie used to make the small initial incision through the organ capsule. Once inside and covered by tissue, the microwave field will effectively seal vessels up to 2 mm in diameter. If the microwave dissecting tool is employed, the soft tissue may be separated and sealed, leaving larger vessels uncut and visible for ligation. In operating the device, the radiation is generally directed outward from the broad side of the device and is concentrated in the area at the front of the loop. The procedure to be used may be described as a slow, steady severing of the tissue using the tip of the device and the coagulating field in the area over the tip and the front of the loop. This technique is simple, and the microwave scalpel and related devices require only minimal training of personnel. Surgeons quickly grasp the essentials of their functioning. Possibly the only obstacle is the minor one that the uninformed user most often attempts to employ the devices in some manner that is based on the familiar RF bovie electrosurgical tool, which functions from outside the tissue. The surgeon must be made aware that because the bovie energy

FIGURE 5.9 Microwave scalpel.

is concentrated at the point of entrance of the spark, it can cut effectively, blowing cells apart. The microwave devices always have their energies absorbed over a larger volume, are ineffective in cutting, and depend on mechanical action for cutting. [Consequently, the surgeons who have employed the microwave devices have found it more convenient to employ a bovie (or an ordinary scalpel) to make the initial cut through the tougher capsule of parenchymal organs.] A viewing of a brief videotape of the microwave devices in use has proved sufficient for instruction. A simple in vitro demonstration can be provided by applying the devices to beef liver or steak that is irrigated with water or saline to provide the cooling normally effected by blood.

5.7.1 Microwave Scalpel System

A block diagram of the microwave scalpel system is shown in Figure 5.10. Power is supplied by a commercial microwave-oven magnetron, capable of providing 300 W of 2.45-GHz power at full voltage. Generator power is reduced for this application to the 145–185 W range, however. After passing through a unidirectional coupler, the energy is carried a special low-loss flexible cable that is connected by a bayonet quick-connector to a short section of $\frac{1}{8}$-in. semirigid coaxial waveguide that passes through the haft and connects in turn to the radiating element tip. Power is turned off and on using a foot-switch run off of a low-voltage supply to control a thyratron switch to turn off the magnetron high voltage. The electronics circuits include a beeper alarm and power lights

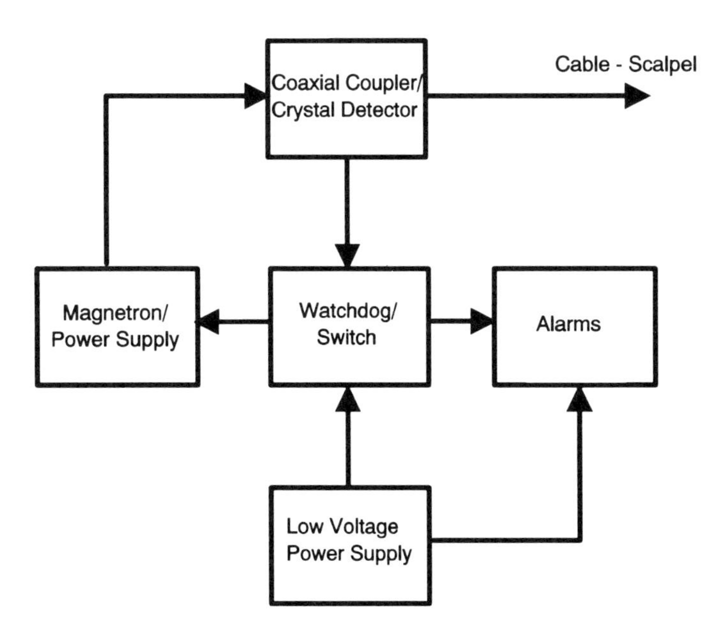

FIGURE 5.10 Microwave scalpel system.

plus a "watchdog" circuit that monitors reflected power using the signal from the directional coupler applied to a second electron switch to protect circuits against overheating. The microwave generator units weigh about 32 lb because of the high-voltage transformers required in the magnetron power supply, but were packaged in compact units measuring only 7 × 10 × 14 in. (see Fig. 5.11). This small package reflected the intention to permit employment of the system in emergency trauma units. This intent has yet to be realized, however, because investigational devices require informed-consent release documents, which cannot be obtained in emergencies.

A particularly critical item in the system is the coaxial cable that connects the scalpel to the generator. It is imperative to the manipulability of the device that this cable be as light and flexible as possible and that the connection be made through a bayonet swivel coupling. It is also necessary that this cable (and its connectors) have high-power handling capabilities and be as low-loss as possible to minimize cable heating. Ordinary cable types cannot meet the requirements; a Goretex 0.19-in.-o.d. coaxial assembly was chosen for this microwave scalpel system. This cable meets the electrical requirements, can be autoclaved, weighs only 14 g/ft, and is very flexible. The insertion loss for a 6 ft cable was measured as less than 1.5 dB in the scalpel system. Even this remarkably low level of absorption is sufficient to create cable heating effects. The power budget is complicated by the fact that part of the energy reaching the scalpel is reflected

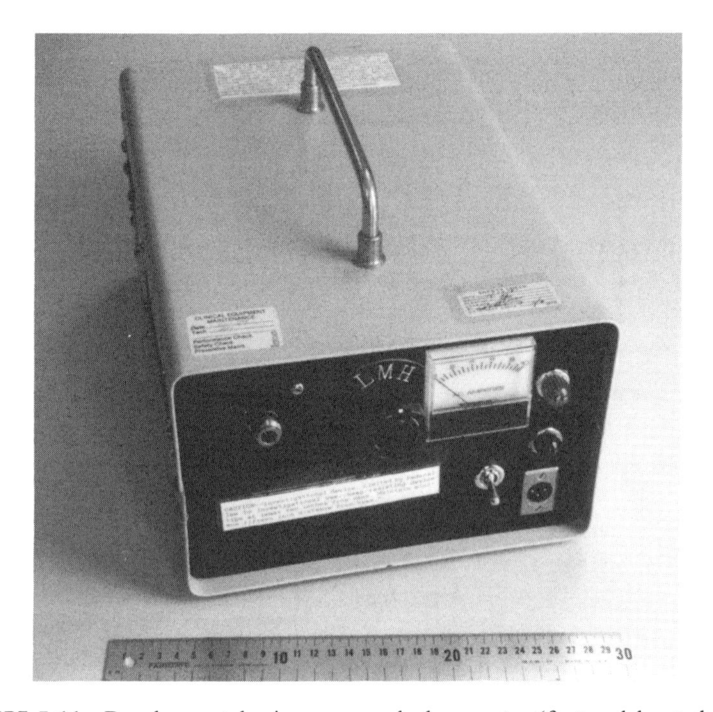

FIGURE 5.11 Developmental microwave scalpel generator (foot pedal not shown).

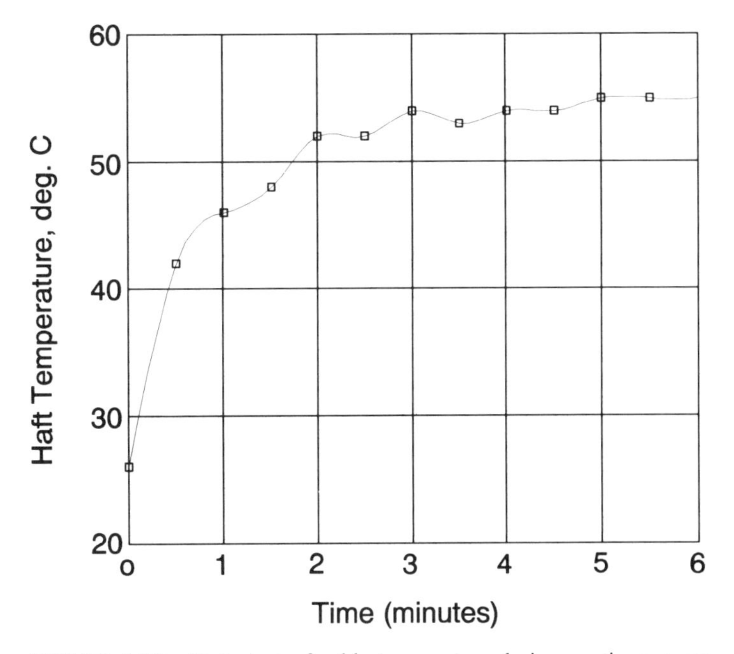

FIGURE 5.12 Static test of cable temperature during continuous use.

back down the coaxial line. It is estimated that nearly 40% of the generator power is absorbed in the cable when the scalpel is in operation in tissue, with the result that after several minutes of use the cable temperature rises to about 50 °C. The results of static tests (Fig. 5.12) confirm the observations made during clinical trials. Common cable types are, of course, unusable in this application.

5.7.2 Animal Trials

The animal trials of the microwave scalpel[21,22] (and many later clinical trials) were designed and performed by Dr. William P. Reed at the University of Maryland School of Medicine, assisted by Dr. Fredrick K. Toy. Because the microwave scalpel was initially conceived of as a device that could be employed in shock trauma units for partial splenectomies of damaged spleens, the initial trials of the microwave scalpel were partial splenectomies using dogs. A midline incision was made, the canine spleen mobilized, and the gastrosplenic ligamentous attachments cleared off at the point of the proposed resection. The major segmental branches of the splenic artery were not ligated, and no clamps were applied to the splenic pedicle. One pole of the spleen was subjected to a sharp surgical trauma, and then using the microwave scalpel, a partial splenectomy of the injured pole was performed. The amputations required only 5–10 min, and the cut surfaces were then dry and free of seepage (see Fig. 5.13). In other animals lacerated areas of the spleen were directly coagulated without splenectomy.

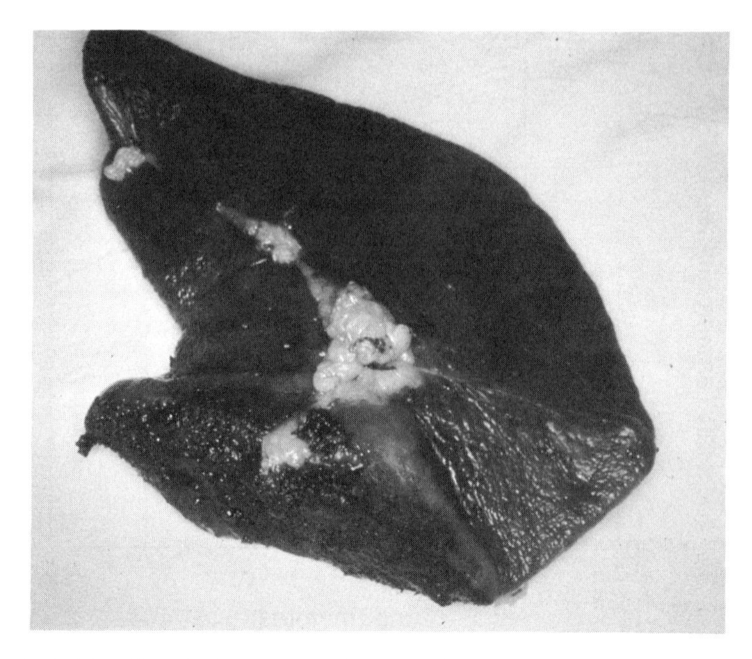

FIGURE 5.13 Canine spleen pole resected using the microwave scalpel.

Blood loss was minor, averaging only 5 ml in one trial. The microwave scalpel performance was compared to that of an RF bovie. Despite repeated attempts, it was not possible to obtain a seepage-free surface using the bovie. Temperature increases in the spleen were measured in one trial and found to be less than 5 °C at 2 cm from the coagulated edge during the procedure (see Fig. 5.14). These results are quite similar to those obtained by Tabuse.

More than 25 dogs were employed in these tests. Four dogs were assessed postoperatively using liver–spleen scans and found to have functional splenic tissue, excepting a sharply marked acellular layer of necrotic tissue about 4 mm thick on the cut surface. The dogs were sacrificed at 2, 3, 7 and 8 weeks; there was no evidence of hemorrhage, infection, abscess, or splenic necrosis. In animals sacrificed at 7 and 8 weeks, the acellular layer was found to be largely absorbed and replaced by a fibrous pseudocapsule.

Besides the canine spleen experiments, liver resections were performed on several pigs. The results of these trials were similar to the canine trials. Experiments were also undertaken to assess any risks posed by the presence of the microwave eschar acting as a possible culture medium for microorganisms in the possible case of splenic repair carried out in the presence of bowel injury.[23] These experiments were carried out using anesthetized rabbits intraperitoneally injected with *E.coli* bacteria one hour before total or partial splenectomy, using either microwave or conventional techniques. These trials showed no significant differences between groups operated on with conventional and microwave surgery;

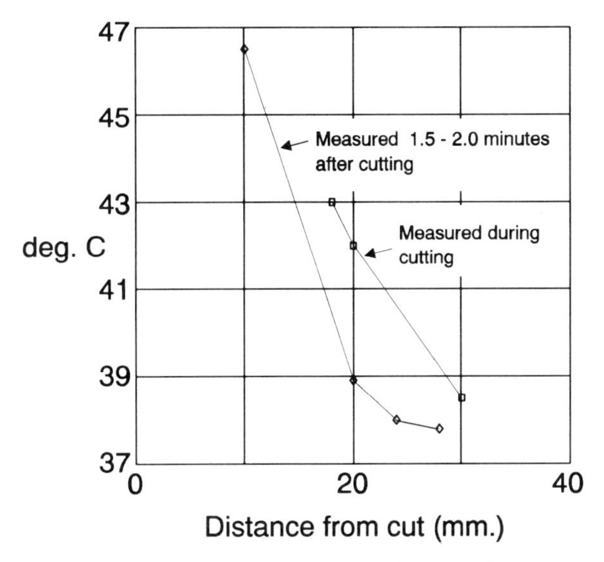

FIGURE 5.14 Temperature increases in tissue during microwave scalpel animal experiments.

all animals showed positive blood cultures, and autopsies showed no abdominal abscesses. Survivors showed neither sepsis nor abscesses. These results led to the conclusion that the microwave eschar presented no risk as culture medium for microorganisms.

Because trauma patients have sometimes sustained simultaneous injuries to the spleen and colon (plus a period of preoperative transportation and resuscitation), an important aim of the program was to determine whether the microwave-produced eschar on the spleen would resist infection in the presence of colonic contamination. The investigation of whether the eschar presents a barrier to bacterial penetration was blocked, however, by a bar placed on the experiments by the University of Maryland Hospital Animal Control Committee, which objected to the proposal to infect dogs before operation without subsequent use of antibiotics. This issue was never resolved. Consequently, although strong evidence exists that the use of microwave coagulation does not increase the risk of sepsis, the conclusion that the thick microwave eschar acts to bar infection is still conjectural.

5.7.3 Microwave Scalpel Devices

During the experimental trials of the microwave scalpel, three modifications of the basic microwave surgical tool were designed to improve the operation and ease of application. Each represents a form of miniaturized loop radiator matched for 2.45-GHz microwave transmission into tissue.

In the original form, the radiator consisted of a single loop encompassed in an ordinary surgical scalpel blade. A modified design consisting of a double-loop radiator was found to be superior (see Figs. 5.9 and 5.15). The double-loop design provides a better electrical match and is more immune to electrical breakdown than is its predecessor. The second of the three modifications also employed this double-loop configuration; it is used as the radiating element in a dull-edged rectangular dissecting tool that is used to separate soft tissue without cutting large vessels (radii $> 2 - 3\,\text{mm}$) that cannot be coagulated by the fields. This tool, which is also shown in Figure 5.15, was found the most useful in liver surgery in which the tissue is readily split without need for a sharp instrument. (As previously mentioned, the energy density in the microwave field is much less than that of either the bovie or the laser scalpel. Consequently, the microwave fields are relatively ineffective for cutting, and the devices depend on mechanical cutting action.) In using a microwave scalpel or dissecting tool, as the surgeon cuts, coagulation is achieved on both sides of the blade. It has been found useful to have the devices coated with Teflon (hence the black appearance in the photographs) to reduce adhesion of char to the blade or dissecting tool. (Only a "soft" Teflon coating is possible since the hard coatings require the use of oven temperatures exceeding the melting point of even the hard solders.) This solution does not eliminate this problem completely; char is built up on the device over several minutes of use, and the surgeon may change the handle or device (taking advantage of the bayonet connector) or simply scrape off the char, as is often

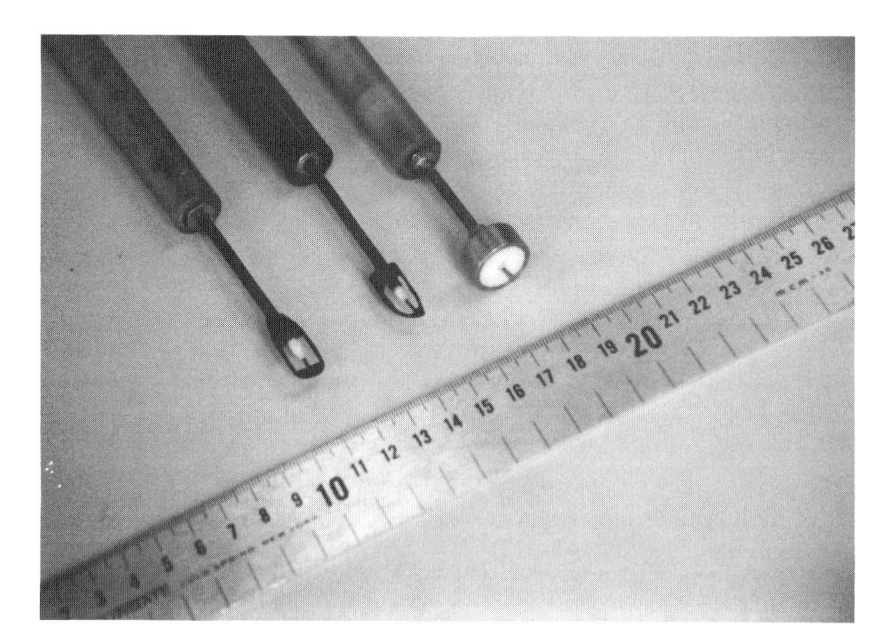

FIGURE 5.15 Microwave dissecting tool, scalpel, and surface coagulating tool.

done with a bovie probe. The device construction is hardy enough to withstand scraping with an ordinary scalpel, for example.

The third new device form is a cylindrical surface coagulator consisting of a loop–cavity radiator, employed for coagulation of flat surface regions that are less conveniently treated by the other devices. The configuration is also shown in Figure 5.15. This applicator has been used, for example, in spot coagulating sites on organ stumps, or on the sides of wedge cuts. Its effect is to produce a coagulated area about 1 cm in diameter.

The design of these devices has been based on simple principles and specifications. The loops in the scalpel and dissecting tool are approximately a wavelength (in tissue) in circumference and are designed to produce a VSWR of < 1.5 when fully immersed in tissue equivalent dielectrics. This modest requirement is achieved without any special matching devices and is sufficient to achieve a power reflection coefficient of less than 4%. Of course, full efficiency is only sporadically attained in use, since the tissue medium surrounding the device is quickly converted to eschar. A brown "mud" is seen bubbling from the area of application; the color is probably the result of the denaturation of hemoglobin. Measurements made during the animal trials showed a wide variation of reflected power as the surgeon used the devices.

5.7.4 Clinical Experience

Clinical trials of the microwave scalpel were conducted at Baystate Medical Center (Springfield, Massachusetts), Mount Sinai Hospital (New York, New York), the University of Rochester Medical Center (Rochester, New York), and the University of Maryland Hospital (Baltimore, Maryland). Twenty-nine patients have been treated at these institutions to date. All procedures could be classified as at least partially successful (Figs. 5.16, 5.17).

The most complete set of clinical trials were performed at Baystate Medical Center by Dr. W. P. Reed.[24] Two preliminary clinical trials of the microwave devices involved two patients, aged 71 and 73. These patients underwent partial splenectomies using the microwave system, before inclusion of the spleen in a resection of the pancreatic tail for malignant tumors. Hemisplenectomy was rapidly accomplished in each case, resulting in a dry resected surface. (Long-term followup for bleeding was not available because the remaining spleen was removed as part of the planned pancreatic procedure.) The results of seven complete later trials at Baystate Medical Center as reported by Reed follow.

The first two cases involved the repair of splenic injury that resulted during colon resection for carcinoma. In the first patient, a 58-year-old male, quick control of rapid bleeding was obtained despite cirrhosis with mild hypertension. Postoperative radionuclide scans of the salvaged spleen showed normal function. In the second patient, a 72-year-old female with a 10-cm carcinoma attached to the abdominal wall and stomach, splenic bleeding was initially controlled by the microwave system; however, a second tear was made in the spleen during later repositioning. Because of the prolonged character of the operation, which

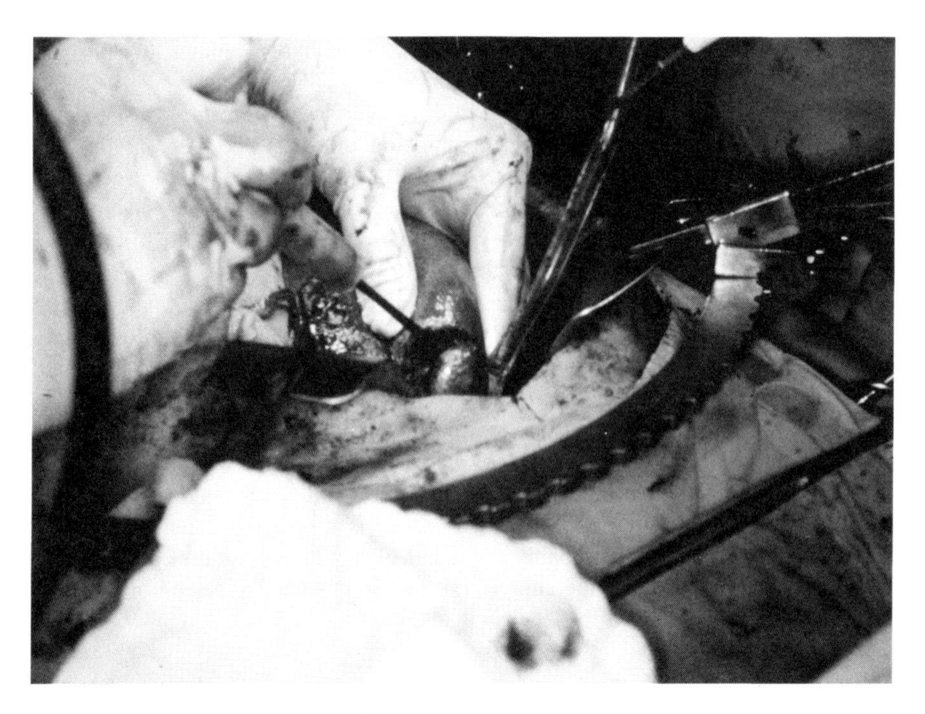

FIGURE 5.16 Microwave scalpel in hepatic resection.

included partial colostomy, partial gastrectomy, and resection of a portion of the abdominal wall, splenectomy was carried out. Both patients are alive and well.

Subsequently, four female patients underwent resection of liver tumors. The first of these patients underwent left lateral segmentectomy for a primary hematoma; hemostasis was excellent despite the presence of cirrhosis. The patient made an uneventful recovery but developed a new tumor 8 months later. This tumor was identified as a new primary, rather than a recurrence of the disease; the patient subsequently committed suicide. The second patient underwent a subsegmental resection of an 8-cm hemangioma on the right lobe of the liver; excellent hemostasis was obtained using the microwave dissecting tool, although it was necessary to ligate two large vessels within the base of the liver wound. Because of the use of the blunt microwave instrument, these vessels were easily identified and ligated before they were injured. The patient made an uneventful recovery. The third and fourth liver patients, aged 71 and 43, respectively, underwent resection of seven and six separate hepatic metastases, respectively. (Metastatic tumors of the colon appear most frequently in the liver, and their surgical removal is a difficult procedure made dangerous by the possibility of uncontrollable bleeding.) Use of the microwave dissecting tool allowed rapid removal of the multiple lesions with minimal blood loss in these cases. The first patient made an uneventful recovery without evidence of fever or other postoperative complications. The second patient also recovered successfully after experiencing a mild postoperative fever.

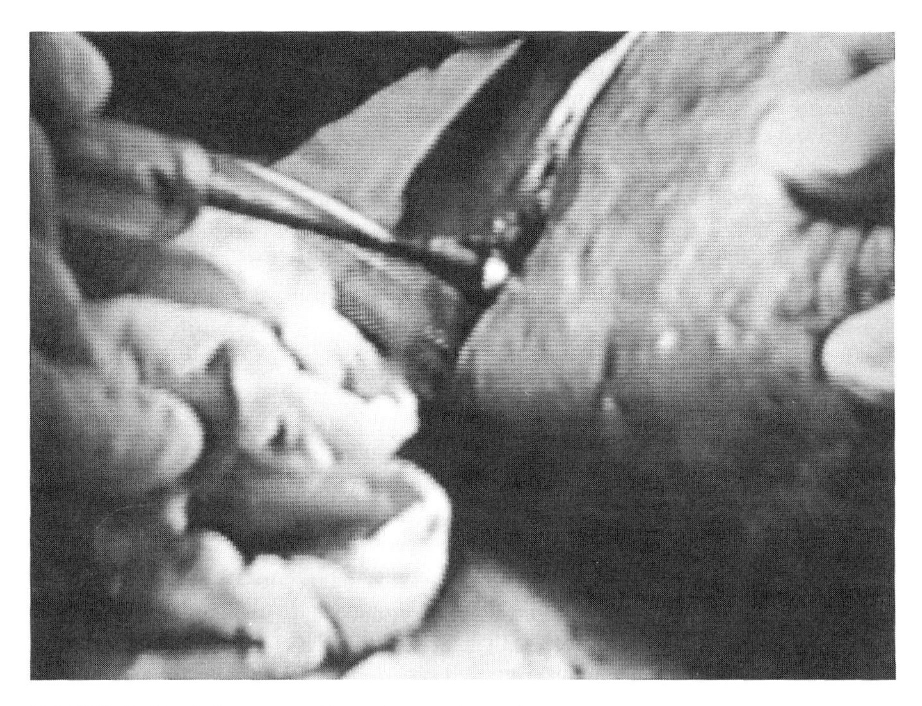

FIGURE 5.17 Microwave dissecting tool employed in resection for splenomegaly (Gaucher's disease).

One patient was treated for massive pelvic hemorrhage that followed abdominal peritoneal resection. This patient was a 95-year-old male who had undergone the resection after failing radiation therapy to control a large rectal cancer; early in the postoperative period, he was noted to have brisk pelvic hemorrhage and was returned to the operating room. It was not possible to control the bleeding with packing or ligation. At this point the microwave system was used, with ultimate control of the hemorrhage. The patient recovered and showed no recurrence of the disease 2 years later.

Besides the two preliminary trials and the seven complete trials listed above, another, incomplete trial of the microwave system was stopped because of a microwave equipment problem, later identified as due to a faulty power switch. This trial involved a 59-year-old patient who underwent resection of four hepatic metastases. The operation was concluded using standard techniques, and the patient remained disease-free 7 months later.

5.8 MICROWAVE STANDARDS AND SAFETY CONSIDERATIONS

Reports appearing in the Soviet literature in the 1960s of many vague nervous system and cardiac symptoms connected to extended exposure to very-low-level electromagnetic fields generated intense scientific and public interest in the possibility that some effect other than simple heating was involved. The scientific studies have continued on a wide scale, producing a large, often confusing scientific literature. Many results obtained (including much of the early Soviet work) have been contradicted in later experiments. No firmly constructed or widely accepted theory of the mechanism of nonthermal effects exists, nor is it understood whether any of the effects reported are harmful or helpful, or simply fall within the range of normal variation of biological parameters. Epidemiologic studies have not produced definite evidence of ill effects. In 1986, an advisory committee to the National Academy of Sciences – National Research Council reported on the results of the large body of work in this field thus far. It concluded that the available evidence was still inconclusive and that judgement must be reserved on the question of health hazards arising from nonthermal effects of longtime microwave exposure. However, as a result of the scientific investigations, a series of standards for electromagnetic field exposures have been adopted in the United States and abroad. Occupational microwave safety exposure standards vary from country to country, but we refer here to the ANSI (American National Standards Institute) C95.1 Committee on Electromagnetic Radiation criterion and to the report of the International Radiation Protection Association (IRPA) produced in collaboration with the World Health Organisation (WHO). The ANSI and IRPA voluntary occupational exposure limit recommendations are based on *specific absorption rates*, that is, on watts absorbed per kilogram of tissue; recommended maximum radiation exposure power density levels are derived from these rates. At the frequency of operation of the microwave scalpel the ANSI and IRPA standards are similar: 4 W/kg averaged over any 6 min and any 1 g of tissue. The corresponding maximum radiation power density level is 5 mW/cm^2 averaged over any 6 min. It may be noted that medical applications of microwaves, such as microwave diathermy and microwave hyperthermia treatments, are exempt from this voluntary standard. Patients receive large exposures during such treatments and because diathermy radiators are designed to operate in air, the leakage fields that have been incident on the operators of this type of equipment have been many times the recommended levels. The safety recommendations for the microwave surgical devices described in the previous section, however, are intended to ensure that even the brief exposure of operators of the microwave surgical instruments falls below the recommended safety limits. The surgeons are informed that if the safety standards are to be obeyed even in the unlikely worst case, their fingers are not to be brought closer than 2 in. to the device during operation.This rule does not impose a significant restriction on the use of the device, but the videotape records of the surgical trials reveal that the rule has sometimes been violated during the trials. It could be argued that any ill effects due to even much higher levels of microwave exposure of the fingers is very unlikely (it would be equivalent to a brief mild diathermy treatment). However, there is no doubt that a slight slip that brought the blade into direct contact with fingers would cause a burn, and avoiding close approach to the blade is a prudent safety precaution.

An obvious risk to the medical staff (and patient) is that of accidental contact with the device while microwave power is on. This type of risk also exists with the use of other electrosurgical devices, of course. Surgical gloves offer only partial protection; clothing, especially loosely fitting or in multiple layers of fabric, offers considerably more protection and reduces the possibility of an accidental direct content microwave burn to other parts of the body. Microwave burns differ from ordinary burns because the heating occurs over an appreciable depth of tissue while the sensory nerves are on the skin surface; consequently, the victim is unable to immediately judge the effect of the heating.

The use of nonexplosive gases in the operating room has greatly reduced the possibility of explosion, but there have been occasional reports of explosions of body gases resulting from the use of the RF electrosurgical devices. This possibility is extremely remote during use of the microwave devices because the devices are not intended for use in colon surgery and do not normally produce a spark. An occasional spark due to ignition of a small fat globule may be observed, however.

Early models of cardiac pacemakers were subject to the effects of microwave fields. These effects occurred as the result of detection of the power modulation of microwave generators by elements in the pacemakers. Pacemakers now in use are shielded and are not subject to such effects. [As a result, the U.S Food and Drug Administration (FDA) has removed the requirement that warning notices be posted to advise pacemaker users of the presence of microwave ovens.] For this reason, as well as because of the low level of the radiated fields, the risk of malfunction of a pacemaker in either patient or staff because of the fields radiated by the surgical microwave devices is regarded as negligible.

Tabuse has not reported any evidence of stray electromagnetic radiation during operation of the needle antenna devices. Presumably such fields are very weak, since the needle radiator is always immersed in tissue during operation. Of course, some leakage from currents reflected back down the outside of the coaxial line is possible. Such effects have been observed in the operation of microwave hyperthermia devices. The microwave scalpel devices, however, are often only partially immersed in tissue. To evaluate the amount low-level of radiation exposure of operating room personnel during use of the microwave scalpel, a series of laboratory tests were carried out. The results of these tests were compared with relevant safety standards, described in this section.

It was clear from the outset that the large mismatch between the loop and free space would prevent the radiation of large energies. Moreover, the small loop : wavelength ratio would also guarantee that no directed (focused) radiation could occur. Thus a thumbnail calculation indicated that the radiation of 2 W from the scalpel loop would lead to power densities of the order of only a fraction of 1 mW/cm^2 at distances at 10 in. Simple measurements made with a Narda radiation probe at distances of 6 and 15 in. from the blade in air confirmed this preliminary conclusion. (These measurements were made with the generator operating at its normal power levels, about 150 W at the loop radiator). The maximum radiated power densities at these distances were found to be about 0.3

and 0.1 mW/cm^2. When the probe was held only 2 in. from the blade (as close as the styrofoam sphere on the probe allows), the measured maximum power density was 2–3 mW/cm^2. These results are in rough agreement with an inverse-square law for the variation of power density with distance (although clearly the distances fall within the near field of the device). As expected from theory, it was observed that the maximum fields were always in the direction perpendicular to the plane of the blade and along the line through the center of the blade loop; the levels fell off rapidly in other directions.

Of course, the necessity of maintaining a sterile field about a human patient prevented any attempts to measure stray radiation during clinical trials. A few attempts to measure stray radiation during an animal surgery trial involved moving the probe about the area of excision at distances of roughly 8–15 in. It indicated very low levels of stray radiation, less than 0.1 mW/cm^2, while the blade was in tissue. Very low levels of stray radiation were also observed in the laboratory when the blade loop completely was immersed in water, or with only the front half of the loop in water, simulating the general conditions often encountered during surgery.

At the suggestion of Dr. H. Lu of the Department of Biophysics at the University of Rochester School of Medicine, an attempt was made to obtain a worst-case condition for stray radiation. It was speculated that this would occur with absorbing material (tissue) on one side of the blade and radiation leaking through a thin layer of tissue on the other side. (This would occur if the surgeon attempted to "pare off" a thin layer of tissue using the scalpel.) This condition was modeled by having the blade placed in water with one side covered by only a thin layer of liquid. It was found that as the thickness of this layer varied from a fraction of a millimeter to several millimeters, the power density on the axis of the loop at a distance of 2 in. from the blade increased from about 2 mW/cm^2 to about 7–8 mW/cm^2 for a layer 2–3 mm thick, falling again to low levels for thicker layers. Our interpretation of this observation was that for a thin layer, the blade is electrically mismatched and the power is for the most part returned to the generator; for thicker layers, the blade is matched but the layer effectively succeeds in absorbing the energy. The stray radiation power density is a maximum for a narrow range of intermediate layer thicknesses. Although this geometry appears unlikely to happen in practice, it could conceivably occur momentarily if the surgeon placed the blade flat against tissue covered by 2 or 3 mm of blood. Of course, this power density decreases rapidly with distance from the blade (presumably it decreases with the inverse square of the distance) and occurs only in the direction along the loop axis. Also, the configuration suggested above would not persist for more than a few seconds since the blood layer would be rapidly coagulated and the mismatched loop would reflect power back to the generator.

The significance of these results is subject to interpretation, but clearly the radiated power densities are very far removed from the levels that are required to produce microwave cataracts. The lens of the eye is sensitive to microwave heating. More than 50 cases of cataracts produced because of microwave exposure have been reported, principally among radar technicians working with ultrahigh-power microwave equipment. Experiments have shown that the generation of cataracts in rabbits by microwaves requires continuous exposure for 30 min to fields of more than 250 mW/cm^2.[25] (Rabbits exposed all day for weeks to power density levels of the order of 1 mW/cm^2 manifested no discernible changes in the character of the eye.) It is also known that because of the geometry of the eye socket, higher fields are required to produce cataracts in humans. Thus, both experiment and theory have indicated several orders of magnitude difference between both the power level and the expected duration of exposure of the surgeon with that required to produce cataracts, assuming that surgeons do not bring the device into near contact with their eyes for an extended time. However, these conditions could be contravened by bringing the radiating scalpel close to the eye. Ordinary eyeglasses cannot provide significant protection. Surgeons are warned that the device should never be brought closer than 15 in. to the eye when active.

In addition, the power densities at the nominal distances to the bodies, arms, and heads of the medical staff are also very low, far below the levels set in current microwave exposure standards. A similar statement also applies to the levels at the position of the surgeon's hand holding the scalpel and even to fingers within a few inches from the device under usual conditions. The worst-case condition described above, however, represents a situation in which the fingers of the surgical staff put into close proximity to the device may be briefly exposed to higher levels. Is such exposure significant? These levels are low compared to leakage levels at similar distances from microwave medical diathermy applicators, and despite the thermal sensitivity of fingers, the levels mentioned are below the level required to produce a sensation of heat. A factor that must also be considered is the protection offered by surgical gloves. A simple experiment in which the plane-wave reflectivity of tissue bolus with and without a covering layer of surgical latex glove material was performed to obtain an indicator of this factor. Our measurements show that surgical gloves will reduce the effective power density level by a factor of about 0.5.

5.9 MICROWAVE METHOD FOR MALE STERILIZATION

An oddity in the history of microwave biomedical devices is the Kihn instrument described in "Method of performing male sterilization".[26] The basic claims of Kihn patent (see Fig. 5.18) refer to "an instrument having a body with two members adapted for closing toward each other and holding there between an anatomical element" (the male vas deferens). The "members" are waveguides, miniaturized by using a high-permittivity dielectric filling. The surgeon places the skin of the scrotum with the vas enclosed between the waveguide, and microwave power is fed through a coaxial cable, to a coax-to-waveguide transition, through one waveguide element to the scrotum and on to the other waveguide element and a coaxial transition, to a detector. Detected transmitter power level is compared

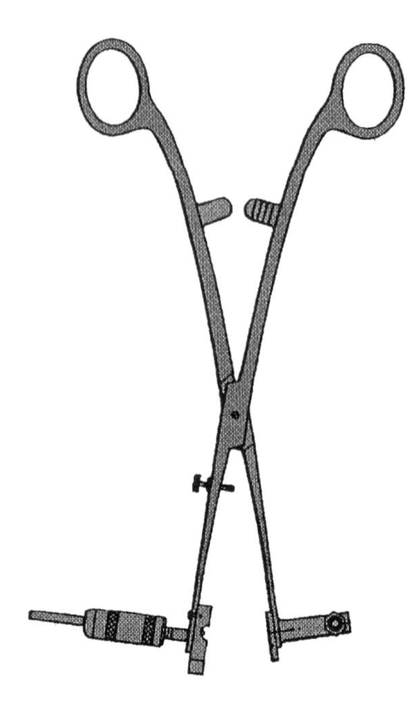

FIGURE 5.18 Microwave male sterilization device.[22]

to generator power and is used as a measure of the microwave radiation used to heat the sperm conduit and destroy its capabilities. The Kihn patent refers to operation at a frequency between 4 and 7 GHz, that is, in the *C*-band of the microwave spectrum, with 6 GHz recommended. The Kihn patent does not provide radiator design details or generator power specifications. It does refer to raising the tissue temperature within the range 50–80 °C, claiming that the vas can be sealed without actually burning the skin. This assertion is based on the claim that the vas have greater microwave absorptivity than does the scrotum. Technical details and clinical data of the results obtained using this system are not available, but we have been told that the Kihn device was sold abroad in considerable quantity for the purpose of human sterilization. Such sales would not be possible under present FDA regulations, which prevent the exportation of equipment that has not been approved for use in the United States.

REFERENCES

1. Licht S: History of therapeutic heat. In Licht S (ed): *Therapeutic Heat and Cold*, 2nd ed. E Licht, New Haven, CT, 1965, Chapter VI.
2. Geddes LA: The beginnings of electromedicine. *IEEE Eng Biol Med Magn* 3:8 1984.
3. LaCourse JR, Vogt MC, Miller WT III, Selikowitz SM: Spectral analysis interpretation of electrosurgical generator nerve and muscle stimulation. *IEEE Trans Biomed Eng* 35:505, 1988.

4. Slager CJ, Schuurbies JCH, Oomen JAF, Bom N: Electrical nerve and muscle stimulation by radio frequency surgery: Role of direct current loops around the active electrode. *IEEE Trans Biomed Eng* 40:182, 1993.
5. van Gemert MJC, Welch AJ: Clinical use of laser-tissue interactions. *IEEE Eng Med Biol Magn* 8:10, 1989.
6. LeCarpentier GL, Motamedi M, McGrath LP, et al: Continuous wave laser ablation of tissue: Analysis of thermal and mechanical events. *IEEE Trans Biomed Eng* 40:188, 1993.
7. Cook MS: Laser photocoagulation of blood vessels. SPIE Vol. 139, *Guided Wave Optical Systems and Devices*. 1978, pp. 34–42.
8. Kosugi Y, Takura K, Huang MT, Okabe T: A technique of tissue coagulation by microwaves. *Med Electron Biomed Eng* (Japan) 11:16, 1973.
9. Taylor LS: Electromagnetic syringe. *IEEE Trans* BME-25:203, 1978.
10. Tabuse K: A new operative procedure in hepatic surgery using a microwave tissue coagulator. *Arch Jpn Surg* 48:160, 1979.
11. Tabuse K: Private communication, March 3, 1993.
12. Tabuse K, Katsumi M, Nagai Y, et al: Microwave tissue coagulation applied clinically in endoscopic surgery. *Endoscopy* 17:139, 1985.
13. Tabuse K, Katsumi M, Kobayashi Y, et al: Hepatectomy using a microwave tissue coagulator. *World J Surg* 9:136, 1985.
14. Tabuse K, Tabuse Y, Kobayashi Y, et al: Hepatectomy using microwave tissue coagulator in liver cancer. *J Microwave Surg* 9:1, 1991.
15. Tabuse K, Katsumi M: Application of microwave tissue coagulator to hepatic surgery —the hemostatic effect on spontaneous rupture of hepatoma and tumor necrosis. *Arch Jpn Chir* 50:571, 1981.
16. Tabuse K, Katsumi M: Microwave tissue coagulation in partial splenectomy for non-parasitic splenic cysts. *Arch Jpn Chir* 50:711, 1981.
17. Tabuse K, Katsumi M: Microwave coagulation therapy of tumors. *J Jpn Soc Cancer Ther* 17:659, 1982.
18. Uda O, Kimura K, Aoki T, et al: An experimental study of a new devised double needle for microwave tissue coagulator. *J Microwave Surg* 9:9, 1991.
19. Taylor LS, Toy K, Reed WP: Microwave coagulating scalpel. *IEEE Trans Biomed Eng* 30:834, 1983.
20. Taylor LS, et al: Development and tests of the microwave coagulating scalpel. *Proc URSI Symposium on Electromagnetic Theory*, Budapest, August 1986, Vol. 1, pp. 198–200.
21. Toy FK, et al: Experimental splenic preservation employing microwave surgical techniques: A preliminary report. *Surgery* 96:117, 1984.
22. Toy FK, et al: Microwave coagulating scalpel for operations on the solid viscus organs. *Curr Surg* 12: 131, 1985.
23. Reed WP, Gershon BA, Taylor LS: Effects of peritoneal contamination on splenic microwave eschar. Unpublished, 1985.
24. Taylor LS, Reed WP: Microwaves in surgery: Methods and results. *Proc IEEE MTT-S International Microwave Symposium*, Boston, MA, June 1991.
25. Cleary SF: Microwave cataractogenesis. *Proc IEEE* 68:49, 1980.
26. Kihn H: Method of performing male sterilization. U.S. Pat. 4,315,510, Feb. 1982.

CHAPTER SIX

Contemporary Ophthalmic Lasers

JOEL M. KRAUSS, M.D., *Mount Sinai Hospital, New York, NY*

6.1 INTRODUCTION

Ophthalmologists have been at the forefront of developing medical uses for new laser technology since the report of the first laser in 1960.[1] Photocoagulation was the earliest therapeutic laser procedure, and remains the most widely employed. There are myriad ocular applications, which have dramatically changed the treatment of eye diseases ranging from diabetic retinopathy to glaucoma. Photodisruption was introduced in the early 1980s. First used to noninvasively treat secondary cataracts, it has also seen an increasing number of applications.

Few recent advances in medical technology have garnered as much public interest as corneal ablation, which may revolutionize the treatment of the most common ocular disease: refractive error. Even as the excimer laser is in the final stages of clinical trials, work continues on the next generation of ablation lasers. The precise tissue removal afforded by these devices may also be applicable to other eye disorders. While less prominent and ubiquitous than their therapeutic counterparts, diagnostic lasers have served an important role in ophthalmic practice, often permitting more accurate and timely treatment with lasers or other modalities.

New instruments and techniques are continually being developed, and few go untested for potential ophthalmic use. Whereas the last decade has seen photodisruption, and increasingly ablation, join photocoagulation as established

New Frontiers in Medical Device Technology, Edited by Arye Rosen and Harel Rosen
ISBN 0-471-59189-0

procedures, there is good reason to believe that the field of ophthalmic lasers is far from mature. This chapter covers principles of ophthalmic laser technology and tissue responses relevant to clinical use, providing historical background but with an emphasis on new technology and areas of current investigation.

6.2 LASER–TISSUE INTERACTIONS

The effect of laser radiation on a particular target depends on the properties of both the laser and the target. The most important laser output parameters are wavelength, duration, and power. Wavelength is a function of the laser cavity's excited medium, which is a gas in argon, krypton, and excimer lasers, a liquid in dye lasers, and a semiconductor in diode lasers. According to the principle of wave–particle duality, radiation is propagated in the form of both waves and discrete quanta, or photons. As such, radiation of a given wavelength is associated with photons of a corresponding energy, such that $E = h\nu = hc/\lambda$, where h = Planck's constant, ν = frequency, c = speed of light, and λ = wavelength. Thus, frequency and energy increase as wavelength decreases. The visible spectrum extends approximately from 380 to 760 nm. The first law of photochemistry (Grotthus–Draper) states that photons must be absorbed by a target in order to initiate a chemical reaction.[2] A chromophore is a molecule, or a portion thereof, that absorbs a photon of a particular energy. Depending on the photon's energy, a chromophore can undergo bond breaking, ionization, or various types of molecular excitation.

The ability of a target, which may or may not be of a homogeneous composition, to absorb radiation is measured by the attenuation in incident radiation after a certain length of the material has been traversed. The absorbance A of a material is defined as $A(d) = \log[I_0/I(d)] = \varepsilon cd$, where I_0 = initial intensity, $I(d)$ = intensity at distance d, ε = absorptivity of the material, and c = molarity of the material. Transmission is that fraction of the incident energy that is not absorbed after traversing a particular target thickness. It is usually written in the form of Beer's law, $T(d) = 10^{-A(d)} = e^{-\alpha d}$, where $\alpha d = 2.3A$ defines the absorption coefficient α. Coefficient α is generally given in units of cm^{-1} and represents the fraction of incident energy that is absorbed per unit length of target material. Absorption length is defined as α^{-1}, or that distance at which $e^{-1} = 0.368$ of incident energy is transmitted, corresponding to 63.2% absorption. The thermal susceptibility of the irradiated tissue is denoted by the thermal relaxation time τ, which gives an indication of the time required for the irradiated tissue to carry away heat energy from the target site. It is wavelength-dependent, and is proportional to $1/4\alpha^2\kappa$, where κ measures the tissue diffusivity in square centimeters per second.

The absorption maximum of a compound is that wavelength in a given portion of the spectrum that has the highest probability of absorption. A plot of absorption versus wavelength yields an absorption spectrum that is characteristic of the chemical composition of that compound. Quantum yield is a measure of the efficiency with which absorbed radiation produces chemical changes, while an action spectrum is a plot of the relative efficiency of the photoreaction versus wavelength.

6.3 PHOTOCOAGULATION

6.3.1 Nonlaser Photocoagulation

The ocular effects of radiant energy have been recognized since at least the time of Socrates, who described solar retinitis after direct observation of an eclipse. Following an eclipse in 1945, Meyer-Schwickerath noted the similarity between damage to the macula (Fig. 6.1) in viewers and that produced by diathermy, and

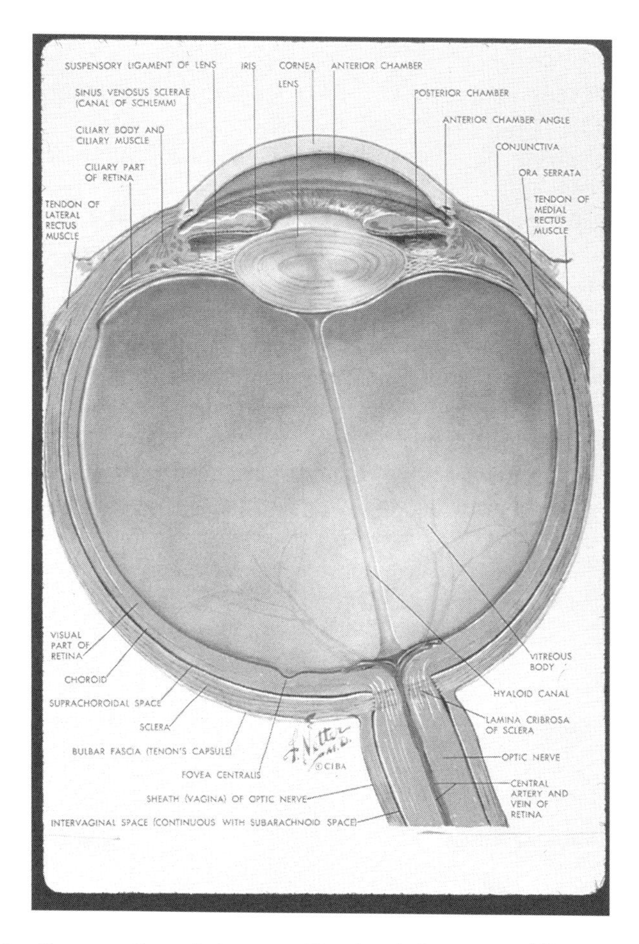

FIGURE 6.1 Cross section of the eye. (Reprinted from Netter[516] with permission of Ciba-Geigy.)

began to study radiation sources and delivery systems for intentional therapeutic photocoagulation.[3] He constructed a device that focused sunlight on the retina, but despite some clinical success, the difficulty in regulating exposure—not to mention the frequent occurrence of cloudy days—made this approach impractical. In 1956 Meyer-Schwickerath settled on the xenon arc, which soon gained widespread acceptance. Although primarily for technical reasons this instrument has been largely replaced by lasers, it is often equally efficacious[4] and remains in use in many areas around the world where lasers are unavailable. Xenon emission consists of all wavelengths between 400 and 1600 nm, so that a full-thickness burn is achieved, without the ability for selective targeting of ocular tissue. The exposures last several seconds, requiring retrobulbar anesthesia to reduce pain and eye movement.

6.3.2 Laser Photocoagulation

In contrast to incandescent sources, a laser emits one or several discrete wavelengths. A high degree of collimation (low divergence) permits focusing to very small spot sizes. The ruby laser produces 694.3 nm radiation in pulse durations in the hundreds of microseconds, and was the subject of early clinical investigation. Although this device was of negligible value for direct vascular treatment, it was effective in controlling proliferative diabetic retinopathy (PDR; see below).[5] L'Esperance introduced the ophthalmic argon laser in 1968,[6] and it wasn't until the value of that instrument was fully realized in the early 1970s that laser photocoagulation became an established procedure. Over the past two decades several other photocoagulation lasers—especially the krypton laser in 1972, the tunable dye laser in 1981, and most recently, the diode laser—have found their way from the lab to the clinic. Many applications have been based on empirical observation, but as laser technology and photobiology advance, selective therapeutic approaches achieve desirable biochemical, cellular, and tissue effects to the exclusion of unwanted damage.

6.3.2.1 Laser Principles

Principal wavelengths for common photocoagulation lasers are listed in Table 6.1. [The carbon dioxide (CO_2) laser, with emission at 10.6 μm in the midinfrared

TABLE 6.1 Principal Wavelengths of Common Photocoagulation Lasers

Laser	Wavelength (nm)
Argon (blue-green)	488.0
Argon (green)	514.5
Frequency-doubled Nd:YAG	532.0
Krypton (yellow)	568.2
Krypton (red)	647.1
Tunable Dye	Variable (most 570–630 nm), depending on dye
Diode	Variable (most 780–850 nm), depending on diode
Nd:YAG	1064.0

(IR), also has coagulative effects, but since it is used primarily to induce vaporization because of its high absorption by water, it is discussed in the section on ablation.] Most such lasers run in a continuous or quasicontinuous mode, so that exposure is controlled by an external shutter. Photocoagulation is generally performed at durations of 100–200 ms, increasing to about 500 ms in cases of ocular media opacity. This should be distinguished from the Q-switched neodymium–yttrium aluminium garnet (Nd:YAG) and other ultra-short pulsed lasers, in which the laser actually produces only very brief bursts of radiation, with resulting average and peak (rather than constant) powers. Thus, whereas output for photodisruption and ablation lasers is listed in units of energy (joules), that of photocoagulation lasers is given in units of power (watts). For a given amount of power, the shorter the time over which it is delivered, the greater the risk of tissue rupture and hemorrhage. Several hundred milliwatts generally suffices for most purposes, although up to 1 W may be necessary. Irradiance is measured in watts per square centimeter and increases as the spot size, to which the laser output is focused, decreases. Spot sizes as small as 50 μm diameter are used to target individual vessels or in the presence of opacity, whereas those of 200 and 500 μm are generally employed for photocoagulation of the macula and peripheral retina, respectively. Smaller spots are more affected by dissipation of heat to surrounding tissue, and hence require higher irradiance to achieve the same central effect as larger spots.

Photocoagulation lasers, as the name implies, exploit tissue absorption and heating.[7] Sophisticated theoretical models have been developed to explain and predict laser-induced thermal damage in the retina,[8–11] which is proportional to the magnitude and duration of the temperature increase (represented by the Arrhenius integral[12]). The temperature rise produced by laser irradiation is a function of time, laser energy and wavelength, and the optical and thermal properties of the absorber. Modest increases of 10–20 °C induce alteration of the genetic apparatus of cells, inactivation of enzymes, and denaturation of proteins and nucleic acids, which lead to necrosis, hemostasis, and coagulation.[13] Immediate effects are visible ophthalmoscopically because of focal increases in necrotic cells and a mechanical disruption of the adjacent neurosensory retina that interferes with its normal transparency.[14] Delayed effects result from inflammation and repair processes. Water vaporization and gas bubble formation, with their attendant secondary mechanical effects, may be seen with greater temperature increases. A moderate increase in temperature under 100 °C is associated with breakage of hydrogen bonds and van der Waals forces, which stabilize the conformation of biologic macromolecules such as proteins, resulting in loss of biologic or structural activity.[15]

6.3.2.2 Argon and Krypton Lasers

Intraocular applications require the transmission of radiation, which occurs to different extents between approximately 380 and 1400 nm. Except for the long-pulsed Nd:YAG and now the diode laser, photocoagulation lasers employ radiation in the visible portion of the electromagnetic spectrum. In this range, important chromophores are melanin [in the retinal pigment epithelium (RPE;

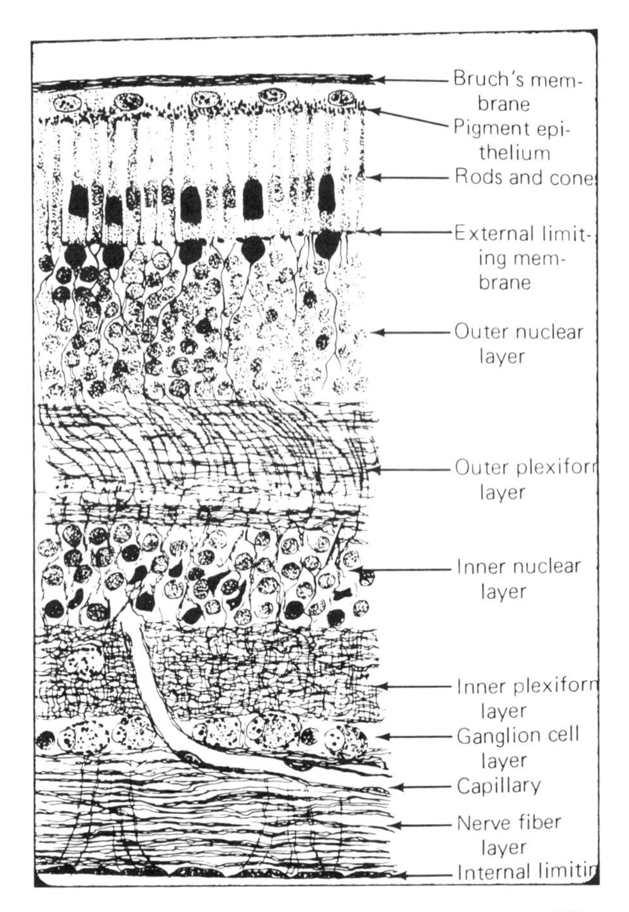

FIGURE 6.2 Cross section of the retina. [Reprinted from Warwick[517] with permission of Chapman & Hall.]

see Fig. 6.2) and iris pigment epithelium, uvea, and trabecular meshwork], hemoglobin (in blood vessels), and xanthophyll (in the inner and outer plexiform layers in the macula, as well as in certain cataracts). Figure 6.3 demonstrates how these chromophores' absorption varies in the region relevant for photocoagulation. Melanin absorption is relatively constant between 400 and 700 nm, although the gradual decrease in absorption at longer wavelengths results in deeper retinal burns.

Argon green light, at 514.5 nm, is minimally absorbed by xanthophyll, but strongly absorbed by both hemoglobin and melanin. Except in the presence of a large retinal vessel, it typically produces a cone-shaped lesion that spares the inner retina,[16] and is appropriate for direct coagulation of retinal vessels, but not for treatment through hemorrhage.[17] Collagen shrinkage in and around vessel walls and hemoglobin heating sufficient to cause thrombi, are thought to be the mechanisms by which photocoagulation seals arteries.[18] Xanthophyll

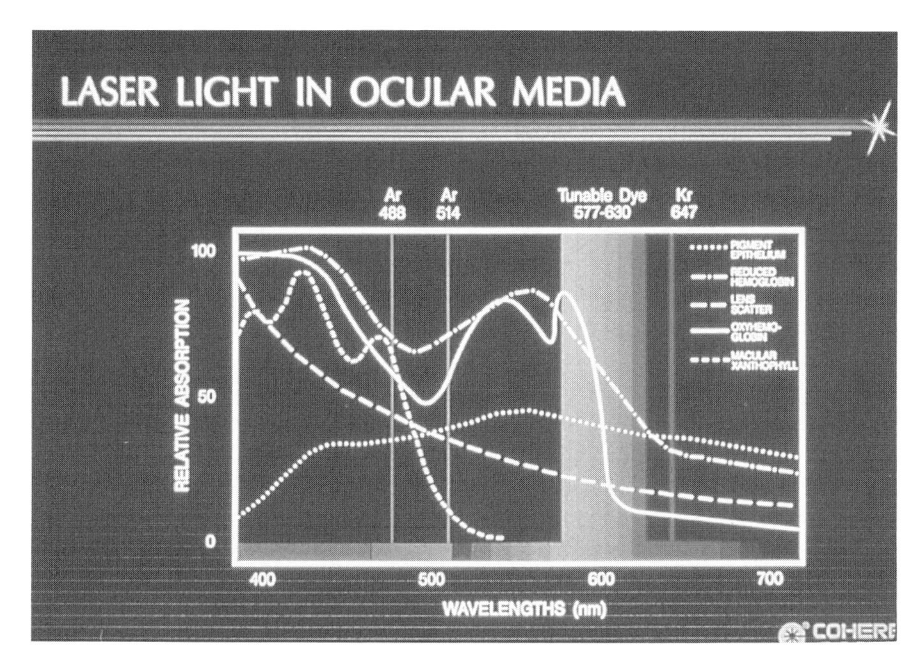

FIGURE 6.3 Relative absorption versus wavelength for various intraocular chromophores. The figure shows that (1) macular xanthophyll has greater absorption for argon blue light than for other laser wavelengths; (2) hemoglobin has excellent blue, green, and yellow light absorption but much poorer red light absorption; (3) light absorption maxima for oxyhemoglobin are located at 542 nm (green) and 577 nm (yellow); (4) reduced (venous) hemoglobin has better krypton red light absorption than does oxygenated (arterial) hemoglobin; and (5) deoxyhemoglobin has roughly the same extinction coefficient for krypton red light as xanthophyll does for argon blue light. (Courtesy of Coherent, Inc., Palo Alto, CA.)

absorbs blue light strongly, but green and especially red and yellow much less so. Photocoagulation at 488 nm thus may damage the retinal nerve fiber layer (RNFL)[19] and is inappropriate for most macular applications.[16,20] There is also concern that use of blue laser light over an extended period may decrease color discrimination in ophthalmologists in a tritan color-confusion axis.[21]

Hemoglobin has strong absorption for all colors with wavelengths shorter than that of red. Red light is thus effective in the presence of vitreous or retinal hemorrhage, through which the light readily passes, while shorter wavelengths are useful for blood vessel closure. Red light is appropriate for macular applications, but unlike green and yellow is unable to directly coagulate vessels in the event of inadvertent hemorrhage, so that it is not useful for vascular pathology. Krypton red's primary absorption is in the RPE and choroid, and it has also proved useful in panretinal photocoagulation (PRP) for PDR.[20] Disadvantages of this wavelength include increased patient discomfort and risk of choroidal hemorrhage or disruption of Bruch's membrane due to the deeper penetration, and the lower efficiency and greater complexity of krypton lasers compared with argon lasers.

If the temperature increase is too great, heat conduction can spread damage to the inner retina.[14]

Yellow light is attractive for retinal photocoagulation, since it is poorly absorbed by xanthophyll, is scattered less than argon wavelengths, is at the peak of oxyhemoglobin absorption, and has the highest oxyhemoglobin to melanin absorption ratio, as well as a high oxyhemoglobin to deoxyhemoglobin absorption ratio.[14] Retinal damage caused by krypton red light is limited mostly to the area near the RPE, whereas krypton yellow light causes less RPE damage, but does result in RNFL and ganglion cell edema and coagulation at the border between the outer plexiform and nuclear layers.[22]

6.3.2.3 Nd:YAG Laser

Radiation at 1064 nm is much less absorbed by melanin than is visible light, so tht it penetrates much deeper than the output of other photocoagulation lasers. Nd:YAG lasers can be configured so that their output is continuous or long-pulsed (up to 20 ms), with predominantly thermal effects. Peyman and associates[23,24] showed that a continuous-wave (CW) Nd:YAG laser produced retinal lesions that were similar in histologic appearance to those created with a krypton red laser, although because of the lower absorption by melanin of the IR wavelength and greater absorption (30%) by the ocular media, approximately 5–10 times more energy was required with the Nd:YAG. The ratio of oxyhemoglobin absorption to that of deoxyhemoglobin is almost 6 times greater at 1064 than at 647 nm, and combined with the IR wavelength's overall lower absorption by hemoglobin, should allow Nd:YAG radiation to maximally penetrate subretinal hemorrhage and treat vascular structures at threshold irradiances.[14] Fankhauser[25] employed 10-ms Nd:YAG pulses for experimental retinal photocoagulation and laser trabeculoplasty (LTP; see below), although the former resulted in fibroblast proliferation that penetrated Bruch's membrane and invaded the retina.[26]

A potassium–titanium–phosphate (KTP) crystal has been used to double the frequency, and thus halve the wavelength, of the Nd:YAG laser. Quasicontinuous output is typically achieved by high repetition (10 kHz) of 1-μs pulses. The resulting "pea-green" output, at 532 nm, is more highly absorbed by the RPE and hemoglobin than is argon green light. As it is nearer to absorption peaks of oxyhemoglobin and deoxyhemoglobin than are other common laser wavelengths, this light requires less power to achieve similar occlusive or obliterative effects on vessels.

6.3.2.4 Tunable Dye Laser

Dye lasers, themselves pumped by other lasers, employ one or more dyes to produce emissions over a wide range of wavelengths.[27] Tunable dye lasers allow the operator to select the desired wavelength within the range of the dye in use. This offers at least the theoretical potential for perfect matching of laser output with tissue absorption. Most photocoagulation needs would be met by a tunable dye laser capable of emitting between 560 and 640 nm, as well as at the standard wavelengths of an argon laser pump.[22] Melanin and hemoglobin would be best targeted with output in the 560–580-nm range. Selective treatment of RPE and choroidal melanin could be achieved between 610 and 640 nm. Orange light at 580–610 nm would allow partial absorption by the RPE and any underlying neovascularization, while permitting sufficient energy to penetrate to the choroid to coagulate any feeder vessels. Oxyhemoglobin absorption decreases dramatically from 590 to 600 nm, enabling highly efficient treatment of vascular abnormalities by suitable wavelength selection in this region. Thus, whereas red light would be preferable in the treatment of subretinal neovascularization (SRNV) in cases of at least normal RPE and choroidal melanin concentration, orange light should be superior for hypopigmented individuals. Orange light is also appropriate for treatment of retinal or vascular tumors, as it achieves strong penetration and coagulation. Treatment may be further customized by employing multiple wavelengths, such as irradiation with red light to obliterate deep feeder vessels, followed by yellow or orange light to directly target SRNV. It should be noted, however, that white lesions appear such because of light scattering, and that once such a lesion develops, it shields underlying tissues from subsequent irradiation.[28]

Romanelli and Puliafito[29] performed a histologic and metabolic comparison of retinal effects in monkeys of a tunable dye laser at 577, 590, and 630 nm. All exposures were performed at 0.1 s and with a 100-μm spot size. Lightly visible white lesions were produced with 100 mW, while more intense ones were created with 200 mW. All lesions were morphologically similar, with the 200 mW lesions displaying more inner retinal damage than the 100 mW ones, regardless of wavelength (Fig. 6.4).

6.3.2.5 Diode Laser

Diode lasers emit radiation with exceptional electrical-to-optical efficiency (around 50%). They are much smaller, less expensive, and more portable than traditional ophthalmic lasers, requiring only a standard electrical outlet to achieve clinically necessary power levels. Such devices are constructed by joining *n*- and *p*-type semiconductors, which serve as electron donors and acceptors, respectively, creating a recombination region at their junction. The size of the bandgap, across which photons jump and that thus establishes the emission wavelength, is determined by the addition (doping) of other atoms. The bandgap in gallium arsenide (GaAs) semiconductors is particularly well suited to producing radiation, with the application of an external electrical potential across the *p*–*n* junction creating an electron population inversion and the parallel GaAs crystal faces serving as semitransparent mirrors. Internally reflected light then stimulates further photon emission in the recombination region, ultimately resulting in output that is coherent and monochromatic, albeit very divergent (in contrast to the collimated output of the argon and many other lasers). Laser power is increased by constructing arrays of many diodes, with coupling between active zones making the total output spatially coherent.

Most diode lasers studied for ophthalmic use contain GaAs crystals doped with aluminum (GaAlAs), and emit between 780 and 850 nm (commercial versions are now generally 810 nm), although any given laser has only a single emission line and is not tunable. Radiation in this range is readily transmitted by the

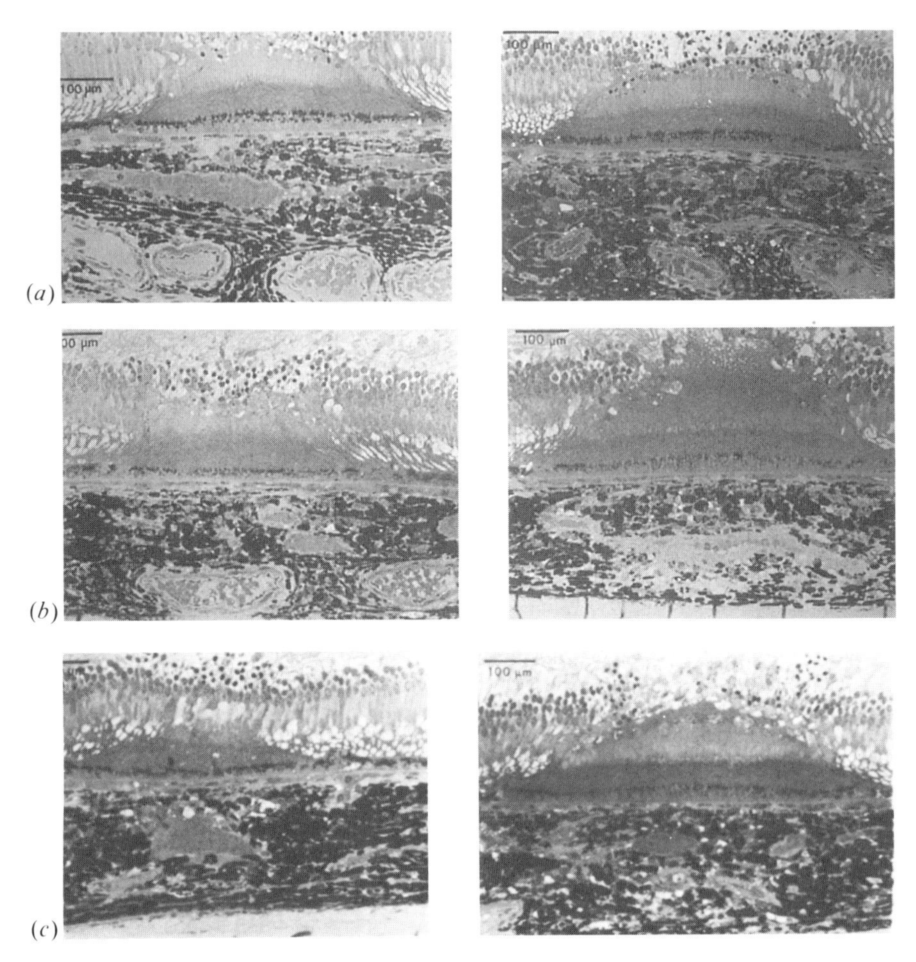

FIGURE 6.4 Lesions produced at 100 mW using (*a*) 577 nm (yellow) light, (*b*) 590 nm (orange) light, (*c*) 630 nm (red) light. Each lesion shows damage to the RPE, choroid, and photoreceptor nuclei and outer segments. (Reprinted from Romanelli[29] with permission of International Ophthalmology Clinics.)

ocular media, and in contrast to some shorter wavelengths, much less energy incident on the retina is absorbed by the RPE or choroid. However, this lower absorption requires higher laser power, whose achievement prevented the diode laser's introduction into clinical use until recently.

Puliafito and associates[30] reported the first therapeutically useful diode laser lesions. Their endophotocoagulation system produced lesions in rabbit retinas that were similar to argon, krypton, and Nd:YAG laser ones on the basis of ophthalmoscopy and fluorescein angiography, but with histologic damage limited to the outer retina. Brancato and Pratesi[31] achieved similar early experimental results, and also reported the first transpupillary diode laser photocoagulation. Using a diode laser endophotocoagulation system in rabbits, Smiddy and Hernandez[32] demonstrated that retinal cell disruption was confined primarily to the outer nuclear layer in mild burns, involved the inner nuclear layer in moderate burns, and involved ganglion cell loss in severe burns. Brancato et al.[33] first coupled a diode laser to a slit-lamp biomicroscope, achieving retinal photocoagulation in rabbits; they subsequently tested this system on a human eye scheduled for enucleation.[34]

Brancato and colleagues[35] conducted a histologic comparison of diode and argon retinal lesions in rabbits. Diode lesions had zones of coagulation necrosis and vacuolation in the RPE, while sensory retina damage was confined to photoreceptor cells and nuclei of the inner nuclear layer. The ganglion cell layer contained many intercellular lacunae, but no alterations were observed in the inner limiting membrane. These changes varied somewhat from those produced by the argon green laser, where some damage occurs in all retinal layers. McHugh and associates[36] demonstrated that, as might be expected, diode laser lesions are histologically most similar to those produced with the krypton red light of 647 nm. Wallow et al.[37] found that moderate diode laser lesions in monkey retinas were comparable to argon ones, but more intense diode laser lesions involved scarring of ciliary nerves in the choroid or sclera, with macrophage invasion and loss of myelin sheaths and axis cylinders.

Radiation at 810 nm readily traverses the sclera, and Jennings et al.[38] successfully performed experimental transscleral retinal photocoagulation in rabbits, transmitting the laser output via a fiberoptic. Treatment at 200 mW produced lesions after exposures of 5–10 s, reflecting a variability experienced with other transscleral wavelengths. The sclera overlying the chorioretinal lesions remained intact, and there was substantially less disruption of the blood–retinal barrier than is seen with cryotherapy (see below), which may be important in reducing the incidence of proliferative vitreoretinopathy following retinal detachment surgery. It has been suggested that the lesion variability may be overcome by gradually titrating exposure power and duration to achieve clinically desirable levels, with an endpoint of gray or gray-white, rather than white, spots.[39] More recent work indicates that altering the diode laser output to bursts of microsecond pulses contained within millisecond envelopes allows more selective and reproducible retinal photocoagulation.[40]

6.3.3 Clinical Applications

It is instructive to consider some currently employed photocoagulation procedures, as well as the mechanisms by which they achieve their effects. Only the more common techniques will be discussed in detail, but because the number of laser-treatable diseases is far greater than the number of fundamentally unique laser modalities, and since many of the diseases have common pathologic components, these will serve as paradigms for those entities (e.g., retinal neovascularization associated with angioid streaks[41] and histoplasmosis[42]) not mentioned here.

6.3.3.1 Diabetic Retinopathy

Most diabetics eventually develop retinopathy, which ranges from background (nonproliferative), with microaneurysms and macular edema, to proliferative, with neovascularization and hemorrhages that may obscure the retina and/or lead to detachment. Early laser therapy of PDR relied on direct and intense targeting of neovascular elements,[6] which often actually exacerbated the problem. It is now accepted that PRP, in which usually several thousand argon blue-green

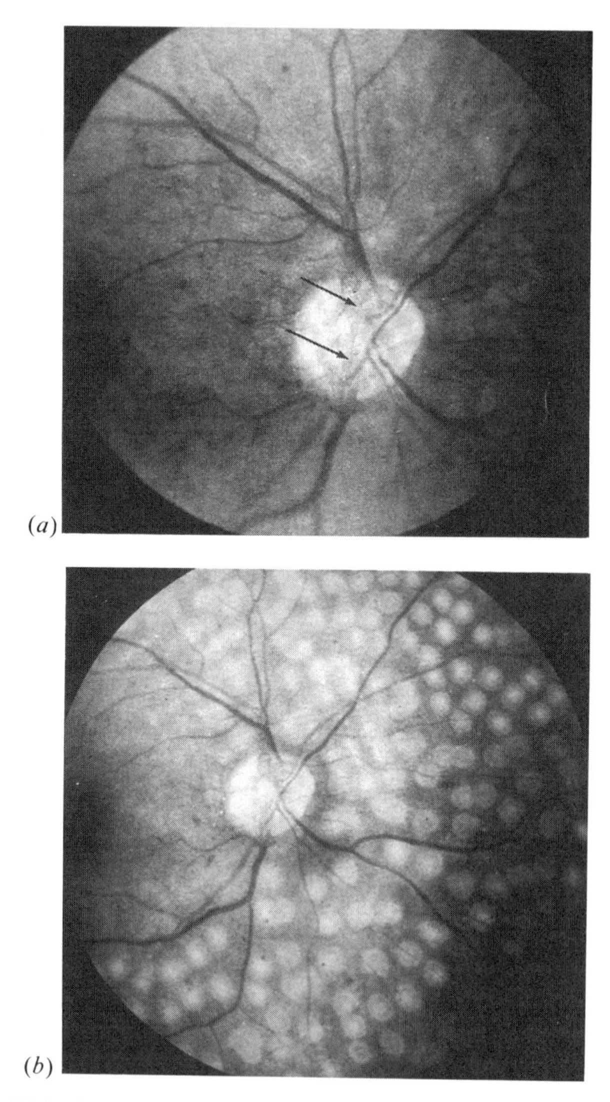

FIGURE 6.5 Diabetic neovascularization extending from the optic nerve before (*a*), immediately after (*b*), and 3 months after (*c*) panretinal photocoagulation with the argon green laser. (Reprinted from L'Esperance[518] with permission of C.V. Mosby.)

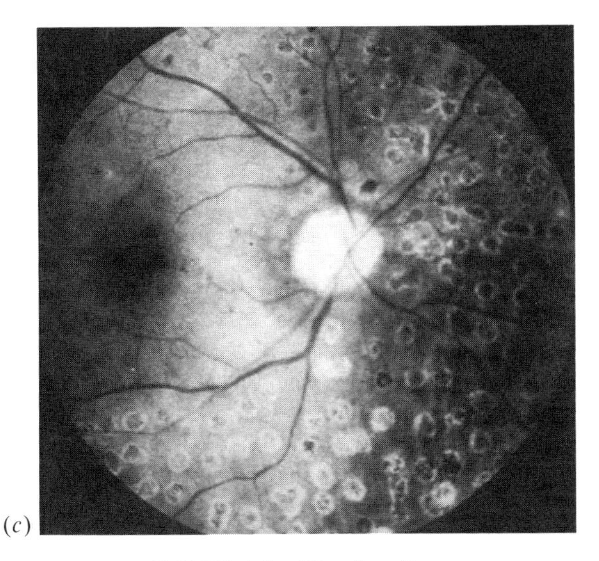

FIGURE 6.5 (*Continued*).

or krypton red burns are placed in the peripheral retina (Fig. 6.5), indirectly improves PDR by reducing the stimulus for neovascularization.[43–45] This follows from destruction of hypoxic retina, especially photoreceptors with their high oxygen requirement,[46] creating tighter adhesions to the choriocapillaris and resulting in decreased vasoproliferative tendencies and better oxygen perfusion to the remaining viable retina.[44,47] RPE cells produce a substance that inhibits neovascularization, which PRP may release.[48] If PRP relies on improved oxygen diffusion from the choroid, the deeper penetration of krypton red light would likely be counterproductive.[49] Considering the amount of tissue damage induced by the laser, it is reasonable to expect that not all the effects are salutary, and complications ranging from transient macular edema and elevated intraocular pressure (IOP) to the development of optic atrophy many years later have been attributed to PRP.[50] Peripheral vision is generally somewhat compromised both qualitatively and quantitatively, with the goals of preventing further deterioration and preserving central vision.[51]

Macular edema is the most common manifestation of diabetic retinopathy, and the localized and diffuse forms are generally treated with focal and grid laser application, respectively.[52] Focal treatment may reduce edema by preventing fluid passage from the subretinal space through the RPE and directly sealing leaking microaneurysms or capillaries. Damage by grid photocoagulation occurs principally in the RPE, with some effect on the photoreceptors and the underlying choriocapillaris.[53] It is unclear exactly how this improves macular edema, but possibilities include reduction of blood flow, increase of inner retinal oxygen, replacement of coagulated RPE cells with new ones,[54] and proliferation of endothelial cells in capillaries and venules overlying the lesions, capable of reinforcing the outer and inner blood–retinal barriers, respectively.[55] The inner retinal effects are believed to result indirectly from targeting of the outer retina; this may explain the superior results sometimes realized with the krypton red, rather than the argon blue-green, laser,[56] although other studies have shown no such difference.[57] Orange dye laser light has not demonstrated any benefit over argon blue-green and may be associated with more patient discomfort.[58]

6.3.3.2 Retinal Vein Occlusion

Treatment of branch (BRVO) or central (CRVO) retinal vein occlusion consists of two main approaches: treating the ocular symptoms or trying to directly address any underlying systemic pathology. Earlier work with diabetic retinopathy showed that laser irradiation can selectively destroy parts of the retina, thus reducing metabolic requirements and the stimulus for retinal or iris neovascularization. While there is a theoretical risk of hemorrhage or other intraocular damage, usually the only side effects experienced by the patient are small, gradually resolving peripheral scotomata. Initial studies on laser treatment of BRVO (Fig. 6.6) offered often conflicting recommendations. Michels and Gass[59] suggested observing patients for at least 1 year before considering laser, Gutman[60] recommended laser only after at least 6 months, and Zweng et al.[61] advised early intervention to forestall complications. Kelley and associates[62] could document no benefit of laser treatment for BRVO with macular edema.

More recently, landmark reports based on extensive multicenter trials have established widely recognized guidelines. The Branch Vein Occlusion Study Group has recommended that since approximately one-third of macular edema cases spontaneously resolve, and considering the frequent early presence of preretinal hemorrhage, which can prevent proper laser treatment and which often resorbs, patients should be carefully followed without treatment for 3–6 months.[63] Recommended laser parameters for grid photocoagulation of macular edema call for making numerous medium white burns of 100 μm diameter surrounding the macula, using argon blue-green light. Although another, smaller study reported that BRVO patients' macular edema did not significantly improve following laser therapy,[64] the Study Group found that the two-thirds of BRVO patients whose macular edema does not spontaneously resolve experience significantly improved visual acuity. The Group only studied BRVO–macular edema patients with visual acuity $\leq 20/40$, and it did not prescribe an optimum time for laser use. A later study suggests that krypton red light may be preferable to argon blue-green for such purposes, given the absorption properties detailed above.[65]

In 1986, the Group[66] published guidelines for treating BRVO complicated by retinal neovascularization, which was present in 36% of patients with large areas of retinal nonperfusion. Employing scatter, or panretinal photocoagulation on those patients with ≥ 5 disc diameters of nonperfusion, they showed decreases of 50% in both retinal neovascularization, and vitreous hemorrhage if neovascularization already existed, compared with untreated patients (40% of whom will develop retinal neovascularization, of whom 60% will eventually sustain vitreous hemorrhage). Their recommendation is to not treat retinal neovascularization, since treatment after it develops—but before the serious complication of vitreous hemorrhage occurs—was found to be equally effective.

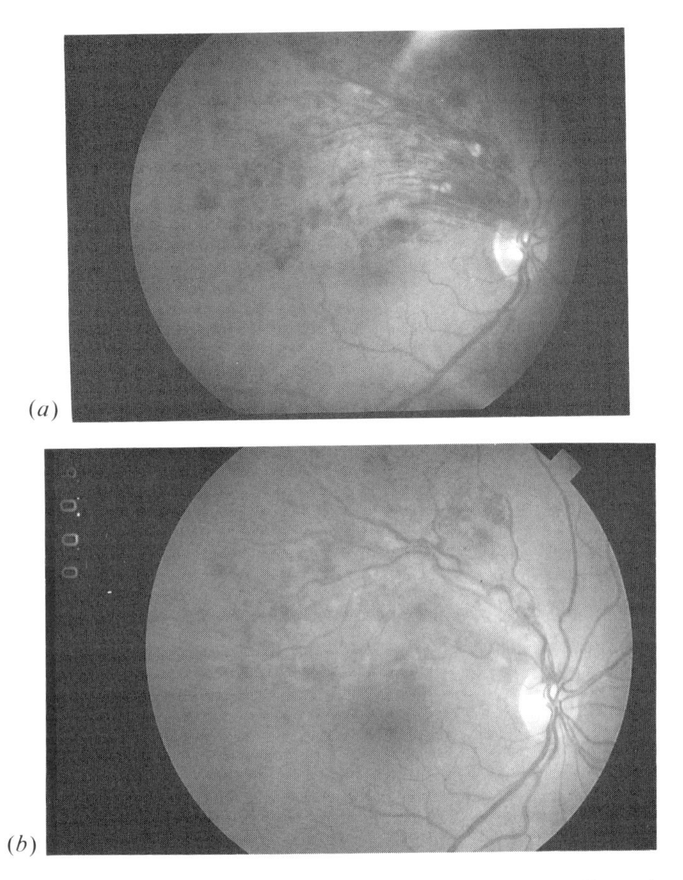

FIGURE 6.6 Branch retinal vein occlusion, with dilated tortuous veins, flame-shaped and blot-like hemorrhages, and cotton wool spots before (*a*) and after (*b*) grid photocoagulation with the krypton yellow laser. (Courtesy of Jason Slakter, M.D.)

No comparable studies have been completed on laser treatment of CRVO. Macular edema in CRVO can be caused by either macular capillary leakage or nonperfusion; in the former case, no treatment exists and vision is irreversibly lost, while in the latter, approximately one-third each will spontaneously recover, remain stable, and have decreased acuity.[67,68] Laser therapy may be useful in such instances but is more likely to play a role in preventing the devastating effects of neovascular glaucoma. Magargal and colleagues,[69] in a nonrandomized, uncontrolled study, found that whereas some 20% of all CRVO patients and 60% of those with extensive ischemia tend to develop neovascular glaucoma, none of those treated prophylactically with panretinal argon laser photocoagulation experienced neovascular glaucoma. Three other studies, using the criterion of ≥ 10 disc diameters of ischemia, also reported the efficacy of preemptive laser treatment for prevention of neovascular glaucoma in severely ischemic CRVO,[70–72] yet at least one investigator still questions the need for such therapy.[67,73] In a 10-year prospective study, Hayreh et al.[73] found that PRP reduced the incidence of iris neovascularization and peripheral visual field loss, but only if treatment was performed within 90 days of the CRVO, and in no case did the laser affect angle neovascularization, neovascular glaucoma, retinal or optic disc neovascularization, vitreous hemorrhage, or acuity. A multicenter CRVO study, similar to the BRVO one, is currently in progress to establish the best therapeutic approach to CRVO.[74]

6.3.3.3 Macular Degeneration

Subretinal neovascularization associated with age-related macular degeneration (AMD) is the most common cause of blindness in the United States and other developed countries. Central, color vision is gradually lost or distorted, especially affecting reading and other tasks requiring fine vision. Laser treatment is likely successful because of heat-induced closure of the new vessels.[75] As is the case with diabetic neovascularization, laser treatment may cause the release of angiogenesis-inhibiting factors, cauterize vessels in the choriocapillaris, and seal breaks in Bruch's membrane that are thought to engender SRNV.[76] Early guidelines for treating SRNV associated with AMD, calling for laser treatment at least 200 μm from the foveal avascular zone (FAZ), were based on photocoagulation with argon blue-green light.[77] A 5-year follow-up has shown that laser treatment decreases by a third the risk of losing 6 or more lines of visual acuity within 5 years.[78]

It was initially felt that SRNV within the FAZ is not amenable to laser therapy, and that attempts at photocoagulation in that region would jeopardize central vision.[79] Krypton red light transmission through blood and xanthophyll and vessel destruction at the level of the choriocapillaris makes it an excellent choice for treatment of juxtafoveal SRNV, assuming that there is sufficient surrounding melanin.[17] Krypton red laser treatment of SRNV with a proximal edge 1–199 μm from the center of the FAZ decreases the 3-year loss of ≥ 6 lines of acuity, but only from 58 to 49%.[80] At 5 years, untreated patients without systemic hypertension with juxtafoveal neovascularization had a risk of 1.82 relative to laser-treated patients of losing at least 6 lines of acuity, whereas hypertensive patients with AMD realized no benefit from photocoagulation.[81]

Reports by the Macular Photocoagulation Study Group[82,83] indicate that patients meeting specific criteria tend to benefit from photocoagulation of subfoveal neovascular lesions, but there is no apparent difference between argon green and krypton red wavelengths (although the latter may be associated with a higher rate of persistence of all AMD-related lesions[84]). Whereas laser treatment causes an immediate decrease of ≥ 6 lines of acuity in 20% of patients, at 2-year follow-up untreated patients are almost twice as likely to have such a decrease. However, the most recent recommendations take initial acuity and lesion size into consideration; patients with small lesions and moderate or poor acuity or medium lesions and poor acuity respond best to laser treatment, whereas those with large lesions and moderate or good acuity respond worst and achieve no discernible benefit from laser therapy through 4-year follow-up.[85]

The high incidence of neovascular recurrence,[86,87] although such recurrent membranes may themselves be treated with laser photocoagulation,[88] reflects the inadequacy of current laser therapy in achieving long-term success. While the underlying disease process is certainly a factor, it would clearly be advantageous to enhance absorption of laser energy within the entire membrane. The proximity of the SRNV to the RPE makes it difficult to determine the relative importance of the vessels' absorption, but it is likely that enhancement of direct laser absorption by the neovascular membranes would increase the destruction of the abnormal vessels (see below).

6.3.3.4 Diode Laser

In early clinical trials, McHugh et al.[89] and Balles et al.[90] demonstrated the potential efficacy of diode laser transpupillary photocoagulation, via a slit-lamp biomicroscope, for the treatment of retinal vascular diseases. The McHugh group encountered no side effects, such as lenticular or corneal opacities, RPE tears, or choroidal hemorrhage. RPE photocoagulation was achieved even through a layer of blood about 150 μm thick. Neovascularization had regressed 6 weeks following treatment of diabetic retinopathy, with visual acuity changes similar to those associated with other lasers. Follow-up of all patients as late as 9 months demonstrated the potential success of diode laser photocoagulation for these conditions.

The Balles group found that 4.5 times more energy, 2.5 times more power, and 1.8 times more irradiance and exposure duration were required to produce diode laser lesions which were clinically comparable to those achieved with the argon laser, all consistent with the much lower absorption by the RPE of the longer wavelength. The deeper diode penetration, even greater than for krypton red light, necessitated retrobulbar anesthesia, but resulted in only four cases of Bruch's membrane rupture or subretinal hemorrhage out of a total of over 9000 treatment exposures. There was excellent penetration through macular edema and serous retinal thickening;[91] transmission through cataracts and hemorrhages was much greater than for shorter wavelength lasers, with less light scattering. It was also noted that since the diode radiation is invisible, there were no complaints of bright flashes. Moreover, the permanent filter blocking the diode wavelength both permits continuous viewing throughout the procedure and eliminates the clicks typically associated with movable shutters, which may startle some patients. However, the diode laser's greater beam divergence did require a larger intraocular focusing cone angle, which limited peripheral treatment. Ulbig et al.[92] used a diode laser to close membranes in seven of nine eyes with parafoveal choroidal neovascularization secondary to AMD or angioid streaks, although four lesions required repeat treatment.

Given the experimental evidence that the diode laser can effectively produce chorioretinal adhesions, with less blood-retinal barrier destruction than either cryotherapy or the argon laser,[93] Haller at al.[94] performed transscleral diode laser retinopexy in conjunction with scleral buckling in a series of patients with rhegmatogenous retinal detachment. Minor complications included a scleral thermal effect and presumed ruptures in Bruch's membrane in 30% of the patients, the latter accompanied by audible "pops." However, these problems were not encountered once the researchers used smaller, gray lesions as their endpoint. Nine of 10 retinas were successfully attached at 6 months, while the tenth redetached at 6 weeks secondary to proliferative vitreoretinopathy but responded to reattachment. A multicenter trial is currently under way to more precisely evaluate this modality. It remains to be seen whether the average of 520 J used by the Haller team is necessary to create retinopexy, and whether such high energy won't cause some hemorrhages in a larger study group.

Endoprobe As clinical experience with the diode laser accumulates, there has been concomitant progress in refining both the laser itself and the means for delivering its output to the eye. This laser's compactness makes it particularly well suited to use with the endoprobe and indirect ophthalmoscope (Fig. 6.7). Numerous applications have been developed for endophotocoagulation since the technique was first introduced in the early 1980s.[95] Recent uses have included treatment of a choroidal bleed in a patient following evacuation of a subretinal clot[96] and ablation of experimental choroidal melanomas in rabbits with high-power argon radiation.[97] Building on early work in animals,[30,32,98] Smiddy[99] reported the first clinical use of diode endolaser photocoagulation, treating patients with PDR, proliferative vitreoretinopathy, complex retinal detachments, and retinal breaks. Exposure parameters were dictated by the particular pathology, always avoiding reaching the whitish lesions typically desired with the argon laser, which with the diode laser has been associated with iatrogenic choroidal folds.[100] It was felt that endophotocoagulation with the diode laser is as clinically efficacious as with the argon laser, but the former offers significant logistic and ergonomic advantages.

In the past few years, considerable advances have been made in probe technology, especially in conjunction with the diode laser, all aimed at maximizing the functions that can be performed by a single probe. Peyman et al.[101,102] developed a couple of such 20-gauge devices, combining fiberoptic diode or argon lasers and aspiration and infusion capabilities, with one also including fiberoptic illumination. These instruments can be used to both drain subretinal fluid and photocoagulate the retinotomy site, thereby obviating repeated forays into the eye with separate devices. Uram[103,104] also constructed a 20-gauge probe incorporating diode laser and illumination fiberoptics, but rather than aspiration/infusion capability, his has a microendoscope with a 70° field of view and recording capability. Using this device, he was able to deliver precisely titrated laser exposures to a specific number of ciliary processes, potentially enabling more effective treatment of neovascular glaucoma. Vitreoretinal endophotocoagulation was similarly facilitated, with the probe providing a clear view even when more anterior structures would obscure the view through an operating microscope, and permitting post-treatment inspection for retinal breaks without resorting to indirect ophthalmoscopy.

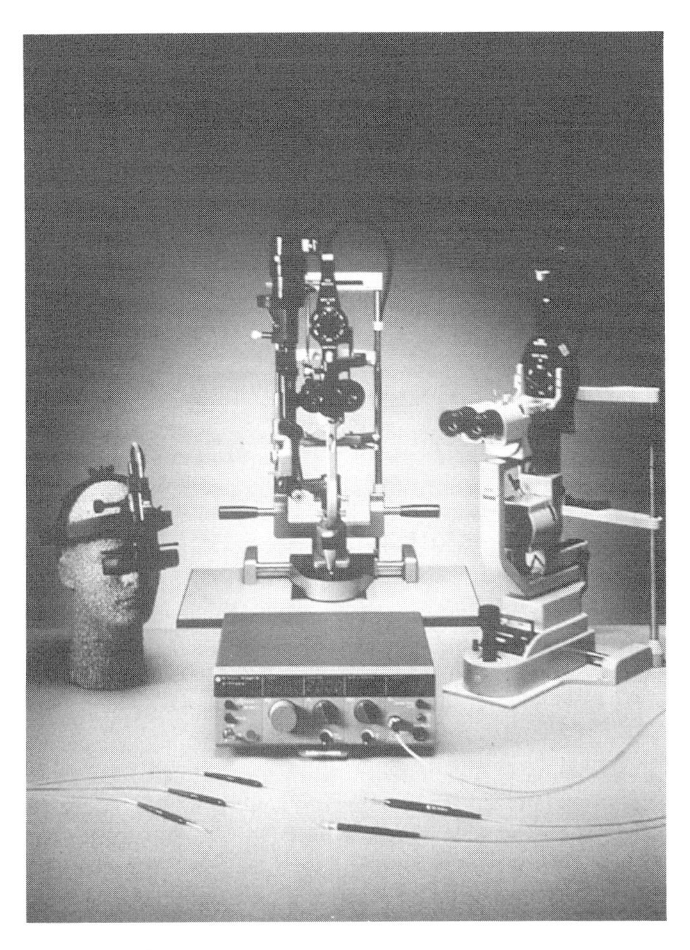

FIGURE 6.7 OcuLight^R SLx diode laser with the three means by which up to 2 W can be delivered to the eye: slit-lamp biomicroscope, endoprobe, and indirect ophthalmoscope. The laser console measures 10 × 30 × 30 cm, weighs 5.5 kg, and runs on either 115 or 230 V without requiring external air or water cooling. (Courtesy of Iris Medical Instruments, Mountain View, CA.)

Laser Indirect Ophthalmoscope The laser indirect ophthalmoscope (LIO) is another instrument that has seen increasing use over the past decade. First described by Mizuno,[105] this laser has all the advantages and disadvantages inherent in indirect ophthalmoscopy, but is indispensable for selected applications. Unlike laser treatment with slit-lamp delivery, the spot size is impossible to standardize, as it depends on the power and position of the handheld and headset lenses, the refractive power of the treated eye, and the presence of any intraocular gas. Macular work is not recommended given the inherent limitations in aiming. However, the field of view is greater than with other laser modalities, and in conjunction with scleral depression, this technique reduces the laser power needed for photocoagulation, probably as a result of the stretched choroid's diminished ability to dissipate heat from the RPE.[106] The LIO is very useful in cases requiring far peripheral treatment, such as retinal tears or peripheral neovascularization, especially those with localized lens opacities or small pupils.[107] It is essential for pneumatic retinopexy reattachment[108] or laser treatment of retinopathy of prematurity (ROP; see below), and has also been reported for treatment of retinoblastoma[109] and choroidal melanoma.[110] Complications include occassional choroidal hemorrhage from too intense exposures, superficial burns of the iris and cornea,[111] and melted haptics [which keep artificial intraocular lenses (IOLs) in position] made of Prolene (which contains copper phthalocyanine dye).[112]

RETINOPATHY OF PREMATURITY This is a potentially devastating condition affecting, to some extent, about two-thirds of infants with birthweight below 1251 g.[113] Cryotherapy has assumed an important role in treating ROP, cutting almost in half the rate of unfavorable outcomes. Nevertheless, it is often ineffective and has been associated with such complications as intraocular hemorrhage, retinal detachment, and scleral trauma,[114] generally because of the pressure with which the probe is applied and the freezing process itself. This pressure, along with the intravenous sedation frequently employed, may also induce episodes of apnea, bradycardia, and oxygen desaturation.[115]

Studies with the argon laser indicated that photocoagulation was at least as effective as cryotherapy, with less trauma to the eye.[116–121] Given its technical convenience, the diode laser has been the subject of most recent studies on photocoagulation of ROP.[122,123] McNamara et al.[124] and Hunter et al.[125] performed randomized trials comparing diode LIO photocoagulation and cryotherapy for the treatment of threshold ROP. In the McNamara study, exposure parameters were 120–600 mW and 0.3 s, with an average of 959 burns placed (Fig. 6.8). Transient vitreous hemorrhages were noted in 3.6% of the laser

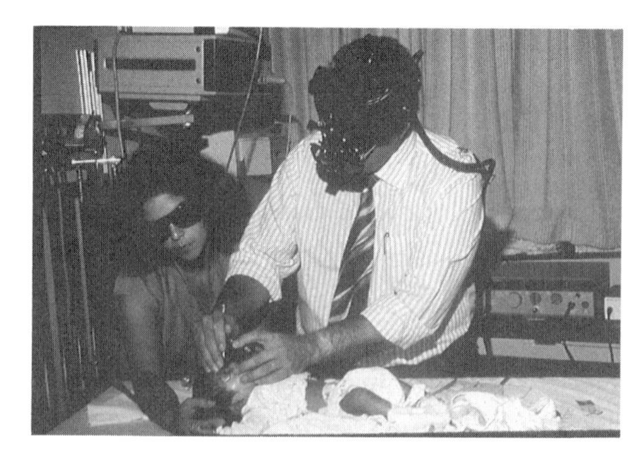

FIGURE 6.8 Diode laser indirect ophthalmoscope photocoagulation for ROP. (Courtesy of Iris Medical Instruments, Mountain View, CA.)

and 12.5% of the cryotherapy eyes. Lid edema, conjunctival hyperemia, and chemosis lasting 1–3 days were seen in all the cryotherapy eyes, while one laser eye showed mild conjunctival hyperemia lasting only several hours. Pain was difficult to assess, but appeared to be comparable to treatment with an argon laser and less than with cryotherapy. In the laser group, 25 of 28 eyes followed for 3 months and all 7 of those followed for 1 year showed regression; the corresponding numbers for the cryotherapy group were 20 of 24 and all 7. Diode laser treatment thus appeared at least as efficacious as cryotherapy, and has the advantage over the argon laser of portability, permitting treatment in neonatal units that might not have access to other lasers. Similar results were obtained by the Hunter group, who noted that treatment through hazy media was easier with the diode laser than with cryotherapy. There was also much less damage to the peripheral fundus. Considering that all preliminary studies have supported the efficacy of photocoagulation and its greater tolerance in the treatment of ROP, Tasman[126] has suggested that future multicenter trials concentrate on establishing the optimal threshold and parameters for laser treatment, rather than on randomized comparisons with cryotherapy.

6.3.3.5 Glaucoma

Angle-closure glaucoma develops when the anterior iris presses against the posterior cornea. This prevents the usual drainage of aqueous humor through the trabecular meshwork and Schlemm's canal into the episcleral space (Fig. 6.1), causing a rise in IOP (22 mmHg is usually considered the maximum normal in humans). The mechanism of an iridectomy or iridotomy in an eye with angle closure is thus clear: producing a passageway (at least 50 μm in diameter[127]) for aqueous to reach the anterior chamber (Fig. 6.9). Given the pigmentation of most irises, photocoagulation offers an attractive alternative to invasive procedures; indeed, Meyer-Schwickerath[128] was able to perforate the tissue with the xenon arc, albeit with considerable heat production and pigment dispersion. The argon laser minimized these complications and made surgical iridectomy virtually obsolete.[129] Although this approach allows simultaneous coagulation of any hemorrhaging vessels, it generally requires dozens of exposures, especially in lightly pigmented irises, and many clinicians now prefer the Q-switched Nd:YAG laser for iridectomy (see below).

Jacobson and associates[130] created diode laser peripheral iridectomies in rabbits that were similar to argon blue-green iridectomies. These authors speculated that the greater transmission through the iris stroma and stronger absorption by the iris pigment epithelium at 810 nm may make the diode laser preferable for this procedure, especially in dark irises. This group[131] also reported the first human iridectomies produced with the diode laser.

Laser trabeculoplasty was first reported by Wise and Witter[132] in 1979, and while it is often successful in controlling elevated IOP in primary and other forms of open-angle glaucoma, its precise mechanism of action remains to be determined. Typically, argon green or diode lasers are used to place circumferential 50-μm burns on the trabecular meshwork, usually around 90°

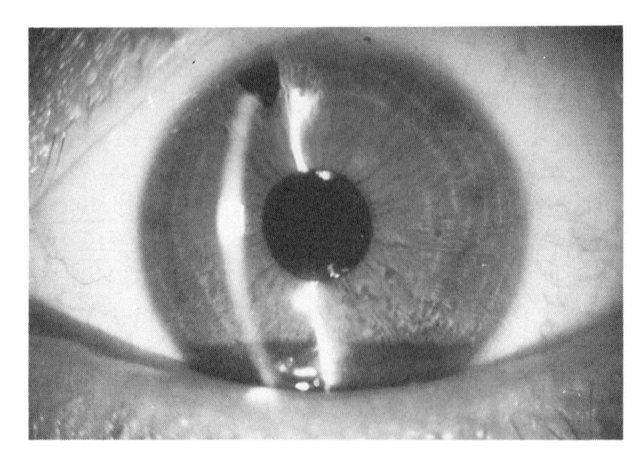

FIGURE 6.9 Iridectomy in an eye with angle closure glaucoma. The opening in the iris, which reestablishes aqueous flow from the posterior chamber through the trabecular meshwork and out of the eye, is ordinarily hidden by the upper eyelid. (Courtesy of Roger Steinert, M.D.)

or 180°, as 360° was found to be associated with more IOP spikes.[100] However, apraclonidine, an α_2-adrenergic agonist, has shown great promise in minimizing postlaser IOP rise, including following treatment of all 360°.[133] There is no significant difference in facilitation of aqueous outflow between argon and krypton wavelengths,[134] whereas the exact location of the laser burns,[135] the percentage of treated trabecular meshwork,[136] and even laser energy,[137] are not correlated with clinical effect. It was initially believed that LTP works by shrinking the superficial collagen of the corneoscleral meshwork, preventing closure of Schlemm's canal by anteriorly displacing the inner trabecular meshwork.[135] More recently, attention has focused on the biochemical effects of LTP,[138] with observation of phagocytic action by trabecular meshwork cells and proliferation of the corneal endothelium onto the trabecular surface, possibly with stimulation of special Schwalbe line's cells capable of producing a phospholipid substance that facilitates aqueous egress through the trabecular meshwork.[139] In any event, it is now believed that mechanical effects of LTP are relatively minor, and that some biochemical changes in and around the trabecular meshwork more likely explain the procedure's clinical efficacy.[140] Although LTP has traditionally been reserved for patients unresponsive to medical therapy alone, it rarely obviates it entirely. LTP has been found to adequately control IOP without surgery or further laser treatment in approximately three-quarters of patients after 1 year and one-half of patients by 5 years, with success rates at 10 years ranging from one-third[141] to one-half.[142] A new multicenter trial has compared outcomes among subjects randomized to initial treatment with either medication or LTP.[143] At 2-year follow-up, 44% of those eyes treated by LTP alone were controlled, whereas only 30% of those receiving only timolol had normal IOP. However, it has been noted that when all single medications were considered, the balance was in favor of medical treatment.[144]

McHugh and coworkers[145] conducted a pilot clinical investigation of diode LTP. Employing parameters of 0.8–1.2 W, 0.2 s, and 100-μm, and placing 50 burns for 180°, they noted that the desired exposure endpoint was a mild blanching of the pigmented portion of the trabecular meshwork. IOP was lowered just as much as with the more established laser (9.6 mmHg at 6 months), with an even greater effect 2–4 weeks after the LTP, possibly as a result of the deeper penetration at 810 nm. There was no IOP spike immediately following treatment. These researchers[146] also showed that the trabecular meshwork damage is histologically similar whether LTP is performed with the diode or argon laser, further reflecting the procedure's independence of wavelength. Indeed, in a direct clinical comparison, Brancato et al.[147] showed that patients undergoing diode LTP retained the same lowering of IOP through 12 months as those individuals treated with argon LTP. Moriarty and associates[148] noted only a small decline in IOP, from 8.4 to 7.9 mmHg, between 12 and 24 months after diode LTP, without the peripheral anterior synechiae seen in some one-third of argon LTP cases.[149]

In patients refractory to medical and surgical modalities designed to increase outflow in order to reduce IOP, cyclodestructive methods for partially destroying the aqueous-producing ciliary body may be necessary. There is a narrow therapeutic window; too much destruction is also not good, as the eye would become hypotonous from inadequate IOP. Cyclodestructive techniques have included diathermy and cryotherapy,[150] the latter of which is still used. In the early 1970s, Lee and Pomerantzeff[151] proposed laser cyclophotocoagulation, but their transpupillary technique proved inadequate. Transscleral radiation with the xenon arc had been described as early as 1961,[152] with use of the ruby laser reported in 1972.[153] Nd:YAG laser cyclophotocoagulation, first described by Wilensky et al.[154] in 1985, has now proved effective in numerous clinical studies and may become the cyclodestructive treatment of choice.[155–158]

Schuman and colleagues[159] performed contact (via fiberoptic) transscleral diode laser cyclophotocoagulation on rabbits, and found IOP lowering and ultrastructural damage comparable to that achieved with the Nd:YAG laser. Work on human cadaver eyes revealed that only some 75% of the energy used with the Nd:YAG laser is needed for the diode laser procedure, consistent with melanin's greater absorption of the latter's wavelength (despite its somewhat lower transmission through the sclera).[160] These findings were in contrast to those of Simmons et al.,[161] who found that the diode laser produces most of its effect in the ciliary body stroma, rather than in the ciliary epithelium as is the case with the Nd:YAG laser, although the clinical significance of this difference is not clear. Initial clinical studies of contact diode laser cyclophotocoagulation by Gaasterland and associates[162] realized average IOP reductions from 36 mmHg prelaser to 23 mmHg at 3-month follow-up, without any hypotony, but with mild surface burns in 40% of patients.

6.3.3.6 Oculoplastic Surgery

Although many of the oculoplastic procedures for which photocoagulation techniques have been attempted are probably better performed by non-laser methods, there are several applications for which lasers appear to offer distinct advantages. Trichiasis, in which eyelashes grow inward and abrade the ocular surface, has been treated with numerous procedures, most effectively with cryotherapy. However, this is often associated with such side effects as corneal ulceration and eyelid edema. As first proposed by Berry,[163] argon laser treatment of trichiasis offers the potential for more selective, less traumatic lash removal. Nevertheless, increasing clinical experience suggests that it may not be applicable to severe cases, and at energy levels sufficiently low to avoid complications, multiple treatments may be necessary.[164] Since blue-green light is well absorbed by hemoglobin and melanin, the argon laser may be used to treat such vascular and pigmented lesions as capillary hemangioma, nevus flammeus, seborrheic keratosis, and telangiectasia.[165] Approximately three-quarters of dark nevi flammeus, known as *port-wine stains*, respond to argon laser therapy.[166] With its superior penetration, radiation from the Nd:YAG laser may be preferable to the argon laser for certain applications, particularly large lesions (Fig. 6.10).[167] The copper vapor laser produces green light at 511 nm and yellow light at 577 nm.[168] The former is strongly absorbed by melanin and can lighten many pigmentary skin lesions, while the latter's absorption by hemoglobin permits eradication of vascular skin lesions. (Oculoplastic uses of the CO_2 and other ablation lasers are discussed below.)

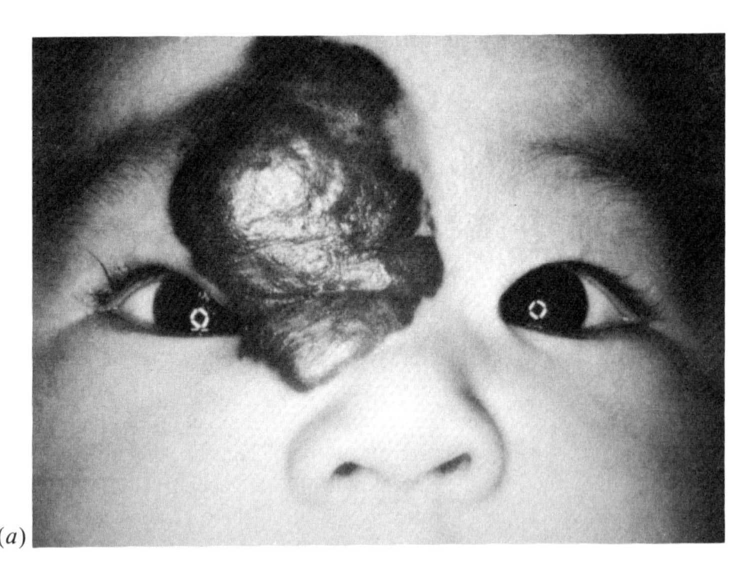

(*a*)

FIGURE 6.10 (*a*) Rapidly growing capillary–cavernous hemangioma of the forehead, upper eyelid, and nose; (*b*) 3 months after Nd:YAG laser photocoagulation and direct injection of steroids; (*c*) following Nd:YAG laser photocoagulation using a sapphire tip, showing good resection of the hemangioma, with improvement of color, contour, and symmetry. (Reprinted from Apfelberg et al.[167] with permission of *Annals of Plastic Surgery*.)

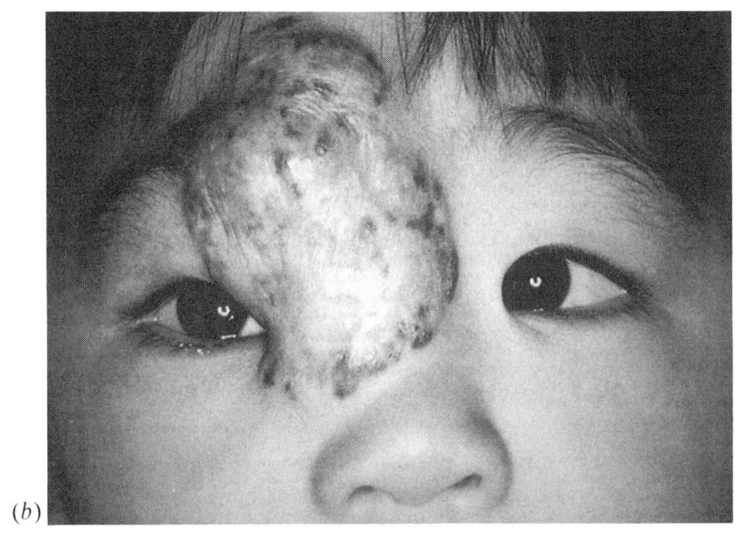

(*b*)

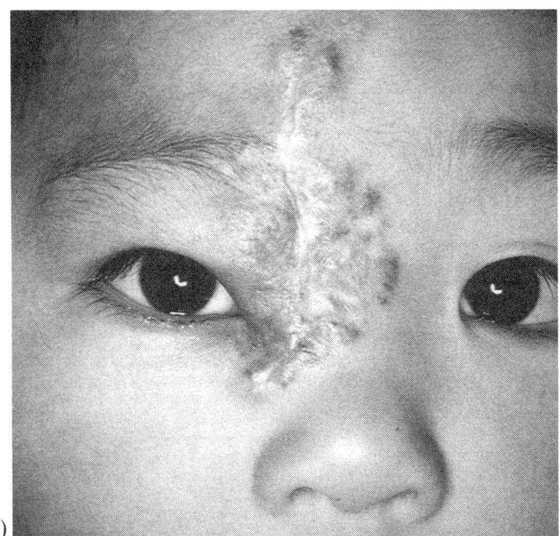

(*c*)

FIGURE 6.10 (*Continued*).

6.3.3.7 Ophthalmic Oncology

Surgery and radiation therapy remain the mainstays for treating eye tumors, but certain cases may be amenable to laser treatment. Retinoblastoma, the most common malignant intraocular tumor in children, can often be destroyed successfully with the argon laser (by eliminating the tumor's blood supply) if detected early,[169] although about one-quarter require additional treatment with other methods. Similarly, small choroidal melanomas may be treated with argon or krypton laser photocoagulation, but the complexity of the protocol and the occasional development of side effects has resulted in a return to radiotherapy in most instances.[170] (Use of exogenous dyes for selective tumor destruction is discussed below.)

6.3.4 Investigational Techniques

Despite the theoretical appeal of highly precise targeting of ocular tissues by appropriate selection of wavelength, it has been suggested that differential absorption is essentially irrelevant beyond threshold lesions.[171] The specificity of laser wavelength and pulse selection greatly exceeds the absorption specificity of the target tissues. The greatest absorption of all photocoagulation wavelengths occurs in the RPE, and aside from the deeper penetration of longer wavelengths due to decreasing melanin absorption, most clinical lesions appear identically gray-white regardless of wavelength. A study of patients with pathologic myopia and SRNV treated with the tunable dye laser revealed no significant anatomic or clinical differences among 577, 590, and 620 nm, although the authors still recommended selection of a specific wavelength according to the precise pathology being treated.[172] Nevertheless, several new investigational approaches do unequivocally enhance the target specificity of ophthalmic laser surgery.

6.3.4.1 Selective Absorption

Anderson and Parrish[173] proposed one method of enhanced absorption, selective photothermolysis, in which selective tissue damage is determined not by precise aiming of the laser beam but by the unique absorption properties of the intended target. If the target has an absorption coefficient at least twice that of the surrounding tissue at a given wavelength, preferential absorption will result in thermal damage localized to the target if the irradiation is performed at a duration similar to or less than the thermal diffusion constant. Brevity of the exposures is determined principally by the size of the target. They used this method to selectively damage blood vessels (3×10^{-7} s, 577 nm) and melanocytes (2×10^{-8} s, 351 nm), although selective photothermolysis is, in principal, applicable even at the subcellular level.

Krauss et al.[174] investigated localization of retinal thermal damage by employing an interferometric technique to project a fringe pattern at various exposure powers and durations. The periodicity of the fringe pattern could be adjusted from macroscopic dimensions to a scale of microns without the need for an imaging plane. Periodicity is more adjustable and unambiguously measurable than spot size, and comparison of tissue response with theoretical models is simplified because the sinusoidal fringe pattern is itself an eigenfunction of the thermal diffusion equation. Tests in rabbits confirmed that exposures at 10 ms, comparable to the retinal thermal relaxation time, resulted in localized damage, whereas those at 100 ms displayed diffusion that belied the fringe pattern of the irradiating beam. Although exposures too short may cause acoustic shock wave and mechanical damage, these findings suggest that, especially in the macula and near blood vessels, treatment at shorter exposure times might result in more localized and effective results. Roider and colleagues[175] employed argon laser

514-nm, 5-μs pulses in rabbits to selectively coagulate the RPE without creating ophthalmoscopically visible lesions and corresponding damage to the adjacent neural retina and choroid.

6.3.4.2 Dye Enhancement

Intravenous injection of exogenous dyes can substantially increase the specificity and effectiveness of certain laser procedures. Hematoporphyrin derivative (HpD), with strong absorption between 625 and 635 nm, is preferentially taken up by neoplasms.[176] Production of singlet oxygen for selective tumor destruction can be induced with red laser light, such as from a dye laser using rhodamine 6G (which emits at 630 nm) or a gold vapor laser (which emits at 628 nm).[177] Liposomal benzoporphyrin derivative (BPD), stimulated by a dye laser at 692 nm, has been used in animal models to treat SRNV[178] and choroidal melanoma.[179]

Indocyanine green (ICG) is a tricarbocyanine dye with an absorption peak at 805 nm. It fluoresces at 835 nm, allowing transmission through overlying blood, exudate, and melanin, and recent advances in IR imaging and digital angiography have made it a powerful tool in defining SRNV.[180,181] The dye's predilection for neovascular nets and their environs makes it a useful exogenous chromophore for diode laser selective photocoagulation and thermal enhancement of membrane closure.[182] In patients receiving ICG-enhanced diode laser photocoagulation of SRNV, minimal deep retinal whitening is noted acutely, followed by chorioretinal scar formation. Puliafito and associates[183] reported a system to perform instantaneous high-resolution ICG digital angiography and ICG-enhanced diode laser photocoagulation of SRNV.

Chloroaluminum sulfonated phthalocyanine (CASPc) is a photoactive dye that generates singlet oxygen on irradiation at 675 nm and fluoresces at 680 nm.[184] Unlike HpD, it is easily prepared as a chemically pure compound and is associated with minimal systemic toxicity and skin sensitization.[185] CASPc has been proposed as a possible photodynamic adjunct to tumor therapy and vessel closure, with less reliance on thermal mechanisms. Bauman et al.[186] demonstrated that dye laser 675 nm irradiation with CASPc, but not irradiation alone, of rabbits with experimental choroidal melanomas achieved significant vessel closure and tumor regression. Ozler et al.[187] found that such tumors regressed in response to CASPc and 675 nm light at 22–60 J/cm^2, only temporarily so at 15–22 J/cm^2, and not at all at less than 15 J/cm^2. All eyes receiving greater than 15 J/cm^2 showed transient corneal edema and conjunctival hyperemia, whereas only those treated with greater than 43 J/cm^2 experienced retinal hemorrhages. Although CASPc work to date has been conducted with dye lasers, it is likely that diode lasers doped with other elements will soon provide sufficient output at that wavelength.[188,189] While CASPc's 675-nm absorption peak is the longest among commercially available photosensitizers, allowing greater penetration of exciting light through tissues, pigment, fluid, and blood, it would be preferable to use an even longer wavelength, with still better penetration and at which diode lasers are currently capable of sufficient energy production. Silicon naphthalocyanines (SlNc) offer promise in this regard, as they have absorption maxima between 770 and 800 nm and a 20% quantum yield of singlet oxygen production.[190] Garrett et al.[191] achieved necrosis of experimental rabbit melanomas with SINc stimulated by a solid state titanium-sapphire laser at 770 nm (see below).

6.3.4.3 *Liposomes*

Zeimer et al.[192] and Khoobehi et al.[193–195] have conducted a series of studies on the intravenous injection of liposomes containing drugs or dye. Low-level irradiation with argon blue-green or dye yellow light is used to achieve heat-induced localized release in the retinal vasculature of those substances from the miniature phospholipid containers with a transition temperature of 41 °C. Potential applications include measurement of blood flow and selective angiography. Work on the latter has shown that with a spot size of 1.5–2.0 mm centered on the optic disc and energy densities of 0.5–3.4 J/cm^2, it is possible to obtain multiple angiograms up to 3 h following dye injection.[195] Potential advantages of this technique include reduction of choroidal fluorescence, which permits optimal viewing of the retinal microcirculation, and clear separation of arterial and venous fluorescence. Selective angiography of a suspected leaking vessel could be documented by specifically targeting it with the laser. Blood flow might be monitored during and after laser treatment of tumors and angiomas. Before any clinical use, however, more extensive tissue damage studies are needed, as are data on the potential toxicity of the liposomes and carboxyfluorescein dye.

6.3.4.4 *Scleral Buckling*

Scleral buckling is the standard procedure for the treatment of rhegmatogenous retinal detachment, but is not without its complications. Looking for a method that would avoid episcleral sutures and large exoplants, Ren and coworkers[196] used the holmium:YAG laser on human cadaver eyes to induce tissue shrinkage and create a buckling effect. Five 250-μs pulses were applied via a fiberoptic probe held some 5 mm from the sclera, to achieve a fluence of 11.3 ± 1.2 J/cm^2, which affected only the outer two-thirds of the sclera; there was no damage to the remaining sclera or the underlying retina. Although their system was successful, they speculated that a CW laser tunable over the 1.8–2.4-μm range would permit more controllable treatment, adjustable to the various absorption characteristics of pathologic sclera. They also suggested the possibility of combining this with laser retinopexy, whose early clinical results were promising.[94]

6.3.4.5 *Tissue Welding*

Using lasers to join tissue without the need for sutures has been an attractive but elusive goal, and remains the subject of periodic studies. Burstein et al.[197] employed the CW hydrogen fluoride (HF) laser to create seals of corneal incisions in porcine cadaver eyes. The welding spot of approximately 0.2 mm diameter was moved across the incision at a rate of 1 mm/min. The fundamental mode, with wavelength 2579 nm and power 30 mW, produced a weld about 100 μm deep which failed at 14 mmHg, while the overtone at 1340 nm and 320 mW created a weld 300 μm deep that withstood pressures up to 34 mmHg. Scleral welding was found to be less successful. Considerably greater resistance was noted by Khadem and coworkers[198] in human cadaver eyes whose corneal incisions had been sealed with a fibrinogen mixture containing a photosensitive singlet oxygen generator, activated by an argon blue-green laser to cross-link a protein solder with stromal collagen. Wolf and associates[199] used the diode laser to convert ICG-enhanced fibrinogen to fibrin in rabbit retinas, and speculated that this may provide a quick chorioretinal adhesion, assisting in the treatment of retinal breaks.

6.4 PHOTODISRUPTION

Photodisruption is the use of high-peak-power ionizing laser pulses to disrupt tissue. Energy is concentrated in space and time to create optical breakdown, or ionization of the target medium, with formation of a plasma, seen as a spark. The use of optical radiation to produce a plasma became possible only after the development of lasers capable of emitting high power through very brief radiation pulses. Although the first lasers were too weak to achieve optical breakdown, in 1962 Hellwarth developed the method of Q-switching, which allowed the creation of very brief but large ruby laser pulses over 10–50 nanoseconds (ns; 10^{-9} s), with maximum powers in the tens of megawatts.[200]

In 1972, Krasnov reported the first use of clinically desirable intraocular photodisruption.[201] To emphasize the relative importance of nonthermal acoustic mechanisms in creating these tissue effects, he used the term "cold laser," which ignores the fact that plasma formation causes very localized temperature increases greater than 10,000 °C. Further work demonstrated that, because of the ruby laser's high-order mode structure (which limits the minimal spot size that can be achieved) it is not the ideal source for a clinically practical ophthalmic photodisruptor. However, the increasing popularity of extracapsular cataract extraction (ECCE; described below) and the pioneering research of Aron-Rosa[202] and Fankhauser[203] and their colleagues with the Nd:YAG laser, combined to make this technique a reality.

6.4.1 Laser Principles

Laser power can be increased by either increasing energy or, more practically, by decreasing the period over which the energy is delivered. The two principal means of compressing the laser output in time to achieve high-peak power are mode-locking and Q-switching. Mode-locking is comparable to the audible summation of musical tones with similar frequencies, known as "beating," which is heard as a periodic surge in intensity. The phase relationships in lasers are synchronized by a shutter near one of the cavity mirrors. For ophthalmic applications the most common shutter is a saturable dye, employed in a process known as *passive mode-locking*. The dye absorbs low-power radiation pulses, but becomes transparent on exposure to high-power ones.

The Q-switch is an intracavity shutter that requires an active medium that allows atoms to remain in the high-energy state for a relatively long time to create high-peak power. Solid-state media such as Nd:YAG are particularly well suited for this process. At the appropriate time the Q-switch shutter is opened, exposing the mirror. Oscillation and stimulated emission follow quickly, with emission of a single brief high-power pulse. Methods of Q-switching include saturable dyes, rotating mirrors, and acoustooptic modulators. Pockel's cell, an electrooptic modulator that is the most common Q-switch, applies voltage across a crystal to vary polarization. Polarity can be rapidly changed by 90°, making the cell either opaque or transparent to the polarized laser beam. The "Q" refers to the quality factor of the laser cavity, which is defined as the energy stored in the cavity divided by the energy lost per cycle. Rapid extraction of high power is accomplished as the Q-switch changes the quality factor of the cavity from a high to a low Q.

Whereas typical mode-locked laser output consists of a train of seven to ten 25-picosecond (ps; 10^{-12} s) pulses, at intervals of 5 ns and contained within a 35–50-ns envelope, Q-switched laser output generally consists of a single 2–30-ns pulse. The total energy required for a single Q-switched pulse and a train of mode-locked pulses is the same, but the peak power necessary to cause avalanche ionization must be 100–1000 times greater for mode-locked than Q-switched lasers.[204,205] Maximum outputs of most ophthalmic models are 10–30 mJ and 4.5 mJ for Q-switched and mode-locked lasers, respectively.

6.4.1.1 Optical Breakdown and Plasma Formation

When a target is heated by absorbing radiant energy, the effect is linearly proportional to the cause. In contrast, nonlinear effects are sudden, all-or-nothing phenomena. Optical breakdown, a nonlinear reaction, occurs when the laser output is sufficiently condensed spatially and temporally to achieve high irradiance. It is manifested by a spark and accompanied by an audible snap, producing dramatic target damage. When focused to a small spot, typically less than 50 μm in diameter, short-pulsed Nd:YAG lasers can produce enough irradiance, usually 10^{10}–10^{11} W/cm^2, to induce optical breakdown, dissociating electrons from their atoms and creating a plasma. Q-switched pulses cause ionization mainly by focal target heating, in a process called *thermionic emission*, whereas mode-locked ones rely primarily on multiphoton absorption.[206] In either case, once the initial free electrons have been generated, plasma expands via electron avalanche or cascade if the irradiance is adequate to cause rapid ionization. Plasma absorbs and scatters incident radiation, thereby shielding underlying structures. Plasma radiation absorption and growth both occur through inverse bremsstrahlung, the process of photon absorption and electron acceleration in the presence of an atom or ion.

6.4.1.2 Mechanisms of Damage

In biologic systems, thermal denaturation of protein and nucleic acids is theoretically confined to a radius of 0.1 mm for a 1-mJ pulse.[207] As such, while high local temperatures exist briefly, total heat energy is low, and significant clinical photocoagulation does not occur.

Several mechanisms combine to generate pressure waves radiating from the zone of optical breakdown, the foremost of which is the rapid plasma expansion that begins as a hypersonic wave.[208,209] A secondary source of hypersonic and sonic waves is stimulated Brillouin scattering, in which the laser light generates the pressure wave that scatters it.[210] The focal heating may lead to vaporization, melting, and thermal expansion, generating acoustic waves.[211] If sufficiently strong, the radiation's electric field will deform a target through electrostriction, which causes simple Brillouin scattering, and radiation pressure induced by momentum transfer from photons to atoms in inverse bremsstrahlung.

The shock wave begins immediately with plasma formation, and expands at a hypersonic velocity of 4 km/s, falling to sonic velocity within 200 μm. The acoustic transient lasts 50 ns at a distance of 300 μm from the focal point, while the pressure falls from 1000 to 100 atm within 1 mm.[212] The next process is cavitation, or vapor-bubble formation. This begins within 50–150 ns after breakdown in water, expands rapidly for the first 20 μs, reaches a maximum size of approximately 0.6 mm at 300 μs, and collapses within 300–650 μs.[208,209] Cavity propagation velocity is about 20 m/s at 300 μm from the breakdown.[213] Cavitation, which is too fast to be seen, should be distinguished from bubble formation. Many shock waves may be generated along the laser beam's path as impurities are encountered.[214] Damage zone size depends on the irradiance and total energy, the plasma's duration, and the mechanical properties (including density, mass, tensile strength, and elasticity) of the target tissue.[215–218]

In recent years, most cataract operations have included the insertion of IOLs, which are generally made of polymethylmethacrylate (PMMA), although older ones may be of glass while newer ones may be of foldable silicone. These lenses can affect the intraocular use of lasers, especially posterior capsulotomy (see below), where damage may take the form of microcracks, melted voids, and large pulverized regions. Unlike the situation in liquids, optical breakdown in PMMA and glass may be associated with self-focusing and self-trapping with both nanosecond- picosecond-range pulses.[219] The damage threshold for glass is approximately 100 times greater than that for PMMA, but once glass damage occurs, it tends to be more extensive.[220] As damage tends to be cumulative, IOLs may be damaged more by bursts of laser shots than by single pulses. Various IOL designs, including the use of spacers to increase the separation between the IOL and the posterior lens capsule, have been created in the attempt to minimize damage from photodisruption.

Since clinical applications of Nd:YAG photodisruption involve energies significantly above retinal damage thresholds, it is important to consider how the retina is protected during these laser procedures. Beam divergence is the angle formed by the cone of light converging on and diverging from the laser system's focal point. The border of the laser beam is described as either the $1/e$ or $1/e^2$ points of the solid angle. Commercial ophthalmic Nd:YAG lasers usually broaden the laser beam with an inverse Galilean telescope and then employ a large-diameter, high-power final focusing lens to achieve the desired combination of cone angle, minimal spot size, and comfortable working distance. As such, for retinal injury to occur during Nd:YAG laser posterior capsulotomy, 96 mJ, some 20 times the energy clinically used, would have to be incident on the cornea.

Plasma formation is a secondary factor in retinal protection during photodisruption in the pupillary plane. It absorbs and scatters incident radiation, thereby diminishing the transmission of radiant energy along the beam path. Plasma shielding assumes a more important role in retinal protection during vitreous photodisruption. Nevertheless, the pressure waves still propagate unattenuated, and may cause retinal or choroidal damage even in the absence of suprathreshold radiation levels.

6.4.2 Instrumentation

Although photodisruption is possible with other lasers and at other wavelengths, including some Nd:YAG harmonics, the fundamental Nd:YAG output at 1064 nm is the only one used in commercial ophthalmic photodisruptors. Most clinical Nd:YAG lasers employ the fundamental TEM_{00} mode, so that the spot size, and consequently the energy required for optical breakdown, can be minimized. Beam divergence is generally 0.5–3.0 milliradians (mrad). The lasers are cooled by ambient air or internally recirculated water, and require only standard 110-V outlets. While 5 mJ is sufficient for most applications, many ophthalmic Q-switched Nd:YAG lasers are capable of producing up to 30 mJ. Higher energies may be needed to cut very dense material and in cases of hazy media, such as corneal edema or blood in the anterior chamber. Mode-locked systems have a maximum output of about 5 mJ per pulse train, but because of the greater control and relative safety of the Q-switch, those models employing mode-locking have largely fallen out of favor.

An aiming beam is required to guide the pulsed, invisible Nd:YAG output. This is achieved with a CW helium-neon (He–Ne) laser that produces 632.8-nm output coaxial with the Nd:YAG's and below the retinal injury threshold. Since high-peak power pulses cannot be satisfactorily transmitted via fiberoptics, ophthalmic Nd:YAG lasers employ fixed mirrors to guide the output to the patient, who is generally seated opposite the surgeon at a specially configured slit-lamp biomicroscope. The larger the solid cone angle, the lower the energy required for optical breakdown and the risk of IOL or retinal damage, but the greater the chances of beam vignetting during some applications. The slit-lamp design limits the angle to approximately 20°, and most systems employ one of 16°.

Although contact lenses are seldom required for simple posterior capsulotomy, they may be helpful to stabilize the eye, prevent blinking, and maintain a regular optical surface. However, to effectively treat the vitreous and some other intraocular structures, specialized instruments with a variety of lenses and mirrors are required (Fig. 6.11).

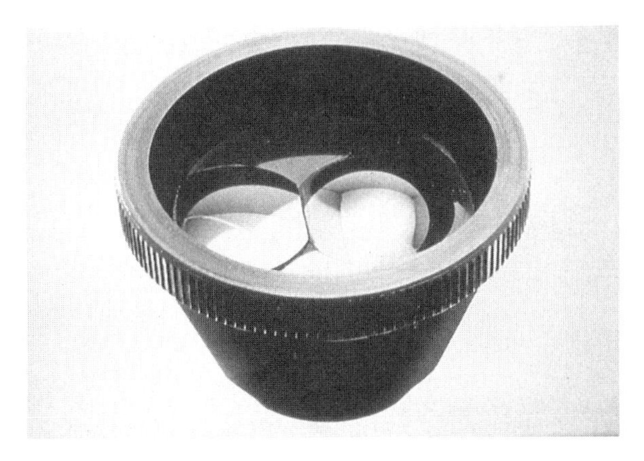

FIGURE 6.11 Goldmann three-mirror contact lens, which permits selective targeting of intraocular structures. (Courtesy of Ocular Instruments, Inc., Bellevue, WA.)

6.4.3 Clinical Applications

6.4.3.1 Posterior Capsulotomy

In the past decade, cataract surgery has largely changed from intracapsular cataract extraction (ICCE), in which the entire lens capsule is removed together with the opaque lens, to extracapsular cataract extraction (ECCE), in which the posterior lens capsule is left in place. This serves to both reduce the incidence of postoperative vitreoretinal complications, such as cystoid macular edema (CME), and to provide support for posterior chamber IOLs. Unfortunately, this membrane often also opacifies, forming a so-called secondary cataract.[221] Until the advent of photodisruption, this membrane was ruptured by introducing a needle into the eye with the patient seated at the slit lamp, with all the attendant risks of any invasive procedure. Laser-assisted removal of primary cataracts remains the subject of considerable research, but is not yet clinically feasible (see below). However, the Nd:YAG laser has proven so successful at sectioning opacified posterior lens capsules that it has completely replaced the traditional surgical approach for the often equally debilitating secondary cataracts.

Although secondary cataracts can substantially diminish vision, their removal doesn't guarantee normal vision. Especially in older patients, concomitant ocular pathology (such as macular degeneration or CME) may impair vision even after capsulotomy. Such instruments as the laser interferometer (see below) and the potential acuity meter can be used to assess best potential acuity even through cataracts, preventing a possibly useless or even deleterious procedure.

Prior to laser capsulotomy, the pupil is dilated to provide the surgeon with maximum visibility of the membrane. Topical anesthesia is necessary only if a contact lens is employed. With most Nd:YAG lasers, a posterior capsule can be opened with pulses of 1–2 mJ. Shots are placed along tension lines, as indicated by capsular wrinkles, to create the most efficient opening (Fig. 6.12).

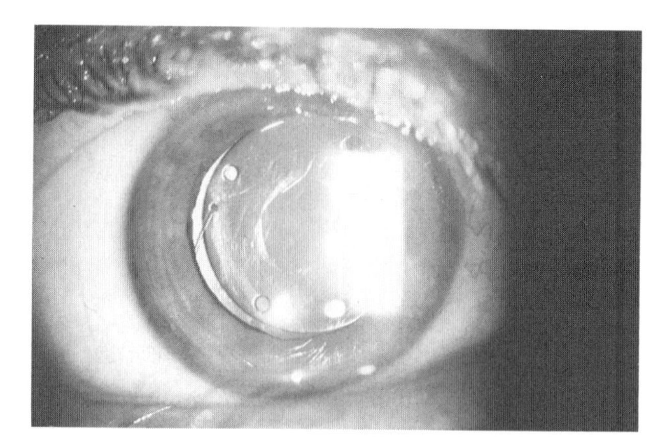

FIGURE 6.12 Secondary cataract after Nd:YAG laser capsulotomy. The IOL edge and haptic are visible. (Courtesy of Roger Steinert, M.D.)

This procedure is associated with a high degree of visual improvement, but is not without occasional complications.[222] By far the most commonly encountered difficulty is a transient rise in IOP,[223] likely caused by impaired aqueous outflow resulting from capsular debris, acute inflammatory cells, and high-molecular-weight protein.[224] IOP is typically checked for several hours following the procedure, with pressure-reducing agents given topically or systemically as needed.[225] Patients with preexisting glaucoma are more susceptible to this IOP elevation, but since it has been encountered in all types of patients, some clinicians advocate prophylactic treatment.[226]

6.4.3.2 Iridectomy

Argon laser iridectomy has largely supplanted the traditional surgical approach, but there remain many instances in which its use is problematic. The argon laser relies on coagulation, vaporization, and necrosis to cut through tissue, and light blue or gray irises may not absorb sufficient energy. Conversely, the strong absorption by dark brown irises may generate a char that impedes further penetration. Since photodisruption does not depend on target pigmentation, the short-pulsed Nd:YAG laser represents an attractive alternative for creating iridectomies. This was verified in clinical studies, which have demonstrated that the Q-switched Nd:YAG laser can create openings in the iris, which unlike some openings created with the argon laser, do not gradually close.[227] Small, self-limited hemorrhages are occasionally encountered with the Nd:YAG laser,[228] but as with all photodisruption procedures, this laser would be unable to coagulate any significant bleeding that might occur. The loss of corneal endothelial cells overlying the treatment site can be minimized with proper technique.[229] To facilitate laser iridectomy, the iris is drawn taught by instilling miotic drops, which constrict the pupil. Openings are often achieved with only a single Nd:YAG laser shot of 4–8 mJ. Nd:YAG laser iridectomy has proved effective in cases in which the argon laser has failed,[230] while use of the two lasers together on dark irises may allow less energy to be employed than with either laser alone.[231]

6.4.3.3 Posterior Segment

While technically more demanding and potentially more dangerous, Nd:YAG laser photodisruption may also be applicable to pathology in the posterior segment. Vitreous membranes sometimes form, and if adherent to the retina, may lead to that tissue's detachment. Experimental vitreous membranes in rabbits have been successfully sectioned with 4-mJ pulses as close as 4 mm to the retina, without retinal injury.[232] Since these membranes may be complex and fibrous, hundreds or thousands of pulses, often in multiple sessions, may be necessary. Aside from the risk of retinal or choroidal hemorrhage, which rises exponentially with proximity to the retina, a lens (crystalline or IOL) may also be damaged if work is performed too close to its posterior surface. However, despite initial concern that photodisruption of the posterior lens capsule or vitreous may cause liquefaction and other vitreous disturbance, Krauss et al.[233] employed MRI and other techniques to demonstrate that this process does not significantly affect the structural integrity of the normal vitreous body.

6.5 ABLATION

The term *ablation* is often used casually to refer to many laser procedures, including photocoagulation. For present purposes, only those methods that involve actual removal of tissue are considered here. As will be seen, these include many promising areas of current investigation, which offer hope for correction of ocular pathology ranging from refractive errors to epiretinal membranes.

The ophthalmic laser development with the most widespread potential applicability, and which has thus received the greatest public attention, is corneal ablation. Photorefractive keratectomy (PRK) with the excimer laser at 193 nm has become a successful clinical procedure some 10 years after the first experimental reports, and will likely soon receive FDA approval. Nevertheless, the technology and techniques are still evolving, and there have been many reports of potential alternatives or successors to the excimer. Moreover, while the cornea remains the most common and suitable target for laser ablation, there is considerable interest in extending the process to intraocular structures.

6.5.1 Cornea

The human cornea consists of five main layers, starting anteriorly: epithelium, Bowman's membrane, stroma, Descemet's membrane, and endothelium (Fig. 6.13). Corneal thickness ranges from about 520 μm centrally to 650 μm peripherally. Water is the largest component of the cornea, representing around three-quarters of its wet weight.[234] The epithelium is approximately 50 μm thick;

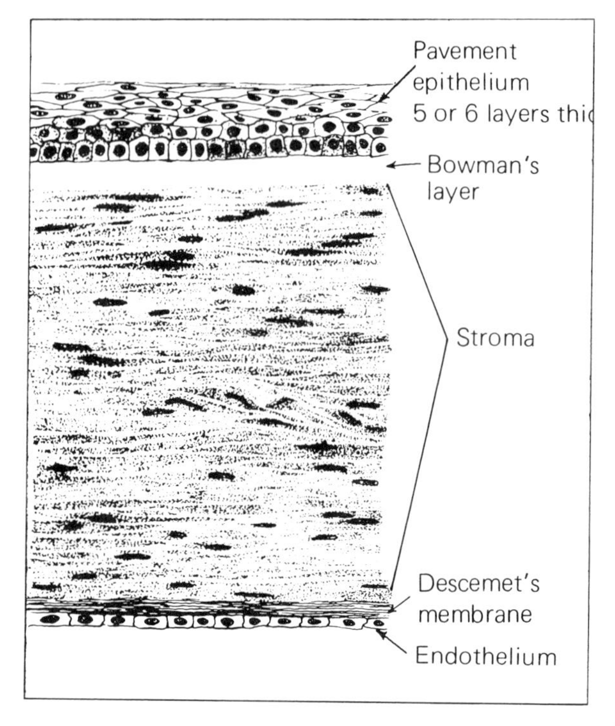

FIGURE 6.13 Cross section of the cornea. [Reprinted from Warwick[517] with permission of Chapman & Hall.]

its basement layer, Bowman's membrane, is acellular and some 12 μm thick. The main solid components of the stroma, which constitutes about 90% of corneal thickness, are collagen, other proteins, and glycosaminoglycans. The endothelium, whose basement layer is Descemet's membrane, is a single layer of cells. These cells lack significant mitotic ability, and damage to them may alter corneal hydration and, potentially, clarity.

The cornea accounts for about two-thirds of the eye's refractive power. Whereas some refractive errors such as astigmatism (irregular curvature) or keratoconus (cone-shaped deformity) are corneal irregularities per se, others such as myopia and hyperopia—most cases of which are secondary to too long and too short axial length, respectively—can be corrected by recontouring the cornea. This is the basis for keratorefractive surgery, which is discussed below.

Early ophthalmic laser applications, particularly retinal photocoagulation, relied on the transmisson by the cornea of the (visible wavelength) light. Although use of the argon laser has been reported for the treatment of such corneal disorders as neovascularization, lipid keratopathy, and adhesions,[235] the cornea had generally not been considered an appropriate target for laser therapy. However, work with the CO_2 laser, and more recently and extensively with the excimer and other new ablation lasers, has dramatically changed this notion.

6.5.1.1 Carbon Dioxide Laser

Ophthalmic investigation of the CO_2 laser followed soon after its development in 1964.[236] The laser emits radiation with a much higher efficiency (some 15%) than argon or krypton lasers, and at a mid-IR wavelength of 10.6 μm. That this radiation is strongly absorbed by water ($\alpha = 950\,\text{cm}^{-1}$) makes the CO_2 laser of potential use in any water-containing tissue, and it is currently employed in many other medical fields. Heat diffusion away from the target area coagulates adjacent vessels and provides hemostasis, which is particularly useful in patients with bleeding diatheses. Along with the lymphostasis afforded by the CO_2 laser, this approach is valuable when treating malignant or severely infected tissues. Water is the most ubiquitous substance in the eye, and the CO_2 laser has been employed, either experimentally or clinically, to treat ophthalmic pathology ranging from the eyelids and adnexa to the vitreous.[237–239] The major disadvantage of the CO_2 laser is that current fiberoptics are not capable of effectively transmitting at this wavelength, necessitating the use of less flexible articulated arms. Also, smoke and steam often develop and must be vented.

Figure 6.14 shows an incision in a bovine cornea produced with the CO_2 laser. There is considerable damage and disorganization to the surrounding tissue. Beckman et al.[240] suggested that the use of a pulsed (60–300 s^{-1}) CO_2 laser might significantly reduce thermal damage to surrounding tissue by allowing less time for heat conduction. This technique enabled them to achieve high peak powers to rapidly vaporize tissue, with relatively low average powers and minimal dissipation of heat. However, there was still a 0.12 mm zone of charred tissue around the incision site, and the edges were considered less sharp than they

FIGURE 6.14 Light micrograph of CO_2 laser (10.6 μm) ablation in a bovine cornea. Dosage parameters: 25 pulses, 0.2 Hz, 6 J/cm^2 per pulse; BaF_2 lens. (Reprinted from Krauss et al.[235] with permission of *Survey of Ophthalmology*.)

would have been with mechanical techniques. Keates et al.[241] proposed an even greater reduction in exposure duration, using a Q-switched CO_2 laser with 500-ns pulses to make experimental corneal incisions. This laser had a repetition rate of $7200\,s^{-1}$, and although the peak output reached 450 W, the average was only 1.6 W. This technique was found to produce more uniform and reproducible lesions than a $90\text{-}s^{-1}$-pulsed mode, with less carbonization. Nevertheless, a study by Peyman et al.[242] in which CO_2 laser burns of various intensities, locations, and patterns were placed on rabbit corneas showed no significant alterations in corneal curvature. Although the notion of using the CO_2 laser as a welding substitute for, or adjunct to, sutures may be alluring (see above), Keates and his colleagues,[243] were unable to achieve adherence with scleral or corneal eye bank tissue. A difficulty of the CO_2 laser remains the degree of tissue shrinkage and vaporization.[241]

6.5.1.2 Excimer Lasers

Ultraviolet Radiation and the Cornea Two far-ultraviolet (UV) wavelengths, 193 and 248 nm, are the excimer emission lines that have been most extensively studied for potential laser surgery of the cornea. As indicated in Fig. 6.15, corneal absorption rapidly increases below 300 nm.[244] Since water does not significantly absorb radiation between 193 and 293 nm,[245] it must be the solid components of the cornea that are responsible for that tissue's absorption in this region and for the mediation of any photophthalmic changes.

The majority of corneal solids are proteins, particularly collagen, which represents about 70% of the stromal dry weight. Protein absorption maxima around 190 nm have been associated with absorption by the C–N peptide

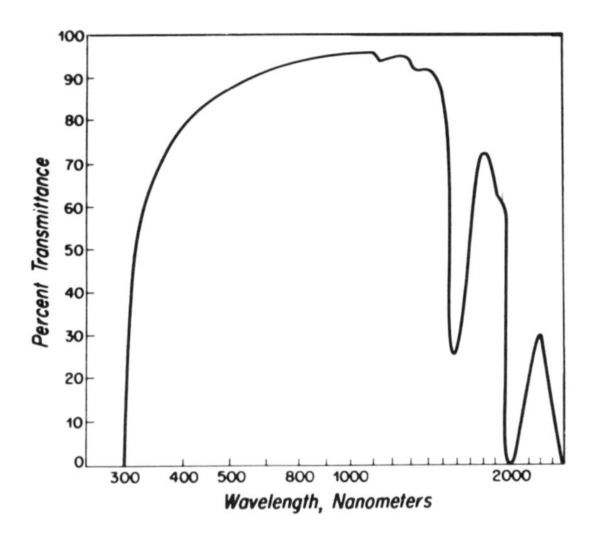

FIGURE 6.15 Transmission of human cornea. (Reprinted from Boettner et al.[295] with permission of *Investigative Ophthalmology and Visual Science*.)

linkage.[246–249] To a first approximation, peptide bonds behave as isolated chromophores.[247] Nucleic acids in the cornea are largely restricted to the epithelium. They absorb strongly at 248 nm, due to a peak around 260 nm corresponding to absorption by the nucleotide bases.[250] Absorption at 193 nm is approximately twice as great as that at 248 nm.[251] The different types of glycosaminoglycans demonstrate absorption spectra that are comparable to one another, with absorption peaks around 190 nm and no significant absorption at 248 nm.[252]

Ultraviolet radiation has been shown to have numerous deleterious effects on cellular activity.[253–255] These are largely mediated by the radiation's effects on DNA, and while the most serious effect is cell killing, others include mutagenesis, carcinogenesis, interference with synthesis of DNA and protein, delay of cell division, and changes in permeability and motility.[249] The action spectrum for UV-radiation-induced mutations parallels the absorption spectrum for DNA,[249] and almost all ultimate carcinogens have been shown to be mutagens.[256] It is well established that UV effects on the cornea are caused in part by absorption of radiation within the nucleoproteins.[257] Since it absorbs strongly and is the most anterior structure, the epithelium is the first corneal layer to be affected by UV radiation.[258]

Early damage studies of the excimer laser indicate that the nucleic acid components of irradiated epithelial cells are the initial absorption sites in the cornea at 248 nm, and that such damage may involve initial DNA (deoxyribonucleic acid) lesions leading to suppressed protein synthesis.[259] On the other hand, the avascularity and low cell content of the stroma make it relatively resistant to UV damage.[260] As expected on the basis of their absorption spectra, glycosaminoglycans are very sensitive to UV radiation.[261] Stromal swelling following UV irradiation may be caused by a breakage of glycosaminoglycans that interconnect adjacent fibers, or by destruction of the endothelium, which has been observed under certain conditions, and which may disrupt normal corneal hydration.[262] Endothelial cells have shown a high rate of unscheduled DNA synthesis when exposed to UV radiation, although the repair capacity is partially responsible for the unexpectedly strong resistance of the endothelium toward UV radiation damage.[263]

Excimer Laser Surgery of the Cornea

LASER PRINCIPLES *Excimers*, or excited dimers, are molecules with bound upper states and weakly bound ground states. The most common excited molecules exhibiting laser action are rare gas excimers, such as F_2 and Xe_2, which emit radiation at 157 and 170 nm, respectively. Such lasers, however, are impractical for clinical or most laboratory uses, not least because oxygen absorption below 190 nm precludes working in room air. The best performance has been demonstrated by excimers formed by the reaction of an excited rare gas atom with a halogen molecule, in which the rare gas atom acts as the corresponding alkali metal and becomes very reactive in the presence of halogen-containing molecules.[264] Such rare-gas monohalides emit radiation as they decay from the bound upper state to the rapidly dissociating ground state. Lasers employing this principle were first developed in 1975[265] and have been valuable sources of UV radiation for research in chemistry, spectroscopy, remote sensing, and dye laser pumping.[266] Different combinations of a rare gas and a halogen gas can be used as the active laser medium to generate a variety of UV wavelengths (Table 6.2).

Excimer lasers, emitting pulses of approximately 10-ns duration, have been employed as a new means of materials processing. Since the early 1980s, excimer lasers have been used to precisely etch submicrometer patterns in a variety of polymer materials.[267,268] Srinivasan[269] has termed this controlled removal of material, in which molecules on the irradiated surface are broken into small volatile fragments, *ablative photodecomposition*. Several theoretical models have been proposed to explain the results of this process.[270–272] UV photons are strong enough to directly break molecular bonds. The driving force for ablative photodecomposition is the energy of the photon in excess of that of the broken chemical bonds, which serves to excite the fragments and ultimately leads to their ablation from the surface.[273] Ablative photodecomposition is thus probably caused by a combination of the high absorption for far-UV radiation possessed by organic polymers, which limits the depth of the radiation's penetration; and the high quantum yield for bond breaking, which results in the formation of numerous fragments in a small volume near the surface.[274] Ablation is thought to result from the intense pressure buildup within this volume.[269,273] The relative photochemical and thermal contributions to excimer laser ablation have been debated, but it appears that there is an increasing thermal effect at longer wavelengths.[275–277]

Excimer lasers offer an intriguing option for ablation and cutting of tissue.[276] They have been used to etch clean and precise patterns in hair and cartilage,[278] arterial tissue,[279] and skin,[280] and are being considered for use in angioplasty and neurosurgery, among other surgical procedures.[281]

REFRACTIVE KERATOPLASTY There are numerous surgical options for altering corneal curvature, but for present purposes it suffices to briefly describe the main procedures and examine some of the areas in which laser techniques may offer advantages.

Radial keratotomy is a still widely practiced procedure in which radial incisions in the cornea reduce its curvature and counteract myopia. Its most serious

TABLE 6.2 Principal Wavelengths of the Rare-Gas Monohalide Excimer Laser

Gas Fill	Wavelength (nm)
Argon fluoride (ArF)	193
Krypton chloride (KrCl)	222
Krypton fluoride (KrF)	248
Xenon chloride (XeCl)	308
Xenon fluoride (XeF)	351

shortcoming is the lack of reproducible incision depth and hence refractive accuracy.[282] With current techniques, incisions are made only one at a time, and it has been reported that corneal dehydration caused by operating microscope lights may thin the cornea up to 10% during the procedure.[283] The goal of maximal radial keratotomy is to bring the incisions as close as possible to Descemet's membrane, which increases the chances for endothelial damage or even perforation. Mean endothelial cell loss may be as high as 10%,[284] while corneal perforation may occur in up to 20% of cases.[285] Although serious complications arising from perforation are uncommon, there have been cases of endophthalmitis following such inadvertent entries into the anterior chamber.[286]

Keratomileusis, keratophakia, and epikeratophakia are all techniques of using a suitably lathed lenticule, obtained from either donor cornea or the patient's own cornea, to alter refractive power. With these procedures, both freezing and unfreezing of the tissue are required, causing swelling of the central corneal stroma, shrinking of the diameter, and keratocyte death, often complicating prediction of final results and compromising visual recovery.[287] While attempts have been made to incorporate these factors into the computer programs that control the cryolathing process, it would clearly be advantageous if the reshaping of the lenticule could be done at room temperature. Direct reprofiling of the whole cornea in situ, if feasible, would obviate lenticule use altogether.

CORNEAL ABLATION In 1983, Trokel and coworkers[288] reported the first use of the excimer laser to achieve precise and controlled etching of the cornea. Using 193-nm radiation, they selectively ablated narrowly defined areas of bovine corneas by employing masks of various designs to restrict the laser energy. Tissue ablation to a depth of 1 μm was achieved with a total energy deposition of approximately 1 J/cm^2. Laser damage was localized to the zone of ablation, with no evidence of thermal effects. Edges of the laser incisions were parallel and straight, and no disorganization of the stromal lamellae or epithelial edge was apparent in the histologic sections.

Puliafito et al.[289] performed a comparative study of excimer laser ablation of the cornea at 193 and 248 nm. Slits of several thicknesses were made in the corneas of freshly enucleated human and bovine eyes, using either a mask or a cylindrical lens. The lowest per pulse fluences at which human corneal ablation was observed were 46 mJ/cm^2 at 193 nm and 58 mJ/cm^2 at 248 nm. Depending on the width of the corneal exposure, 193 nm ablations were either trough-like or slit-like (Fig. 6.16a). Transmission electron microscopy revealed a zone of damaged stroma approximately 0.1–0.3 μm thick on the edges of the ablation, with preservation of corneal fine structures beyond this region. Such edge effects were thought to represent a thin band of thermal denaturation or photoablated material that was not completely ejected and that adhered to the wall of the incision. Scanning electron microscopy gave further evidence of the sharp demarcation of the cuts (Fig. 6.16b). In contrast, ablation at 248 nm produced incisions with a region of adjacent stromal damage at least 2.5 μm wide (Fig. 6.17). In addition to the wider zone of damage at 248 nm, the stroma was

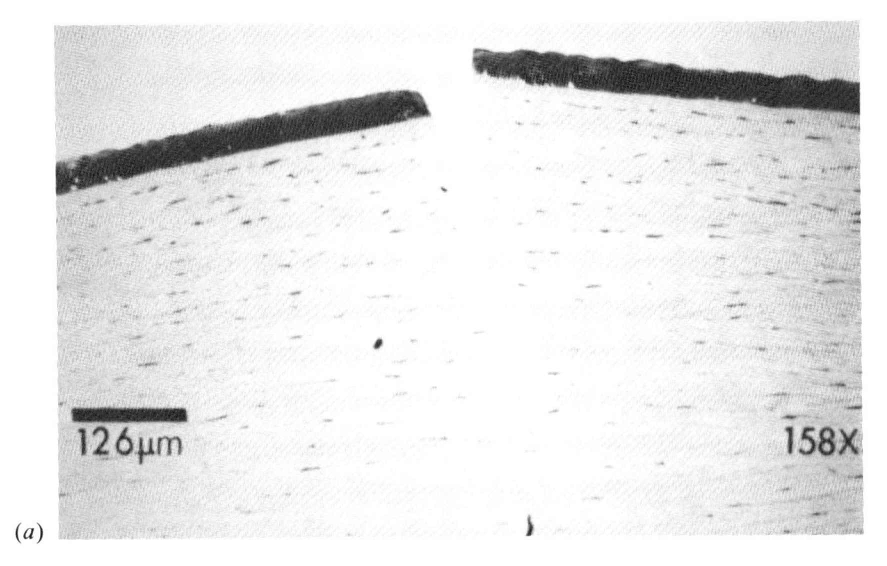

(*a*)

(*b*)

FIGURE 6.16 Slit-like corneal ablation performed with excimer laser at 193 nm. Dosage parameters: 20,000 pulses, 50 Hz, 125 mJ/cm^2 per pulse, 10-μm mask. (*a*) Light micrograph; (*b*) scanning electron micrograph, showing sharp cleavage of the corneal epithelium (arrow) and stroma (arrowhead). (Reprinted from Puliafito et al.[289] with permission of *Ophthalmology*.)

also markedly disrupted. Residual thermal damage produced by 248-nm excimer laser pulses is similar to that produced by CO_2 laser pulses of 2-μs duration.[290] This is consistent with the respective absorption levels measured at the two wavelengths, $\alpha = 2700\ \text{cm}^{-1}$ at 193 nm and $\alpha = 210\ \text{cm}^{-1}$ at 248 nm. Kerr-Muir and associates[291] found that excimer laser ablation yields tissue at least 10 times

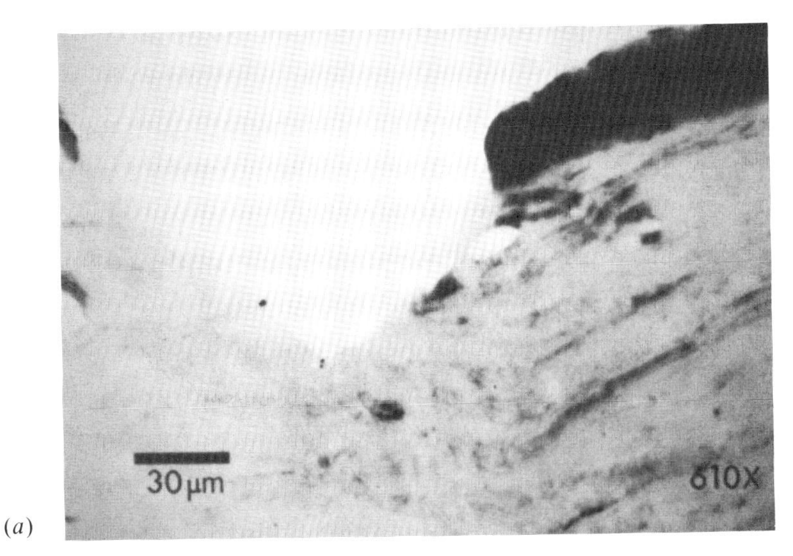

(*a*)

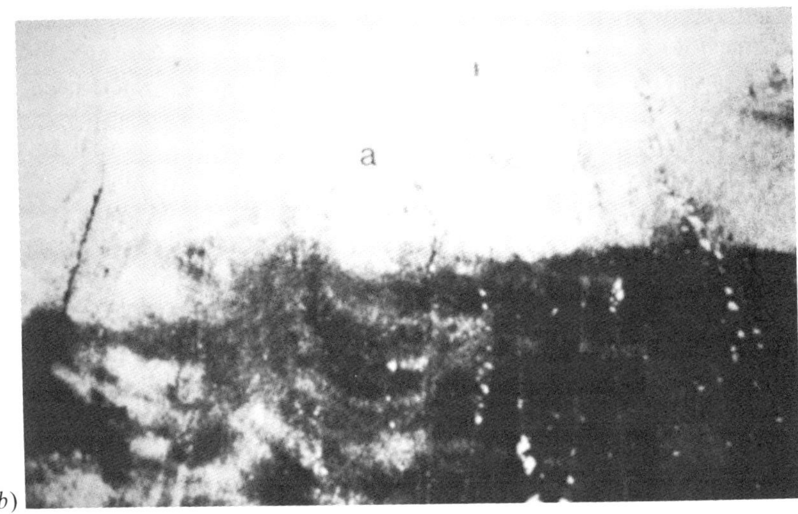

(*b*)

FIGURE 6.17 Corneal ablation performed with excimer laser at 248 nm. Dosage parameters: 1050 pulses, 50 Hz, 190 mJ/cm^2 per pulse, quartz lens. (*a*) Light micrograph – note irregular edge and disorganization of stromal collagen adjacent to the ablation; (*b*) transmission electron micrograph. The corneal stroma adjacent to the region of ablation (a) shows a broad zone of disorganization measuring at least 2.5 μm wide. (Reprinted from Puliafito et al.[289] with permission of *Ophthalmology*.)

as smooth as does conventional diamond knife surgery, and further described the pseudomembrane which seals laser-treated surfaces and appears to minimize postoperative scarring.

A number of factors may explain the minimal damage to adjacent tissue observed with 193-nm corneal ablation. The absorption length of 193-nm radiation in stroma is only 3.7 μm, so that incident energy is deposited in a relatively small volume of tissue. Moreover, the incident laser radiation is short-pulsed (10–20 ns), so that heat diffusion beyond the irradiated region is minimized. Indeed, calorimetric studies suggest that these laser pulses have a much shorter penetration depth than the simple absorption coefficient would indicate, and that Beer's law may not hold for ablation.[292] As previously noted, other investigators have demonstrated decreased thermal effects at shorter exposure times using non-UV wavelengths. Non-CO_2 laser alternatives to the excimer are discussed below, but it is possible that ablative photodecomposition at 193 nm may be fundamentally different from ablation using IR radiation. This may be related to differences in photon energy, and the type and location of target chromophores (i.e., tissue water at IR wavelengths vs. organic biomolecules at 193 nm). The photon energy of 6.4 eV (electron volts) for 193-nm radiation is more than sufficient to cleave peptide bonds (3.0 eV) or the adjacent carbon–carbon bonds (3.5 eV) of the polypeptide chains, especially collagen. It is possible that the high quantum yield of peptide bond cleavage at 193 nm, along with the high absorption of glycosaminoglycans, which are intertwined throughout the stromal collagen moiety, are responsible for the superior tissue removal and lower fluence requirement associated with 193-nm ablation.

Krueger and his associates[293, 294] further investigated some of the quantitative aspects of excimer laser ablation of the cornea. They found that plots of etch depth per pulse versus fluence per pulse generate roughly sigmoidal curves in whose steep parts there is an approximately logarithmic relationship, as had already been observed in simpler polymer materials. Inflection points in these plots correspond to minima of graphs of fluence required to remove a given depth of tissue versus fluence per pulse. These values—about 200 mJ/cm^2 at 193 nm, 1000 mJ/cm^2 at 248 nm, and 1500 mJ/cm^2 at 308 nm—represent the per pulse fluences for the most efficient ablation, where the largest portion of incident energy is converted to tissue removal. Probably 308-nm radiation is not useful for such work, since corneal transmission is high at this wavelength[295] and can lead to damage of intraocular structures. A study by Peyman et al.[296] demonstrated that exposure of the cornea to 308-nm excimer laser radiation produced a combined ablative and coagulative effect, with corneal necrosis, stromal opacification, and endothelial cell damage. The tissue depths removed per excimer laser pulse were 0.45 and 6.25 μm at the preceding maximum efficiency levels for 193- and 248-nm radiation, respectively. Since 193-nm radiation removes considerably less tissue per pulse, control of etch depth is more precise at this wavelength. Energy in excess of these optimum values may produce undesired side effects such as heat or shock waves. Whether these levels produce the best clinical results, however, remained to be determined.[297–299] Recent clinical applications have employed fluences in the 160–180-mJ/cm^2 per pulse range, and Campos and coworkers[300] found that even lower fluences may prolong laser hardware lifetime, allow larger ablation zones (thereby minimizing duplex optical images[301]), and produce a thinner pseudomembrane, although the clinical ramifications of the last are uncertain.

Many issues remained to be resolved before excimer laser corneal ablation could be attempted clinically. Corneal smoothness may also be affected by the inherent inhomogeneity of the excimer beam, which as described below requires elaborate techniques to minimize, and by the somewhat asymmetric distribution of the corneal components and variation with age of its UV absorption.[295]

Dehm et al.[302] reported that 193-nm excimer laser incisions to 90% of corneal depth produce endothelial alterations similar to those seen underlying diamond knife incisions of comparable depth. These include minimal disruption of cell junctions, cellular edema, and a ridge that corresponds to the site of the incision. Stress waves generated during ablation were thought to be analogous to the mechanical injury associated with conventional surgery. No endothelial cell loss was observed with scanning electron microscopy for 193-nm ablation, while severe endothelial cell damage and loss were observed for 248-nm ablation performed at identical fluences. Differences in endothelial damage may be due to differences in absorption length, damage mechanisms, and recoil forces associated with ablation at these two wavelengths. Zabel et al.[303] found that although the pressure near the back of the cornea reaches approximately 100 atm during ablation of the superficial stroma, endothelial disruption occurs only after some 85–90% of the corneal thickness has been ablated.

Since the excimer laser employs UV radiation, possible mutagenesis and carcinogenesis are obvious concerns and were the subject of several early studies. Nuss et al.[304] found that unscheduled DNA synthesis, a measure of excision repair of pyrimidine dimers, was not increased in rabbit corneas following 193-nm irradiation at 400 mJ/cm^2 per pulse, as compared with diamond knife controls. In contrast, 248 nm did produce a statistically significant increase in such repair. Kochevar[305] indicated that the fact that excimer laser radiation at 193 nm causes less cytotoxicity than predicted by the DNA absorption spectrum may be due to absorption at that wavelength by protein present between the cell surface and the nucleus, or to induction by photons reaching the nucleus of DNA photoproducts that are either not cytotoxic or readily repaired by the cells. DNA-damaging effects resulting in cytotoxicity were least at 193 nm, greatest at 248 nm, and intermediate at 308 nm. It has also been shown that 193-nm irradiation of the cornea at a fluence of 60 mJ/cm^2 per pulse causes a fluorescence emission between 295 and 425 nm.[306] Although this includes wavelengths known to be cataractogenic, the percentage of incident energy reaching the crystalline lens is only approximately 10^{-5} that incident on the cornea, and is thus clinically insignificant.

There are two principal categories of clinical application of excimer laser corneal ablation: PRK and phototherapeutic keratectomy (PTK). While the goals and technical details differ, they both rely on a comprehensive understanding of excimer laser–tissue interaction and corneal wound healing.

PHOTOREFRACTIVE KERATECTOMY Preliminary experimental work on animal studies[307–326] preceded clinical trials[327–346] of excimer laser ablation. Several researchers performed radial keratotomies with the excimer laser,[307,308,312,317] but most concentrated on wide-area corneal ablation.[235,309,315,322,323] Radiation at 193 nm consistently yielded smoother surfaces than did that at 248 nm, and use of the longer wavelength was discontinued. Even at the shorter wavelength, however, wound healing was often associated with some degree of opacification and regression in refractive results, especially with deeper ablation. Moreover, there was visual and histologic variability in wound healing.

Unlike the emissions of many lasers, such as the argon and krypton lasers, that of the ordinary excimer laser is largely non-Gaussian and irregular. Modifications are necessary for ophthalmic applications if the beam's homogeneity is to be commensurate with the fundamentally precise laser–tissue ablation, allowing macroscopic as well as microscopic smoothness; L'Esperance[313] described one such technique using prisms to create a "top hat" beam configuration. With the exception of astigmatism, correction of refractive errors requires radially symmetric alteration of the corneal curvature. This is most commonly achieved with a computer-controlled iris diaphragm.[322] Ordinarily this would be useful only for myopic corrections; however, a rotating diaphragm has been described that can create spherical negative and positive—as well as cylindrical—corrections, to correct myopia, hyperopia, and astigmatism, respectively.[311]

Conventional radial keratotomy spares the optical axis. Although refractive keratoplasty necessarily violates this region, only the stroma is actually lathed. However, in order for the excimer laser to realize its full potential for refractive surgery, the corneal curvature must be altered in situ. This means that for laser keratomileusis, not only must the optical axis be treated, but this must involve disruption, if not obliteration, of what has traditionally been viewed as the inviolable Bowman's membrane. However, one advantage of this greater invasiveness is the ability to correct much higher myopia than that treatable with radial keratotomy.[326] The pseudomembrane covering the ablation surface, which consists of a dense exterior part 20–100 nm wide and a less dense interior portion 60–200 nm wide, may act as a substitute for Bowman's membrane, by supporting the establishment of a regular basal epithelial layer until a true basement membrane, albeit slightly more undulating than the original, reappears.[315]

Munnerlyn et al.[316] calculated the ablation diameters and depths necessary to correct spherical refractive errors, assuming that the epithelium will regrow with a uniform thickness and produce a new corneal curvature determined by the new stromal curvature. They showed that flattening to correct myopia is determined by the equation $t_0 \approx -S^2D/8(n-1)$, where t_0 is the required ablation depth, $D = (n-1)/(1/R_2 - 1/R_1)$; n is the corneal index of refraction, 1.377; R_1 and R_2 are the initial and final corneal radii of curvature, respectively; and S is the diameter of the ablation zone. This is often approximated as t_0 = dioptric correction $\times$ $S^2/3$, and indicates that the required ablation depth per diopter increases with increasing optical zone, going, for example, from 3.0 μm at 3 mm, to 5.3 μm at 4 mm, and to 8.3 μm at 5 mm. Hyperopic correction is achieved by sparing the visual axis and removing tissue peripherally, in an amount nearly equal to that required for the same magnitude of myopia.

A team led by L'Esperance and Taylor performed the first PRK trial on humans in the United States.[327,328] Initial FDA investigational device exemptions limited work to blind eyes or those scheduled for enucleation, and results were consistent with the preceding animal work. Using a delivery system with enlarging apertures for myopic correction, this group treated eyes with parameters of 10 Hz and 80–125 mJ/cm^2 per pulse, creating ablations 3–5 mm in diameter and 30–150 μm in depth. With the patient in the supine position, the eye to be treated was stabilized with a vacuum ring and retrobulbar injection of lidocaine. Prior to the start of ablation, the epithelium was mechanically debrided with a blade to about 1 mm beyond the intended ablation zone. Following the procedure, the eye was treated with balanced salt solution and antibiotic ointment, and then taped closed for 2–3 days. Patients reported either no or minimal postoperative pain, the latter relieved with oral analgesics. Reepithelialization occurred in all eyes within 3 days, with no subsequent erosions. All eyes not enucleated by 1 week showed some inflammatory response, albeit without any corneal leukocyte reaction; whether this resulted from the ablation per se, or was secondary to the suction ring or epithelium removal, could not be determined. The early inflammatory response was only partially responsive to local corticosteroids, and there was no apparent effect on the amount of clinical haze, but the study group was quite small (10 patients). As had been observed in animals, there was a gradual filling in of the ablation zone, with residual refractive change two-thirds that originally obtained. Slit-lamp examination initially revealed mild superficial edematous haze in the ablated area, which progressed after 2 weeks to a mild speckled interface haze between the epithelium and stroma. Specular microscopy at 3 months showed no loss of endothelial cells. Increased collagen and ground substance was observed at 4 months, and these researchers suggested that the use of pharmacologic means of maintaining the communication between the epithelium and the underlying keratocytes, in the absence of Bowman's membrane, may reduce new collagen formation and retain the initial ablation results. In a similar study, McDonald et al.[329] found that regression in refractive correction was proportional to the amount initially attempted, and that, unlike many conventional keratorefractive procedures, excimer laser ablation did not cause any astigmatism.

Seiler and associates,[330] working in Germany, published early results of PRK on both blind and sighted humans. Using parameters of 180 mJ/cm^2 per pulse and 10 Hz, and a computer-controlled diaphragm dilating in step widths of 5 μm, they ablated 4.5-mm-diameter discs for correction of up to −6 D (diopters), gradually decreasing to 3.5 mm discs for −10 D so as to avoid unnecessarily deep keratectomies. Postoperative treatment consisted of gradually tapered topical antibiotics and steroids. All patients reported considerable discomfort and foreign-body sensation, and were provided with systemic analgesics. Glare was common initially in the sighted eyes, but subsided after 2 weeks. There were no diurnal fluctuations in vision, although one patient complained of persistent halos at night (prompting these investigators to eschew 3.5-mm ablation zones in future work). Manifest refraction showed the typical initial overcorrection and subsequent mild reversal, and although there was a temporary reduction in best-corrected visual acuity, at 1 month it was back to the original level in most patients, and was even increased by 1 line in a few subjects. That stromal remodeling is the likely cause of refractive regression was suggested by the relative stability in those eyes undergoing small corrections, in which Bowman's membrane was not completely ablated. IOP was unaffected in all but one patient, in whom reduction in steroid frequency lowered the pressure from a high of 21 mmHg to 18. After 3 months, 12 of 13 sighted eyes had corrections within 1 D of those intended, while at 6 months only 10 did. Although transient subepithelial haze occurred in almost all eyes, it resolved to clinically insignificant levels after 6 months.

Brancato and colleagues[332] reported the results of over 1000 myopic patients who underwent excimer laser PRK in an Italian multicenter study. Patients with at least 10 D of myopia were treated in a two-step process, in which the initial ablation attempted more correction in a smaller diameter than did the second one immediately following it, permitting shallower ablations than with the one-step method. All patients received topical corticosteroids, for a minimum of 2.5 months following the procedure. The percentage of eyes within 1 D of attempted correction at 12 months ranged from 71.2% for those with initial errors less than −6 D to 28.2% for those with initial errors between −10 and −25 D, with mean residual errors of -0.52 ± 1.04 D and -1.86 ± 3.47 D, respectively. There were no severe postoperative complications, but 2.4% of patients lost two or more lines of best-corrected visual acuity; at least half of these cases were attributable to other, concomitant ocular pathology. Recent reports of 2-[333] and 3-year[334] follow-ups of PRK patients suggest that minimal refractive changes occur after the first year.

In a Swedish study, Tengroth and coworkers[335] experienced residual refractive errors at 12 months of 0.31 ± 0.52 D and -0.42 ± 1.05 D in subjects with initial errors of −1.25 to −2.90 D and −5.00 to −7.50 D, respectively, confirming the greater tendency toward myopic shift at higher attempted corrections observed in other studies. All patients received steroids, with 12-month refractions of 0.17 ± 0.47 and -0.09 ± 0.29 D in those treated for 3 months and 5 weeks, respectively, indicating that longer drug therapy reduces early regression in the refractive results. Subepithelial haze developed in all patients, peaking at 4–5 months postoperatively and declining to a mean of 0.75+ on a scale to 5+ by 1 year. Both myopic shift and haze could be diminished by later use of steroids, but longer treatment was required, and less effect was achieved compared with earlier institution. Gartry et al.[336] found no significant difference in haze or refractive regression between patients treated with steroids for 3 months and those who received a placebo, and recommended against routine pharmacologic intervention, although they didn't follow patients beyond 6 months. Stevens et al.[337] found higher complication rates with no significant benefits in those patients receiving steroids for 6 months compared with those treated for only 3 weeks.

Liu and associates[338] analyzed numerous factors in blind and sighted humans potentially affecting the outcome of PRK. They found that the larger the attempted correction, the greater the residual myopia. Refraction after 1 month is not predictive of that at 3 or 6 months, whereas refraction at 2 and 3 months is highly correlated with results at 6 months. There was no correlation between patient age and either rate of healing or development of corneal haze, while there was some evidence that older patients tend to have more accurate refractive and visual results at 6 months. Some correlation may exist between corneal haze and ablation depth or diameter, but refractive and visual results were entirely unrelated to the rate of healing. Significant regression of refractive results occurred only for attempted corrections of greater than 5 D.

Seiler et al.[342] followed 193 PRK eyes for up to 2 years. In contrast to radial keratotomy, PRK resulted in less induced astigmatism and progressive hyperopia. Complications were rare in patients with less than −6 D of myopia, but higher myopes were much more likely to experience steroid-induced rises in IOP, scars, loss in glare vision, under- and overcorrection, and continued regression. These researchers suggested, however, that these effects might be mitigated by using larger ablation zones. Sher et al.[326] performed PRK on a series of highly myopic patients (up to −14.5 D), with ablation depths to 230 μm and diameters to 6 mm, and achieved mean refractions at 6 months of -0.90 ± 2.13 D, although the long-term effect on corneal mechanical stability of such deep ablation remains to be determined. Heitzmann and coworkers[343] employed a multizone (4-, 5-, and 6-mm) protocol for correction of myopia greater than −8 D, which allowed shallower ablations but still resulted in considerable regression and haze. They suggested that such patients might be better treated by multiple ablation sessions (as has also been recommended by Förster[344]) or by some combination of PRK and radial keratotomy.

Binder and coworkers[345] found decreases in eye-bank eyes' central and peripheral endothelial cell counts from 2035–3051 to 940–1836/mm^2 following PRK. Central islands, or areas of localized steepening, probably caused by a combination of technical factors involving the laser beam and the corneal surface, have been noted in some patients.[346] Although they tend to improve with time, they may cause a decrease in best corrected visual acuity. Loss of corneal sensation may complicate any refractive procedure, but Campos et al.[339] showed that postoperative recovery is less in those patients treated for high myopia, undoubtedly because of the larger ablation volume.

It has been shown that PRK yields a relatively small central area of uniform refractive power, surrounded by a large transition zone of linearly decreasing power.[347] Moreira et al.[348] have proposed deliberately attempting a multifocal refractive effect to compensate for presbyopia, although such an approach may be complicated by monocular diplopia or a decrease in image contrast.

ASTIGMATISM The large variety of surgical procedures developed for the correction of natural and postoperative astigmatism is testimony to the inability of any single approach to effectively treat all patients. The excimer laser may prove valuable in this effort, as its output can be harnessed for a diversity of techniques. In 1988, Seiler et al.[349] used the excimer laser to produce linear corneal T-excisions for the correction of astigmatism in a series of blind and sighted patients in Germany. Radiation was directed at metal-foil-coated PMMA contact lenses containing tangential slits 4.5 mm × 150 μm, resulting in cylindrical corrections of up to 4.16 D. Postoperative care was as for this group's myopia work,[330] although topical antibiotics and steroids were used for only 1 week after the procedure. Astigmatism initially fluctuated, but tended to stabilize within 2 weeks. Epithelial regrowth occurred within 3 days, but even after a month there was considerable variation in the extent to which the entire keratectomy was filled with an epithelial plug. Visual acuity was stable after 1 week, while some patients noted glare for up to 8 weeks.

One attraction of the excimer laser is that corneal excisions aren't limited to patterns produced with conventional surgery, and McDonnell and coworkers[350] proposed the use of the excimer laser to create toric ablations for the correction of cylindrical errors. In contrast to the linear excision method, in which tissue removal ranged from 51 to 93% of corneal thickness,[349] this approach allows more superficial areas to be ablated, using an expanding slit with flattening in the meridian of expansion, perpendicular to the slit itself. Preliminary tests were conducted on rabbits, and although there was good correlation with intended corrections, keratometry at 12 weeks revealed corrections about only half of those expected. Initial clinical use of this technique, including for the combined treatment of myopia and astigmatism, termed *photoastigmatic refractive keratectomy* (PARK), also showed significant regression.[351,352]

TREPHINATION Of course, the greatest promise for the excimer laser is in situ reprofiling, but the technology may also prove useful for precise trephination, such as in obtaining tissue for transplants. Conventional trephination is often associated with significant residual astigmatism, and requires suction or some other means of direct stabilization along with contact cutting, introducing further distortion. In 1987, Lieurance and associates[353] reported using the excimer laser to remove lenticules from human eye-bank eyes, resulting in a normal epithelium, intact Bowman's membrane, normal stroma, and a smoother surface than is achieved with cryolathing. Serdarevic et al.[354] employed a rotating-slit delivery system[314] for trephination of donor and recipient human eye-bank and rabbit corneas. Fluence was 110 mJ/cm^2 per pulse, and operating at a rate of 15–20 Hz, the laser required from 30 s to a few minutes to achieve perforation, depending on corneal thickness and eye stability. Trephination diameter ranged from 4 to 8 mm, according to the distance from the spherical lens in this configuration. Buttons and recipient beds produced with the excimer laser were freer of distortion and damage to adjacent tissue, including loss of fewer endothelial cells, when compared with those created by free-hand or suction trephines. Deposits at the wound edges never measured more than 0.08 μm, minimizing the disparity between donor and recipient wound configuration. Healing through 3 months was comparable between the laser and mechanical trephination groups.

Lang et al.[355] proposed using the excimer laser with appropriate corneal masks for elliptical corneal transplants, conforming to the natural elliptical meniscus shape of the cornea, facilitating suture placement, and enhancing graft stability.

Gabay and his associates[356] used the excimer laser to prepare plano donor lenticules for clinical use, by placing a mechanically obtained lenticule in a special mold and ablating the excess tissue. Surface topography was superior to that of a hand-cut lenticule, and postoperative recovery was uneventful. While the laser's cost is considerable, the actual lenticule preparation was performed at a fraction of the expense of conventional techniques, and the group recommended using this method for processing optical power epikeratophakia tissue by providing a concave mold base for hyperopia and a convex one for myopia.

PHOTOTHERAPEUTIC KERATECTOMY Although the great attention, both among researchers and the popular press, which the excimer laser has garnered in recent years is due to its potential for PRK, PTK is at least as efficacious. In 1985, Serdarevic et al.[357] reported that 193-nm excimer laser ablation, but not that at 248 nm, was successful in completely eliminating and sterilizing experimental *Candida albicans* corneal infections in rabbits. The underlying stroma was unaffected, although since removal of tissue may reduce tensile strength and alter refractive power, this was deemed advisable only in instances of antifungal therapy failure. Dausch and Schröder[358] used excimer laser ablation to treat patients with malignant melanomas of the conjunctiva, pterygia, recurrent erosions, persistent epithelial defects following keratoplasty, and herpes infection. Steinert and Puliafito[359] used the excimer laser to remove a 1-mm zone from a patient with longstanding keratoconus who developed an apical fibroblastic nodule. At 1 week the surface contour was smooth, albeit with minor subepithelial anterior stromal haze, and although visual acuity remained at 20/25, the patient reported a subjective sense of improved vision and increased contact lens tolerance.

Sher et al.[360] conducted a trial of PTK for such conditions as anterior stromal and superficial scarring following infection or trauma, anterior corneal dystrophy, recurrent erosion, and band keratopathy. Most patients received peribulbar anesthesia and underwent mechanical removal of the epithelium. Corneal scarring was reduced in most subjects, while approximately half had improved visual acuity. Reepithelialization occurred within 5 days, without significant scarring. Postoperative treatment consisted of application of a patch with or without a disposable contact lens, and slowly tapered antibiotics and steroids. These researchers proposed a combination of myopic ablation followed by secondary hyperopic steepening to minimize the hyperopic shift encountered in about half the patients. (Induced hyperopia, which increases with ablation depth, may also be lessened by shifting the ablation zone and employing different spot sizes to modify the transition zone.[361]) They also concluded that removal of the epithelium is undesirable for PTK, as the intact tissue can serve as a modulator of ablation. In another series of PTK patients,[362] these researchers treated a variety of corneal scars and found that at 6 months postoperatively, 49% of subjects had visual acuity improvement of at least 2 lines, 36% had change of less than 2 lines, and 15% had worsening of at least 2 lines.

Kornmehl et al.[363] compared several masking fluids intended to shield deeper tissues while exposing surface irregularities during PTK. Dextran 70, with high absorption at 193 nm and moderate viscosity, was found to produce the least surface irregularity, followed by carboxymethylcellulose and saline; corneas ablated without fluid had the greatest irregularity. Compared with surgical superficial keratectomy, PTK has the potential to result in considerably less astigmatism and more predictable corneal power.[364]

6.5.1.3 Midinfrared Lasers

Although the vast majority of PRK research has been concentrated on the excimer laser at 193 nm, there have been periodic reports of attempts at employing other lasers and wavelengths to achieve the same precise ablation. Many of these studies have used IR lasers, a motivation for which was provided by early concern about the excimer laser's toxic gases and possible UV mutagenesis, the latter without any evidence, as well as the inability to transmit its output via fiberoptics. In 1986, Loertscher and associates[365] created corneal incisions in eye-bank eyes with a pulsed HF gas laser, which produced a combination of emissions from 2.74 to 2.96 μm and was operated at fluences of 0.7–2.3 J/cm^2 per pulse and a rate of 10 Hz. Water has an absorption peak around 2.9 μm, which makes the HF laser theoretically superior to the CO_2 laser in limiting thermal damage to adjacent tissue. The 200-ns pulse duration is considerably shorter than the 1.7-μs relaxation time of HF-irradiated water, further reducing heat spread. However, while damage to deep portions of the HF incisions was limited to 1–2 μm, that in the shallow parts was some 10–15 μm wide and was similar to damage associated with the excimer laser operating at 248 nm.[289] It was believed that greater penetration by some of the shorter HF emission lines, along with difficulty in focusing the laser output on the cornea, may have contributed to the larger than expected thermal effects.

Similar results were reported by Thompson et al.,[366] who compared HF, 193-nm excimer, and 2.94-μm solid-state erbium:YAG (Er:YAG) lasers in experimental corneal trephination. Excimer excisions were the sharpest, and while the mid-IR lasers required less time to penetrate the tissue, they also produced a 10–15-μm zone of adjacent stromal damage and wounds about 2.5 times larger than those made with metal scalpels. Peyman et al.[367] used an Er:YAG laser, emitting 200-μs pulses, to ablate 3.5-mm discs in rabbit corneas. Thermal damage extended up to 40 μm from the margins of the ablation zone, yet the endothelium was unchanged by ablations as deep as 320 μm. There was faint corneal light scattering in most animals, which progressively cleared. It was believed that the multimodal nature of that laser's output may have produced the less-than-optimal results. Tsubota[368] transmitted 400-μs Er:YAG pulses via a 200-μm-diameter fiberoptic, achieving fluences of 636–954 mJ/cm^2 per pulse. Ablation of the cornea and various intraocular structures yielded thermal damage results consistent with previous findings. Seiler et al.[369] found that Q-switched 80-μs Er:YAG pulses reduce thermal damage to 1–2 μm.

Stern et al.[370] studied corneal incisions produced by a Raman-shifted Nd:YAG laser, operating at 2.8 and 2.92 μm and emitting 8-ns pulses. At 2.8 μm, there was an ablation threshold of 250 mJ/cm^2 per pulse, and etch depth per pulse increased sigmoidally from 0.15 μm at 390 mJ/cm^2 per pulse to 3.8 μm at 2200 mJ/cm^2 per pulse. The considerable difference in ablation threshold and etch rate between the 193-nm excimer and this mid-IR laser undoubtedly reflects their fundamentally different mechanisms for tissue removal, namely, photodecomposition and heating, respectively. In contrast to results with the excimer laser, thermal damage in this case was highly dependent on fluence, ranging from 1.5 μm at 600 mJ/cm^2 per pulse to 10 μm at 2200 mJ/cm^2 per pulse, better than achieved with the HF laser but still inferior to the 0.3-μm zone of damage surrounding excimer ablations.[289] As with the HF laser, thermal damage surrounding the incisions decreased from top to bottom. It was expected that results at 2.92 μm should have been even better than those at 2.8 μm, and their similarity to those at the shorter wavelength was attributed to large pulse-to-pulse fluence variations at 2.92 μm.

6.5.1.4 Visible and Nonexcimer Ultraviolet Lasers

Another study by the Stern group[371] examined several short-pulsed lasers in the visible portion of the spectrum for production of corneal incisions in enucleated bovine eyes. They demonstrated that ablation threshold energy is proportional to the square root of the pulse duration, going from 2.5 μJ at 100 femtoseconds (fs; 10^{-15} s) with a colliding-pulse mode-locked ring dye laser, to 500 μJ at 8 ns with a frequency-doubled Q-switched Nd:YAG laser. Ablation with ns lasers at visible wavelengths proved impractical, as excision morphology and control of ablation depth were poor. Collagen denaturation and disorganization were severe at high energies with both the picosecond- and femtosecond-range lasers, but 30-ps pulses near the ablation threshold produced almost as little collateral damage as seen with 193-nm excimer excisions, with nearly identical precision of ablation depth. It is important to note that these were single picosecond pulses, and that the mode-locked train common in some commercial picosecond Nd:YAG lasers would likely have a much higher ablation threshold and thus cause far greater collateral tissue damage. Similar results were achieved with the femtosecond laser, although occasional shock-wave damage raised concerns about potential damage to the endothelium. It was believed that some nonlinear process aside from optical breakdown may account for the increased ablation efficiency of picosecond and shorter pulses. These researchers concluded that ultra-short-pulsed lasers at visible and near-IR wavelengths may be an alternative to the excimer laser for corneal surgery, but more likely, may offer unique advantages for vitreous surgery (see below).

6.5.1.5 Laser Thermokeratoplasty

While excimer laser PRK for mild-to-moderate myopia is relatively predictable and successful, that for high myopia and especially hyperopia is less so. An old idea that has been recently resurrected for the alteration of corneal curvature without removing any tissue is thermokeratoplasty. It was originally proposed as

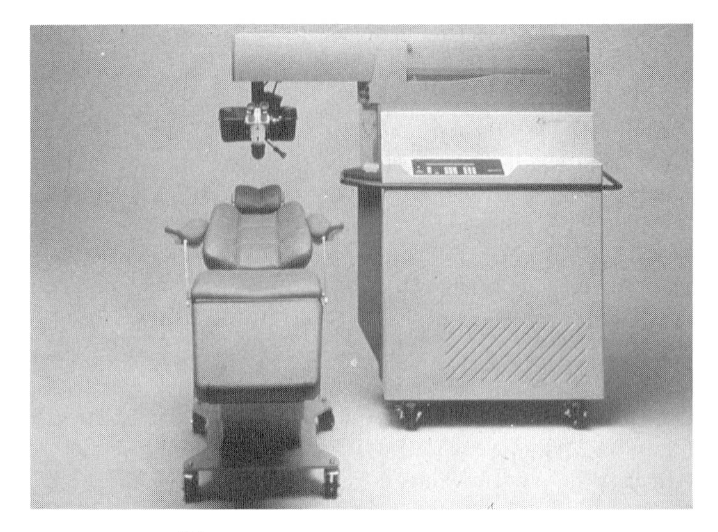

FIGURE 6.18 OmniMed™ laser refractive workstation, combining excimer and holmium: YAG lasers for the correction of myopia and hyperopia, respectively. The manufacturer also offers the emphasis™ Erodible Mask for hyperopia correction using the excimer laser. (Courtesy of Summit Technology, Waltham, MA.)

a treatment for keratoconus,[372] and despite such early complications as transient refractive results, recurrent erosions, scars, and necrosis, Seiler et al.[373,374] used the pulsed solid-state holmium:YAG laser at 2.06 μm to create intrastromal, cone-shaped coagulations to correct hyperopia. Patients reportedly have no complaints of pain with this technique, termed *laser thermokeratoplasty*, and the epithelium regrows in 24–48 h. Although still in early clinical trials, at present this is the most widely accepted means of laser therapy for hyperopia.[375] At least one manufacturer currently offers a single unit containing both excimer and holmium lasers (Fig. 6.18). Holmium laser scleroplasty has been suggested as a means of reducing both axial myopia[376] and postoperative astigmatism.[377] There has also been a report of a micropulsed diode laser being employed to achieve 7 D of steepening in porcine cadaver corneas to whose surfaces indocyanine green had been added.[378]

6.5.2 Noncorneal Applications

6.5.2.1 Eyelids and Adnexa

Disorders of the conjunctiva and ocular adnexa are especially well suited to treatment with the CO_2 laser. Clinical applications have included removal of neoplasms near the punctum,[379] superficial lid margin tissue in tarsorrhaphy,[380] plexiform neurofibromas of the lid and orbit,[381] squamous conjunctival papillomas,[382] and orbital lymphangiomas,[383] and repair of ectropion.[384] In the treatment of nasolacrimal duct obstruction, Gonnering et al.[385] avoided cutaneous scars and minimized recovery time and postoperative pain by performing dacryocystorhinostomies (DCRs) and conjunctivodacryocystorhinostomies with either CO_2 or frequency-doubled Nd:YAG lasers. Mittelman and Apfelberg[386] found no differences in pain, edema, or healing between the CO_2 laser and conventional techniques in blepharoplasty. More recently, Woog and associates[387] employed the holmium:YAG laser to perform endonasal DCRs, achieving excellent intraoperative hemostasis without medial canthal scarring. Long-term ostium patency was 82%. Troutman et al.[388] have suggested excimer or other ablation for precise transconjunctival weakening or strengthening of extraocular muscles.

6.5.2.2 Crystalline Lens

After the cornea, the ocular structure that has attracted the most attention for possible ablation is the crystalline lens. In 1986, Nanevicz et al.[389] published a comprehensive report on parameters for bovine lens ablation using various excimer wavelengths. Using a rate of 20 Hz for 4 min, and fluences up to 1280 mJ/cm^2 per pulse, they found that the ablation thresholds were 110, 265, and 160 mJ/cm^2 per pulse at 193, 248, and 308 nm, respectively, whereas no ablation at 351 nm was achieved within the tested fluence range. Once the threshold was surpassed, the greatest ablation rate was observed for 248 nm. Ablation craters were smoothest at 193 nm, while those produced with 248-nm radiation showed vacuolation and greater disruption in the surrounding tissue, and those at 308 nm were of intermediate smoothness. Absorption coefficients were calculated to be 1360, 410, 122, and 36 cm^{-1} at 193, 248, 308, and 351 nm, respectively. The fact that ablation at 308 nm was only slightly less than that seen at 193 nm, despite the tremendous difference in absorption coefficients, suggested that 308-nm radiation may produce UV-absorbing chromophores, enhancing ablation at that wavelength.

Ultraviolet absorption by the lens is much more dependent on age than is that of the cornea; indeed, Bath and associates[390] showed that human cataractous lenses have lower ablation thresholds than those reported in normal bovine eyes. One of the main advantages of 308 over 193 nm for lens ablation is the ability of the former to be transmitted via fiberoptics, and the very high absorption of the latter by chloride ions in saline, which would complicate intraocular applications. However, Keates and his colleagues[391] found that exposure of the eye to 308 nm produces fluorescence that is potentially harmful to the retina and other intraocular structures, whereas absorption at 193 nm by chloride and saline is not sufficiently large to preclude the use of that wavelength for photoablative cataract surgery if an appropriate delivery system can be devised.

6.5.2.3 Glaucoma

For any filtering procedure aimed at creating a fistula between the anterior chamber and the episcleral space, precise incision and minimization of trauma and wound healing are of paramount importance if patency is to be maintained. Procedures may be performed either endoscopically or gonioscopically through the anterior chamber (ab interno), or externally with or without conjunctival dissection (ab externo). In 1979, Beckman and Fuller[237] created scleral dissections and filters with the CO_2 laser in patients with neovascular and open- and closed-angle glaucoma. L'Esperance et al.[392] achieved considerable success in treating patients with neovascular glaucoma by performing CO_2 laser trabeculostomies and sclerotrabeculostomies. More recently, photoablative lasers have been considered as alternatives to the thermal lasers previously examined for sclerostomy.[393] Berlin et al.[394] coupled the 308-nm output of an excimer to an optical fiber, and reported scleral perforation in eye-bank specimens using 80–100 pulses at 35 mJ/cm^2 per pulse and 20 Hz. Seiler and coworkers[395] identified the juxtacanalicular portion of the trabecular meshwork as the primary site of outflow resistance by using the excimer laser at 193 nm to create partial trabeculectomies; increased aqueous outflow immediately stopped the ablation process. While this approach obviates entering the anterior chamber, it does require prior episcleral dissection. Allan and associates[396] avoided dissection during 193-nm excimer laser sclerostomy by employing an open mask to plicate the conjunctiva before ablation. They suggested that this technique would avoid the secondary thermal and mechanical damage caused by contact endoscopic systems. An en face air

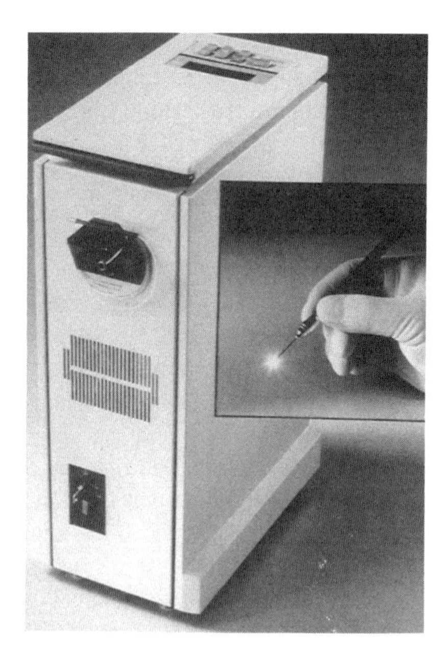

FIGURE 6.19 gLASE™ 210 thulium holmium chromium–yttrium aluminum garnet (THC:YAG) laser, which produces 2.1-μm radiation at 5 Hz, 300-μs pulses, and up to 350 mJ, transmitted via the SUN-LITE™ probe [a 200-μm fiberoptic in a casing with outer diameter of 712 μm (22-gauge)]. It is currently employed for sclerostomies, but is under early investigation for thermokeratoplasty treatment of hyperopia. (Courtesy of Sunrise Technologies, Fremont, CA.)

jet to prevent aqueous outflow enabled ablation to continue until sufficiently large fistulas were created. Margolis et al.[397] created filtering blebs in bovine and eye-bank sclera with both the Er:YAG (250 μs) and holmium:YAG (300–500-μs) lasers, coupled to optical fibers, without significant disruption of adjacent structures.

On the basis of preliminary eye-bank studies, Eaton et al.[398] suggested that the thulium–holmium–chromium (THC)-doped YAG laser at 2.01 μm (Fig. 6.19) may be preferable to the holmium, since the former requires fewer pulses and less energy to create sclerostomies, presumably causing less collateral damage. Initial clinical work by Iwach and colleagues[399] with THC:YAG ab externo laser sclerostomy (requiring a total of 1.4–7.2 J) in patients with intractable glaucoma (in whom IOP in the low teens is often desirable) yielded a success rate of 66% at 12 months and 57% at 30 months. McAllister and coworkers[400] obtained similar results, and suggested strategies for avoiding the most frequent complication, iris plugging of the sclerostomy. Employing this laser, Namazi et al.[401] found that the antimetabolites 5-fluorouracil, and especially mitomycin C, significantly increase the duration of laser-created filtration channels.

6.5.2.4 *Vitreous and Retina*

Although retinal photocoagulation was the first medical use of lasers, practical applications of ablation in the posterior segment have been far more elusive. The CO_2 laser at 10.6 μm has been employed in animals to cut vitreous bands[239] and drain subretinal fluid,[402] but cannot be used close to the retina, is large and inefficient, and requires an articulated arm. Experimental vitreous membranes in rabbits were cut with the excimer laser at 308 nm,[403] but no additional work in this area has been reported. Margolis et al.[404] used a fiberoptic-coupled, 250-μs pulsed Er:YAG laser to cut experimental vitreous membranes in rabbits, at distances of 500–3600 μm from the retina. All attempts at cutting the membranes were successful, although 53% resulted in some form of retinal lesion, either hemorrhages or nonhemorrhagic burns. Nevertheless, the 200–300-μm size of the latter was not considered a contraindication to clinical application of the technique if the membranes occurred in extramacular sites. Lin et al.[405] showed that the Er:YAG output creates a bubble at the tip of the fiber through which it is transmitted, which can cause thermal and mechanical tissue damage. To minimize the size and movement of this bubble, and thus the resulting damage to the retina, they suggested reducing the pulse energy below 0.5 mJ and using a shielded tip. These steps may have their own adverse consequences, namely, requiring a higher repetition rate, and decreasing some of the advantages of laser ablation over mechanical cutting for removal of membranes tightly adherent to the retinal surface, respectively; investigation of these issues is ongoing. Use of pulses in the picosecond domain just above the ablation threshold may significantly reduce untoward thermal and mechanical effects, allowing safe cutting very near the retina.

Borirakchanyavat et al.[406] attempted transection of experimental vitreous membranes using a 250-μs pulsed holmium:YAG laser at 2.12 μm. When an optical fiber was used alone, only thin membranes could be sectioned at energy levels or repetition rates that permitted work near the retina. This problem was solved by encasing the fiber in a retina-shielding pick, which allowed almost three-quarters of membranes, as close as 0.5 mm to the retina, to be completely transected. The laser directly caused one nonhemorrhagic retinal burn, while the pick caused two retinal injuries, including one with a small hemorrhage. Cutting precision was histologically comparable to that of the CO_2 laser, but an order of magnitude less than that of the Er:YAG laser.[404] One technical advantage of holmium:YAG radiation is that it is much more easily transmitted by existing fiberoptics.

Selective RPE damage has been reported with the Q-switched Nd:YAG[407] and micropulsed diode[408] and argon[409] lasers. A significant impediment to the intraocular use of 193-nm excimer laser radiation is the technical difficulty in transmitting it with the necessary accuracy. Lewis and coworkers[410] constructed a guide consisting of a fused-silica lens of 1000 mm focal length and a rapidly tapered stainless-steel tube whose outer diameter was that of an 18-gauge needle, with an inner diameter of 120 μm. They used this instrument to ablate bovine cadaver retinas and rabbit retinas in vivo following lensectomy and vitrectomy, employing pulses of 0.5–1.2 J/cm^2 and 120 μm, at 30–100 Hz. Once a low-pressure stream of air was used to displace the thin layer of adherent fluid from the area to be treated, they were able to achieve retinal ablation with the precision typically associated with the cornea. Possible applications include ablation of epiretinal membranes—as even a slight amount of underlying fluid would protect the retina—and the creation of precise retinal incisions.

6.6 DIAGNOSTIC LASERS

6.6.1 Scanning Laser Ophthalmoscope

As first described by Webb and colleagues,[411,412] the scanning laser ophthalmoscope (SLO) produces video images of the retina. A low-power laser beam is scanned horizontally and vertically across the retina, creating a raster pattern that is used to map the retina or take highly localized measurements. The SLO also has excellent depth resolution and can produce tomographic images.[413] This is particularly valuable in creating three-dimensional images of the optic disc,[414] since increased cupping is a frequent antecedent of visual field loss in glaucoma. Edema, scars, and macular holes are among other retinal conditions that have been assessed with this technique.[415,416]

Scanning laser ophthalmoscope image contrast can be further enhanced by employing confocal optics.[417] Figure 6.20*a* shows a confocal SLO specifically designed to analyze the RNFL. It uses a polarization detector to produce a thickness map of the RNFL, to document and follow glaucomatous changes. Normal RNFLs have a symmetric, hourglass-shaped distribution, which is gradually eroded as glaucoma progresses (Fig. 6.20*b*). The thickness of the form birefringent RNFL determines the degree to which the low-power diode laser

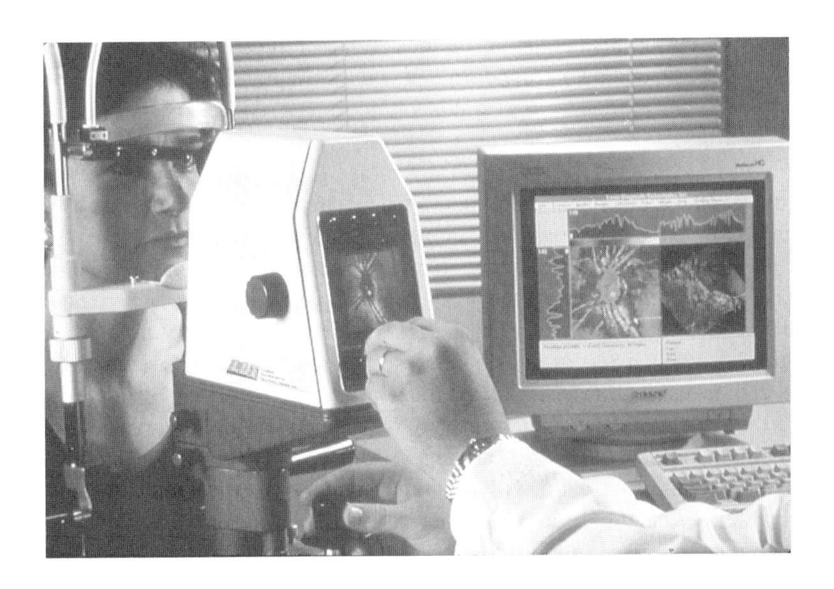

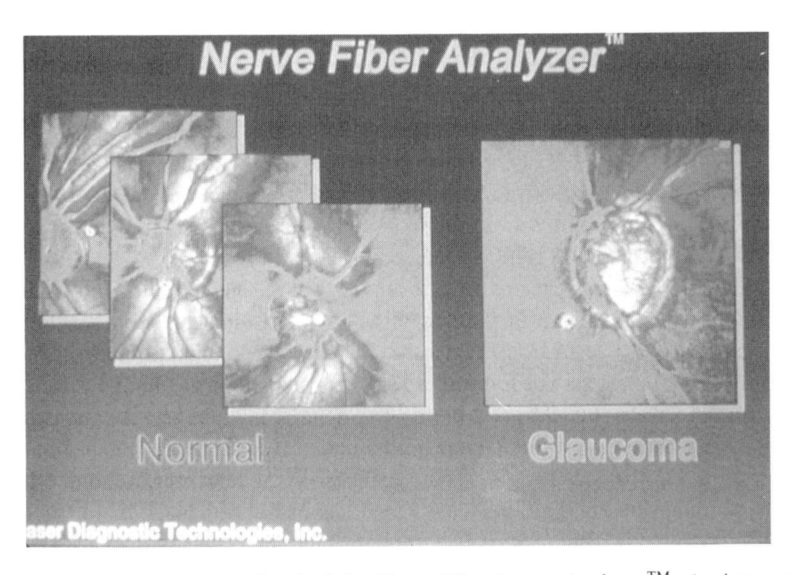

FIGURE 6.20 (*a*) Instrument head of the Nerve Fiber Layer Analyzer™. An integrated monitor allows the physician to simultaneously view the retina and maintain eye contact with the patient. (*b*) At left are nerve fiber layer thickness maps of three normal eyes, with thick superior and inferior arcuate bundles. At right is a map of a glaucomatous eye, with thinning superior arcuate bundle and absent inferior arcuate bundle. (Courtesy of Laser Diagnostic Technologies, Inc., San Diego, CA.)

beam is polarized.[418] This device obtains readings in less than 1 s, doesn't require pupil dilation, may be used in patients with cataracts, and avoids the flashes associated with conventional fundus photography. Since the RNFL ordinarily doesn't exceed 150 μm in thickness, yet the optical depth resolution of the human eye is around 200 μm, the potential advantage of this method, with its resolution of 15–20 μm, over standard photographic techniques is apparent.[419]

Figure 6.21 shows how a tomographic SLO is employed to perform topographic analysis of the retina. This device, which relies on specular reflection of laser light from the retina's internal limiting membrane, can be used to determine optic nerve head volume[420] and map the macula and tumors. It assesses contour, in contrast to the instrument described above, which measures thickness.

Visual evoked potentials can be measured more precisely with the SLO.[421] Visual function testing is performed by varying the SLO laser beam intensity to create patterns in the raster. The examiner sees the same pattern, localized to a specific portion of the retina, which the patient perceives, allowing the locus of fixation to be correlated with retinal pathology.[422] Angiography is another process that has been aided by the SLO. Less fluorescein than is required with conventional angiography is excited by an argon blue laser, yielding more and higher resolution images.[423] A method for stereoscopic video angiography has been described.[424] The SLO has been particularly important for ICG angiography (see above), in which the low-fluorescence dye is excited by a diode laser to more

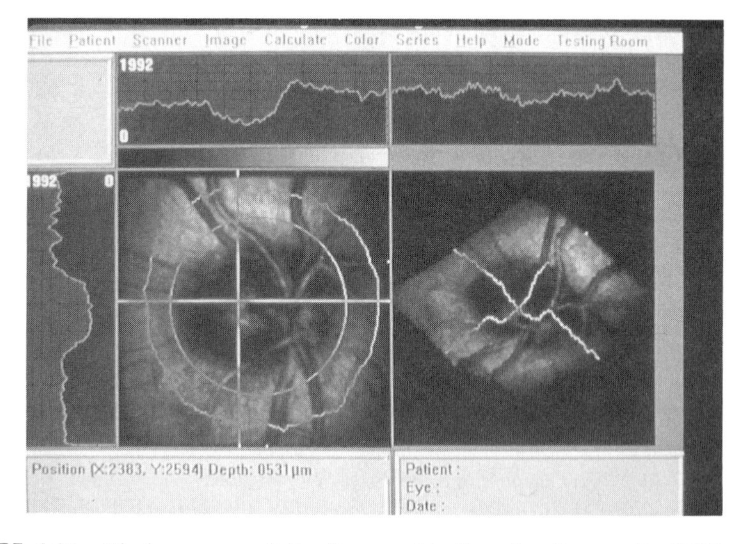

FIGURE 6.21 Display screen of the Topographic Scanning System TopSS™. At left is an intensity image of an optic nerve head. Above and to the left of this image are horizontal and vertical cross sections, respectively, through the optic nerve head. To the right is a three-dimensional view of the same optic nerve head. The viewing angle can be adjusted by the physician. A peripapillary retinal contour of nerve fiber layer height along the circle superimposed around the optic nerve head is displayed in the upper right-hand corner of the screen. (Courtesy of Laser Diagnostic Technologies, Inc., San Diego, CA.)

precisely delineate SRNV membranes.[425, 426] An alternate technique—combining slit-lamp and He–Ne 543-nm laser illumination—has been described to enhance visualization of fine vitreoretinal structures.[427]

6.6.1.1 Laser Flare and Cell Meter

Anterior segment inflammation, which may result from numerous types of intraocular pathology, causes breakdown of the blood–aqueous barrier and is manifested by increases in aqueous protein and cells. Fluorophotometry was the first method to objectively quantify this process but is relatively time-consuming and requires intravenous injection of dye.[428,429] In 1988, Sawa et al.[430] reported the development of a slit-lamp-based instrument that employs a low-power He–Ne or diode laser to detect the amount of protein—to which flare, produced by the Tyndall phenomenon, is linearly related over a wide range of protein concentrations—and the number of cells in the anterior chamber [Laser Flare Cell Meter (LFCM), Kowa Acculas, Inc., San Jose, CA (Fig. 6.22)]. Flare measurements are corrected for background scatter, are reproducible to within ~8%,[431] and take less than 1 s.[432] These readings may be converted directly to protein concentration if the type of the protein is known, but not in cases of heterogeneous scatterers such as low-molecular-weight albumin and high-molecular-weight globulins. Studies have shown that flare has a diurnal variation[433] and tends to increase with age.[434] The LFCM has been used clinically to examine inflammation associated with uveitis of various etiologies,[435] retinal detachment,[436] cataract removal and IOL insertion,[437] posterior capsulotomy,[438] and LTP.[439] Cytomegalovirus (CMV) retinitis is a frequent complication in AIDS (acquired immune deficiency syndrome) patients, and this instrument's ability to

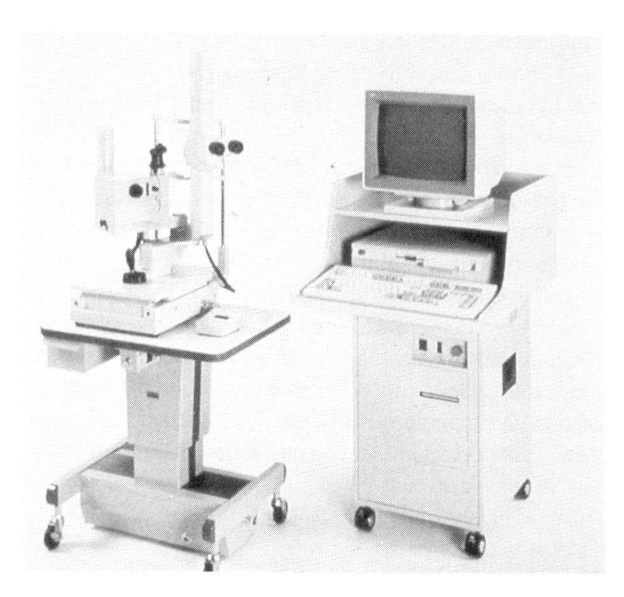

FIGURE 6.22 Kowa FC-1000 laser flare cell meter, showing slit-lamp-based system and computer. (Courtesy of Kowa Acculas, Inc., San Jose, CA.)

detect slight increases in aqueous flare before any retinal changes are visible allows earlier institution of treatment.[440]

To count cells, the laser beam is repeatedly scanned over a 0.075-mm^3 volume, with peaks further analyzed so that only white blood cells are registered. Cell count is less reproducible than flare measurements, especially with relatively few particles, while a spurious cell count in the clinical absence of cells may be caused by agglutination of proteins in very high concentrations.[441] There is also less correlation with direct slit-lamp measurements, which are made over larger volumes and longer periods, so that the clinical value of this application is less certain than that for flare.

6.6.2 Laser Interferometer

The interference pattern produced with the laser interferometer is largely independent of the eye's various optical components. It offers the potential to assess visual acuity by projecting the pattern onto the retina and determining the greatest periodicity (corresponding to the closest fringe spacing) at which the subject is able to resolve discrete fringes.[442] To be of much practical value, the examiner must know the periodicity, which is estimated by using standard values for the refractive indices of the cornea and lens, and ultrasound to measure the eye's axial length. The laser interferometer has been used to determine potential visual acuity in patients undergoing cataract extraction,[443] and therapy for amblyopia[444] and uveitis,[445] especially in the presence of concomitant retinal or neurologic pathology. Laser interferometry has generally been found to be more accurate than previous techniques,[446] although this may not be the case in the presence of maculopathy.[447] While it has been suggested that this instrument may be employed even in the presence of dense cataracts or turbid media, these factors necessarily diminish the fringe pattern contrast. To avoid underestimating postoperative visual acuity, correction for any significant optical obstruction is necessary.[448]

Recently, laser interferometric techniques have been described for precise depth measurements of the cornea, retina, and other ocular structures. Both Fabry Perot and Michelson systems have been developed, and offer the promise of highly accurate one- and two-dimensional measurements.[449–455] Use of the laser Doppler principle allows considerably faster readings than are achievable by interpretation of static interferometric images. Selection of IOL refractive power is based largely on the eye's axial length, which the noncontact laser interferometric approach may determine more precisely and safely than the standard ultrasound method.[456] Although patient fixation is required with the laser, further technical enhancements should reduce scan time to under 1 s. The potential use of such systems in excimer laser surgery is described below.

6.6.3 Laser Doppler Velocimeter

The laser Doppler velocimeter (LDV)[457] was developed over 20 years ago but is probably the least common type of ophthalmic diagnostic laser; it has been used almost exclusively for research purposes. It typically combines a low-power He–Ne or diode laser with a fundus camera or slit-lamp biomicroscope, permitting simultaneous viewing and projection of an approximately 100-μm-diameter laser spot onto the desired vessel. Light of frequency ν is scattered by red blood cells, and autodyne optical mixing spectroscopy is employed to measure the resulting frequency shift of $\Delta\nu$ relative to the light scattered from the wall of the vessel containing the cells. Since there is a range of cell velocities, the LDV produces a spectrum indicating how many cells are moving at each velocity. This method only measures relative velocity, but absolute velocity can be determined by adding a second detector.[458] Readings can be taken in patients with poor fixation or those who require particularly long study by using rapid calculation algorithms or techniques to compensate for eye movement.[459] The LDV has shown normal retinal blood flow to be about 80 μL/min, which increases with antiglaucoma medication,[460] hyperglycemia,[461] and transition from dark adaptation to light,[462] and decreases after treatment with insulin,[463] in cases of optic atrophy,[464] and in PDR.[465]

6.7 FUTURE DEVELOPMENTS

The diode laser is revolutionary more for economic and ergonomic reasons than because it provides some fundamentally new laser–tissue interaction. Aided by advances in endoprobes and indirect ophthalmoscopy, it is already being used to treat a variety of vascular diseases, and may become the treatment of choice for ROP. Its low cost and easy portability will permit treatment in remote areas, including many third-world nations. Just as GaAlAs diode technology has improved in the past several years so that sufficient power is now available for many applications, the future holds the promise that with new materials, useful diode output will be achieved at visible wavelengths, perhaps supplanting the much bulkier and less efficient argon, krypton, and dye lasers. At present, diodes are beginning to replace flashlamps and other lasers as pumping sources for visible wavelengths.[466] If sufficient energy can be generated by Q-switched diode lasers, they might be used in place of Nd:YAG lasers for photodisruption. Lesion variability – with the diode or other lasers – is unavoidable, but early work with a dynamic confocal reflectometer offers the possibility of reproducible retinal photocoagulation independent of pigmentation, media opacities, or optical aberrations.[467]

As phase III clinical trials proceed at numerous academic centers, excimer laser ablation algorithms are still being refined. Recent work suggests that aspheric PRK may yield better optical homogeneity than the standard spherical approach[468] (perhaps by minimizing the refractive transition zone at the ablation periphery) and that a series of several concentric ablations may allow shallower tissue removal to achieve greater corrections, similar to a Fresnel lens.[469]

Erodible masks have been proposed as an alternative to the usual apertures or diaphragms, theoretically permitting correction of myopia, hyperopia, and astigmatism (including irregular forms), but initial animal work in this area has been successful primarily for myopia.[470] The unqualified success of PTK notwithstanding, the widespread acceptance and use of the excimer laser will likely depend on the successful correction of refractive errors. Individual variation in wound healing may be an insurmountable biologic reality, but overall control of wound healing appears necessary for maintenance of optimal refractive and visual results. Studies of corticosteroids and the antimetabolite mitomycin C[471,472] suggest that the former is more successful in safely modifying the results of PRK. Lohmann and associates[473] found that tear plasmin levels increase after ablation, so that inhibitors of plasmin and plasminogen activator, perhaps in conjunction with steroids, might be used to modify postlaser healing. Moreover, individuals with initially high plasmin levels experienced more regression and may not be candidates for PRK in its current form. Topical interferon-alpha 2b, also alone or as an adjunct to steroid therapy, has also been identified as a potential agent in minimizing corneal haze,[474] as have antioxidants[475] and inhibitors of glycosaminoglycan formation.[476] The topical nonsteroidal antiinflammatory drug (NSAID) diclofenac has been shown to significantly reduce pain in the 2–3 days following ablation.[477]

It is also necessary to determine whether, if regression cannot be entirely controlled, repeat ablation is possible, as this is not allowed under current FDA investigational guidelines. (Studies of patients undergoing repeat PRK because of initial undercorrection, scarring, or regression indicate that the additional laser treatment is successful,[478,489] as is PRK following undercorrection with radial keratotomy.[480]) Thompson et al.[481] proposed circumventing variable corneal wound healing following PRK by performing excimer laser ablation of (as yet unidentified) acellular synthetic epikeratoplasty lenticules, which would be attached to the cornea in a technique termed *laser adjustable synthetic epikeratoplasty* (LASE).

Even if standard excimer laser PRK proves unreliable for treatment of high myopia, this laser may still be employed for room-temperature keratomileusis, preserving Bowman's membrane and permitting ablation based on the visual center even if such doesn't correspond to the center of the lenticule (a luxury not now afforded by the cryolathe). This could be done on the posterior stromal surface of the lenticule, or on the anterior surface of the exposed stromal bed; the latter is termed *in situ keratomileusis*.[482]

Correction of hyperopia needs to be more carefully studied, since although it is theoretically similar to that of myopia, in practice epithelium or stroma has tended to fill in the ablated annulus, negating the refractive effects. If it proves feasible, PRK for hyperopia might be combined with either conventional or laser cataract surgery (see below). Various ablation strategies have been proposed,[483] with at least one attempted—with moderate success at 3-month follow-up—on blind eyes in a phase I FDA study.[484]

Clinical experience is insufficient to discount all potential side effects, but results to date are encouraging. While there may be an early decrease in contrast sensitivity following PRK, this has been found to resolve within a year of surgery.[485] There is also debate within the medical community about whether excimer laser ablation constitutes surgery, and some optometrists have been lobbying for the right to use the technique.

While there are many engineering and treatment nuances among the major ophthalmic excimer lasers, the systems are much more alike than not. Significant technical hurdles had to be overcome before the excimer laser was ever used clinically, but recent enhancements and modifications have largely been variations on the same theme. Different strategies are continually being studied to optimize refractive outcome and minimize healing complications. A significant improvement would be the addition of real-time eye-tracking[486] and tissue depth monitoring; possible approaches to the latter include femtosecond laser optical ranging[487] and optical coherence domain reflectometry.[451] However, clinical considerations aside, the excimer's considerable size, cost, and energy requirement represent significant impediments to its widespread use. There is also concern about exposure to fluorine gas. LaserSight, Inc. (Orlando, FL) has introduced the "miniexcimer" (Fig. 6.23), which at $\sim$ 70 kg weighs much less than conventional ophthalmic excimer lasers. It operates from a 110-V outlet, is air-cooled, and uses an excimer gas mixture containing only 0.19% fluorine. Since this device can generate sufficient fluences for ablation over spot sizes only 1.0–1.5 mm in diameter, it utilizes a computer-controlled scanning system to treat the cornea. Preliminary clinical trials for correction of myopia were conducted in China in early 1993, reportedly with good results, with phase I FDA studies scheduled for later. Studies on hyperopia correction are also planned.

Work on laser modification of corneal curvature is proceeding so rapidly that even before the excimer laser has received FDA approval, potential replacements

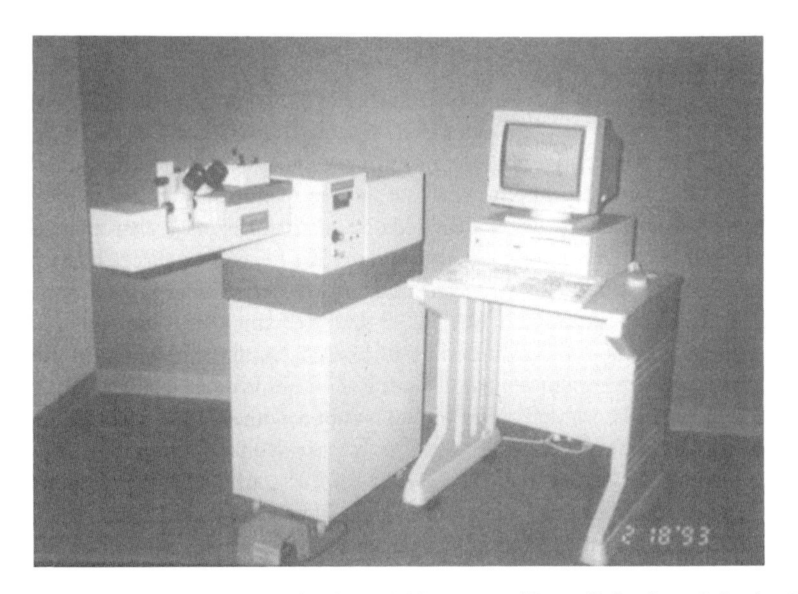

FIGURE 6.23 Compak-200 "miniexcimer." (Courtesy of LaserSight, Inc., Orlando, FL.)

are being studied. Early worries about possible mutagenesis appear to have subsided (except for radiation around 248 nm, which has not been used clinically), and despite the achievement of IR laser ablation, researchers have not been able to duplicate the precision of the argon fluoride excimer laser. As such, several investigators are studying solid-state, UV alternatives to the excimer. Researchers from the Bascom Palmer Eye Institute are collaborating with LaserSight, Inc. on the use of the Nd:YAG fifth harmonic at 213 nm (achieved with a combination of cesium–dihydrogen–arsenate and barium borate crystals[488]), 250 mJ/cm^2, 10 ns, and 10 Hz for possible ablation.[489] Since sufficient energy for ablation can be achieved only with a small spot size, tissue removal is effected by scanning a 0.5–1.0-mm quasi-Gaussian beam across the cornea in a precise pattern to achieve the desired profile.[490] While technically difficult, such scanning may permit more gradual and customized ablations. Preliminary results in cadaver and rabbit eyes revealed ablations comparable to those of the argon fluoride excimer laser, with damage areas less than 1 μm wide, no step-like transition zones, and mild residual subepithelial haze at 3 months, but with poor correlation between intended and achieved refractive changes.[491,492] Although commercial availability is at least several years away, another goal of this group is the development of a laser capable of generating several harmonics at various pulse durations, potentially also permitting photocoagulation and photodisruption.[493] Researchers at Tufts University and Schwartz Electro-Optics (Concord, MA) are investigating ablation with the titanium–sapphire laser, using barium borate crystals to frequency-quadruple its 10 ns, 10 Hz output to 205–225 nm.[494] Although it is very inefficient, the free-electron laser is capable of generating a wide range of IR wavelengths, and allows independent variation of wavelength, energy, and pulse duration. Very preliminary studies achieved corneal ablation,[495,496] including central steepening for hyperopia correction,[497] but it remains to be seen whether this is preferable to any of the solid-state IR lasers.

Studies have shown the importance of the epithelium in moderating repair, and reducing corneal haze and reversal of refractive changes following excimer ablation.[320] Although, in conjunction with steroids or antimetabolites, the cornea appears to heal with little or no residual scarring, the long-term implications of this procedure remain to be determined. Some researchers have thus proposed selective intrastromal ablation without damage to the anterior layers as an alternative to excimer laser anterior ablation,[380] although the best such laser remains to be determined.

Zysset et al.[498] provided early guidance for intrastromal ablation by demonstrating that for suprathreshold Nd:YAG laser corneal lesions, the damage zone radius is proportional to the cube root of the pulse energy. Whereas picosecond- and nanosecond-range pulses of equal energy cause approximately the same damage, since the former tend to be on the order of microjoules and the latter on the order of millijoules, in practice the use of low-energy picosecond pulses can significantly reduce collateral tissue damage. In contrast to nanosecond exposures, picosecond ones were not associated with significant shock-wave propagation or cavitation bubble expansion in the primary interaction zone. However, given the low energy of individual picosecond pulses, they concluded that in order to achieve effective cutting or ablation, it would be necessary to employ many such exposures at moderate-to-high repetition rates.

At least two companies now produce short-pulsed lasers capable of intrastromal ablation and other applications. The Phoenix Model 2500 Ophthalmic Laser Workstation (Phoenix Laser Systems, Inc., Fremont, CA) employs a frequency-doubled, 532-nm Nd:YAG laser with 8-ns, 40–350-μJ pulses. It has a nominal spot size of 5 μm, with three-dimensional accuracy consistent with a 15–20-μm treatment area, and incorporates an eye-tracking system to eliminate effects of even subtle movements. Early attempts at intrastromal ablation (which the company terms *transepithelial stromal keratectomy*) with this laser in enucleated cat eyes yielded vacuoles 30–80 μm in diameter, surrounded by compressed collagen anteriorly and edematous collagen posteriorly, without significant endothelial damage (Fig. 6.24).[499,500] The vacuoles collapse as the fluid inside is absorbed, thereby altering corneal curvature. Work is ongoing to determine the ablation scanning patterns necessary for correction of various types of refractive error. The only procedure for which the laser has already been approved is posterior capsulotomy, but other applications in various stages of investigation, for which the laser's precision may provide unique advantages, include nuclear photofragmentation (see below), sclerostomy, PTK, radial keratotomy, iridectomy, and focal retinal ablation.

The Nd:YLF (yttrium lithium fluoride) Eye Laser System™ (Intelligent Surgical Lasers, Inc., San Diego, CA) produces 1053-nm pulses of energy 20–

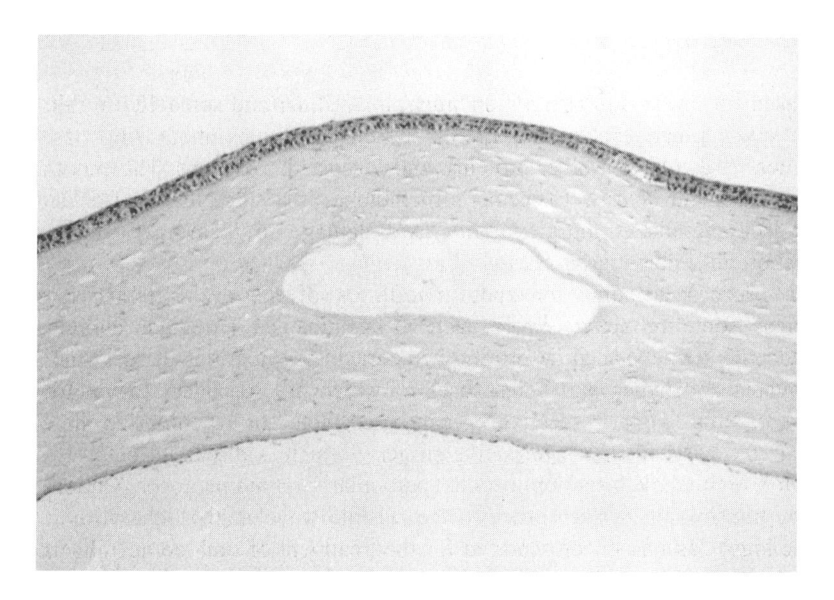

FIGURE 6.24 Histologic view following intrastromal ablation of a porcine cornea with the Phoenix Model 2500 Ophthalmic Laser Workstation. (Courtesy of Phoenix Laser Systems, Inc., Fremont, CA.)

350 μJ, spot size ≤ 7 μm, duration ≤ 60 ps, and repetition rate 10–1000 Hz. It consists of a stable mode-locked oscillator laser operating at 80 MHz whose output is a continuous train of 1-nJ pulses, which a high-gain, Q-switched regenerative amplifier then converts to 1-mJ pulses.[501] Nd:YLF has a larger fluorescence bandwidth than Nd:YAG, which allows pulses of shorter duration to be generated. Using an earlier Intelligent Surgical Lasers, Inc. model, Frueh and colleagues[502] created intrastromal transverse excisions in rabbit corneas, which initially induced approximately 5 D of flattening, without intrastromal scarring. Steroids were not employed, and the changes regressed by 6 weeks. Remmel et al.[503] demonstrated that a continuous layer of stroma could be removed in human cadaver eyes. Also using this system in human cadaver eyes, at 80–140 μJ, 50 ps, and 1 kHz, Niemz et al.[504] scanned the laser focus in a spiral pattern to create intrastromal vacuoles approximately 40 μm in diameter and 20 μm deep, 100 μm below Bowman's membrane, without damaging it or the epithelium. It has also been demonstrated that this process causes no acute endothelial damage.[505] However, Brown and associates[506] found that ablation depth in human and rabbit cadaver corneas was related to energy but not to programmed depth. Collagen disorganization ranged from 3–8 μm with 100-μJ pulses, and 10–20 μm at 300 μJ. The ablation effect varied with the beam path direction and plasma threshold changes within the cornea. Moreover, there was ablation of tissue anterior to the desired target, lamellar separation by plasma–gas formation without ablation, and inconsistent ablation, indicating the need for further studies and refinement of this technique. The Nd:YLF laser has also been used to create sclerectomies in human cadaver eyes,[507] sclerostomies in rabbits,[508] and iridotomies in patients,[509] all with a high degree of precision. As of mid-1993 the Intelligent Surgical Lasers, Inc., laser had received FDA clearance only for posterior capsulotomy, but was in phase I trials for vitreolysis and in phase II trials for cataract fragmentation. There was an open iridotomy study, while investigational device exemptions were filed for intrastromal ablation and internal and external sclerostomy.

Taboada and associates[510] described a technique they termed *intrastromal photokeratectomy*, in which Q-switched Nd:YAG laser pulses at 100 ns and 500–2000 Hz were applied to the cornea via a contact probe. This resulted in vacuoles approximately 100 μm in diameter, localizable within 20 μm, and with a transition zone between normal tissue of 0.5 μm. The track of vacuoles, which can be created in radial, wide-area, or other patterns, optically disappeared within 2 days, as treated tissue dissolved into a diffusive liquid, yet the refractive effects in the rabbit and monkey eyes persisted for the 5-month duration of the study. There was no immediate or long-term damage to either the epithelium or endothelium. It remains to be seen whether intrastromal ablation with any laser can approach the precision of the excimer, and whether Descemet's membrane isn't more likely to push forward than is Bowman's membrane backward, but potential advantages include increased patient comfort and no need to patch or instill antibiotics, due to retention of the epithelium, as well as lower cost and avoidance of any UV radiation.

Novatec Laser Systems, Inc. (San Diego, CA) indicates that its LightBlade™ laser is capable of corneal surface, intrastromal, and intraocular ablation, and includes an eye-tracking system, but clinical trials of the as-yet-unspecified "deep UV solid state" device have yet to begin.

Whereas laser removal of primary cataracts remains a popular misconception, reality may be catching up with myth. Photophacofragmentation is the process of employing pulsed laser energy to soften a cataractous lens, to facilitate aspiration at the time of surgery; studies employing the Nd:YAG laser have been reported.[511,512] It has been shown that the threshold for excimer laser ablation of the lens capsule is significantly higher than that for the nucleus and cortex.[513] This raises the possibility of ablation, such as with the excimer, or the Nd:YLF, the frequency-doubled Nd:YAG, or some other very-short-pulsed laser,[514] through a tiny opening in the anterior capsule, via an anterior or pars plana approach. This could preserve the zonules and potentially allow the use of injectable polymers to reconstitute the lens following aspiration of the ablated remnants, perhaps even restoring accommodation.[515]

REFERENCES

1. Maiman TH: Stimulated optical radiation in ruby. *Nature* 187:493, 1960, p. 1051.
2. Longsworth JW: Photophysics. In Regan JD, Parrish JA (eds): *The Science of Photomedicine*. Plenum Press, New York, 1982, p. 43.
3. Meyer-Schwickerath G: Koagulation der Netzhaut mit Sonnenlicht. *Ber Deutsch Ophth Gesellsch* 55:256, 1949.
4. Blankenship GW: Fifteen-year argon laser and xenon photocoagulation results of Bascom Palmer Eye Institute's patients participating in the diabetic retinopathy study. *Ophthalmology* 98:125, 1991.
5. Beetham WP, Aiello LM, Balodimos MC, et al: Ruby laser photocoagulation of early diabetic neovascular retinopathy. *Arch Ophthalmol* 83:261, 1970.
6. L'Esperance FA: An ophthalmic argon laser photocoagulation system: Design, construction, and laboratory investigations. *Trans Am Ophthalmol Soc* 66:827, 1968.
7. Mellerio J: The thermal nature of retinal laser photocoagulation. *Exp Eye Res* 5:242, 1966.
8. Clarke AM, Geeraets WJ, Ham WT: An equilibrium thermal model for retinal injury from optical sources. *Appl Optics* 8:1951, 1969.
9. Mainster MA, White TJ, Allen RG: Spectral dependence of retinal damage produced by intense light sources. *J Opt Soc Am* 60:848, 1970.
10. Roulier A: Calculation of the thermal effect generated in the retina by photocoagulation. *Graefes Arch Clin Exp Ophthalmol* 181:281, 1971.
11. Birngruber R, Hillenkamp F, Gabel VP: Theoretical investigations of laser thermal retinal injury. *Health Phys* 48:781, 1985.

12. Birngruber R: Thermal modeling in biologic tissues. In Hillenkamp F, Pratesi R, Sacchi CA (eds): *Lasers in Biology and Medicine*. Plenum Press, New York, 1980, pp. 77–97.

13. Hillenkamp F: Interaction between radiation and biological systems. In Hillenkamp F, Pratesi R, Sacchi CA (eds): *Lasers in Biology and Medicine*. Plenum Press, New York, 1980, pp. 37–68.

14. Mainster MA: Wavelength selection in macular photocoagulation: Tissue optics, thermal effects, and laser systems. *Ophthalmology* 93:952, 1986.

15. Lapanje S: *Physicochemical Aspects of Protein Denaturation*. Wiley, New York, 1978.

16. Trempe CL, Mainster MA, Pomerantzeff O, et al: Macular photocoagulation: Optimal wavelength selection. *Ophthalmology* 89:721, 1982.

17. Swartz M: Histology of macular photocoagulation. *Ophthalmology* 93:959, 1986.

18. Boergen KP, Birngruber R, Hillenkamp F: Laser-induced endovascular thrombosis as a possibility of selective vessel closure. *Ophthalmic Res* 13:139, 1981.

19. Apple DJ, Wyhinny GJ, Goldberg MF, et al: Experimental argon laser photocoagulation. I. Effects on retinal nerve fiber layer. *Arch Ophthalmol* 94:137, 1976.

20. Marshall J, Bird AC: A comparative histopathological study of argon and krypton laser irradiations of the human retina. *Br J Ophthalmol* 63:657, 1979.

21. Arden GB, Berninger T, Hogg CR, et al: A survey of color discrimination in German ophthalmologists. Changes associated with the use of lasers and operating microscopes. *Ophthalmology* 98:567, 1991.

22. Peyman GA, Raichand M, Zeimer RC: Ocular effects of various laser wavelengths. *Surv Ophthalmol* 28:391, 1984.

23. Peyman GA, Conway MD, House B: Transpupillary CW YAG laser coagulation: A comparison with argon green and krypton red lasers. *Ophthalmology* 90:992, 1983.

24. Peyman GA, Larson B: Effects of the CW YAG laser on the human iris and retina. *Ophthalmology* 91:1034, 1984.

25. Fankhauser F: The Q-switched laser: principles and clinical results. In Trokel S (ed): *YAG Laser Ophthalmic Microsurgery*. Appleton-Century-Crofts, Norwalk, CT, 1983, pp. 128–131.

26. Van der Zypen E, Fankhauser F, Loertscher HP: Retinal and choroidal repair following low power argon and Nd:YAG laser irradiation. *Doc Ophthalmol Proc Ser* 36:61, 1984.

27. L'Esperance FA: Trans-spectral organic dye laser photocoagulation. *Trans Am Ophthalmol Soc* 83:82, 1985.

28. Halldórsson T, Langerholc J, Senatori L: Thermal action of laser irradiation in biological material monitored by egg-white coagulation. *Appl Optics* 20:822, 1981.

29. Romanelli JF, Puliafito CA: Metabolic studies of dye laser retinal photocoagulation. *Int Ophthalmol Clin* 30:95, 1990.

30. Puliafito CA, Deutsch TF, Boll J, et al: Semiconductor laser endophotocoagulation of the retina. *Arch Ophthalmol* 105:424, 1987.

31. Brancato R, Pratesi R: Applications of diode lasers in ophthalmology. *Lasers Ophthalmol* 1:119, 1987.

32. Smiddy WE, Hernandez E: Histopathologic results of retinal diode laser photocoagulation in rabbit eyes. *Arch Ophthalmol* 110:693, 1992.

33. Brancato R, Pratesi R, Leoni G, et al: Retinal photocoagulation with diode laser operating from a slit lamp microscope. *Lasers Light Ophthalmol* 2:73, 1988.

34. Brancato R, Pratesi R, Leoni G, et al: Semiconductor diode laser photocoagulation of human malignant melanoma. *Am J Ophthalmol* 107:295, 1989.

35. Brancato R, Pratesi R, Leoni G, et al: Histopathology of diode and argon laser lesions in rabbit retina: A comparative study. *Invest Ophthalmol Vis Sci* 30:1504, 1989.

36. McHugh JDA, Marshall J, Capon M, et al: Transpupillary photocoagulation in the eyes of rabbit and human using a diode laser. *Lasers Light Ophthalmol* 2:125, 1988.

37. Wallow IHL, Sponsel WE, Stevens TS: Clinicopathologic correlation of diode laser burns in monkeys. *Arch Ophthalmol* 109:648, 1991.

38. Jennings T, Fuller T, Vukich JA, et al: Transscleral contact retinal photocoagulation with an 810-nm semiconductor diode laser. *Ophthalmic Surg* 21:492, 1990.

39. Fankhauser F, Kwasniewska S, Henchoz P-D, et al: Versatility of the cw-Nd:YAG and diode lasers in ocular surgery. *Ophthalmic Surg* 24:225, 1993.

40. Trokel S, Wapner F, Schubert H: Contact transscleral retinal photocoagulation using a micropulsed diode laser. *Invest Ophthalmol Vis Sci* 33(Suppl):1311, 1992.

41. Lim JI, Bressler NM, Marsh MJ, et al: Laser treatment of choroidal neovascularization in patients with angioid streaks. *Am J Ophthalmol* 116:414, 1993.

42. Macular Photocoagulation Study Group: Krypton laser photocoagulation for neovascular lesions of ocular histoplasmosis. Results of a randomized clinical trial. *Arch Ophthalmol* 105:1499, 1987.

43. The Diabetic Retinopathy Study Research Group: photocoagulation treatment of proliferative diabetic retinopathy: Clinical application of diabetic retinopathy study (DRS) findings. DRS Report Number 8. *Ophthalmology* 88:583, 1981.

44. Early Treatment Diabetic Retinopathy Study Group: Results from the early treatment diabetic retinopathy study. *Ophthalmology* 98(Suppl 5):739, 1991.

45. The Krypton Argon Regression Neovascularization Study Research Group: Randomized comparison of krypton versus argon scatter photocoagulation for diabetic disc neovascularization. The krypton argon regression neovascularization study report number 1. *Ophthalmology* 100:1655, 1993.

46. Wolbarsht ML, Landers MB: The rationale of photocoagulation therapy for proliferative diabetic retinopathy: A review and a model. *Ophthalmic Surg* 11:235, 1980.

47. Weiter JJ, Zuckerman R: The influence of the photoreceptor-RPE complex on the inner retina: An explanation for the beneficial effects of photocoagulation. *Ophthalmology* 87:1133, 1980.

48. Glaser BM, Campochiaro PA, Davis JL: Retinal pigment epithelial cells release an inhibitor of neovascularization. *Arch Ophthalmol* 103:1870, 1985.

49. Singerman LJ: Red krypton laser therapy of macular and retinal vascular diseases. *Retina* 2:15, 1982.

50. Liang JC, Goldberg MF: Treatment of diabetic retinopathy. *Diabetes* 29:841, 1980.

51. Prendiville PL, McDonnell PJ: Complications of laser surgery. *Int Ophthalmol Clin* 32:179, 1992.

52. Early Treatment Diabetic Retinopathy Study Research Group: Photocoagulation for diabetic macular edema: Early treatment diabetic retinopathy study report number 1. *Arch Ophthalmol* 103:1796, 1985.

53. Wilson DJ, Finkelstein D, Quigley HA, et al: Macular grid photocoagulation: An experimental study on the primate retina. *Arch Ophthalmol* 106:100, 1988.

54. Bresnick GH: Diabetic maculopathy: A critical review highlighting diffuse macular edema. *Ophthalmology* 90:1301, 1983.

55. Okisaka S, Kuwabara T, Aiello LM: The effects of laser photocoagulation in the retinal capillaries. *Am J Ophthalmol* 80:591, 1975.

56. Clover GM: The effects of argon and krypton photocoagulation on the retina: Implications for the inner and outer blood retinal barriers. In Gitter KA, Schatz H, Yannuzzi LA (eds): *Laser Photocoagulation of Retinal Disease*. Pacific Medical Press, San Francisco, 1988, pp. 11–17.

57. Olk RJ: Argon-green (514 nm) versus krypton red (647 nm) modified grid laser photocoagulation for diffuse diabetic macular edema. *Ophthalmology* 97:1101, 1990.

58. Seiberth V, Schatanek S, Alexandridis E: Panretinal photocoagulation in diabetic retinopathy: Argon versus dye laser coagulation. *Graefe's Arch Clin Exp Ophthalmol* 231:318, 1993.

59. Michels RG, Gass JDM: The natural course of retinal branch vein obstruction. *Trans Am Acad Ophthalmol Otolaryngol* 78:166, 1974.

60. Gutman FA: Macular edema in branch retinal vein occlusion: Prognosis and management. *Trans Am Acad Ophthalmol Otolaryngol* 83:488, 1977.

61. Zweng HCH, Fahrenbruch RC, Little HL: Argon laser photocoagulation in the treatment of retinal vein occlusions. *Mod Probl Ophthalmol* 12:261, 1974.

62. Kelley JS, Patz A, Schatz H: Management of retinal branch vein occlusion: The role of argon laser photocoagulation. *Ann Ophthalmol* 6:1123, 1974.

63. Branch Vein Occlusion Study Group: Argon laser photocoagulation for macular edema in branch vein occlusion. *Am J Ophthalmol* 98:271, 1984.

64. Shilling JS, Jones CA: Retinal branch vein occlusion: A study of argon laser photocoagulation in the treatment of macular oedema. *Br J Ophthalmol* 68:196, 1984.

65. Kremer I, Hartman B, Siegal R, et al: Static and kinetic perimetry results of krypton red laser treatment for macular edema complicating branch vein occlusion. *Ann Ophthalmol* 22:193, 1990.

66. Branch Vein Occlusion Study Group: Argon laser scatter photocoagulation for prevention of neovascularization and vitreous hemorrhage in branch vein occlusion: A randomized clinical trial. *Arch Ophthalmol* 104:34, 1986.

67. Hayreh SS: Classification of central retinal vein occlusion. *Ophthalmology* 90:458, 1983.

68. Finkelstein D: Laser treatment of branch and central retinal vein occlusion. *Int Ophthalmol Clin* 30:84, 1990.

69. Magargal LE, Brown GC, Augsburger JJ, et al: Neovascular glaucoma following central retinal vein obstruction. *Ophthalmology* 88:1095, 1981.

70. May DR, Klein ML, Peyman GA, et al: Xenon arc panretinal photocoagulation for central retinal vein occlusion: A randomised prospective study. *Br J Ophthalmol* 63:725, 1979.

71. Laaitkainen L, Kohner EM, Khoury D, et al: Panretinal photocoagulation in central retinal vein occlusion: A randomised controlled clinical study. *Br J Ophthalmol* 61:741, 1977.

72. Magargal LE, Brown GC, Augsburger JJ, et al: Efficacy of panretinal photocoagulation in preventing neovascular glaucoma following ischemic central retinal vein obstruction. *Ophthalmology* 89:780, 1982.

73. Hayreh SS, Klugman MR, Podhajsky P, et al: Argon laser photocoagulation in ischemic central retinal vein occlusion: A 10-year prospective study. *Graefe's Arch Clin Exp Ophthalmol* 228:281, 1990.

74. Clarkson JG: Photocoagulation for ischemic central retinal vein occlusion. *Arch Ophthalmol* 109:1218, 1991.

75. Bird AC, Grey RHB: Photocoagulation of disciform macular lesions with krypton laser. *Br J Ophthalmol* 63:669, 1979.

76. McMeel JW, Avila MP, Jalkh AE: Subretinal neovascularization in senile macular degeneration. *Trans New Orleans Acad Ophthalmol* 291, 1983.

77. Macular Photocoagulation Study Group: Argon laser photocoagulation for neovascular maculopathy: Three-year results from randomized clinical trials. *Arch Ophthalmol* 104:694, 1986.

78. Macular Photocoagulation Study Group: Argon laser photocoagulation for neovascular maculopathy: Five-year results from randomized clinical trials. *Arch Ophthalmol* 109:1109, 1991.

79. Bressler SB, Bressler NM, Fine SL, et al: Natural course of choroidal neovascular membranes within the foveal avascular zone in senile macular degeneration. *Am J Ophthalmol* 93:157, 1982.

80. Macular Photocoagulation Study Group: Krypton laser photocoagulation for neovascular lesions of age-related macular degeneration: Results of a randomized clinical trial. *Arch Ophthalmol* 108:816, 1990.

81. Macular Photocoagulation Study Group: Laser photocoagulation for juxtafoveal choroidal neovascularization. Five-year results from randomized clinical trials. *Arch Ophthalmol* 112:500, 1994.

82. Macular Photocoagulation Study Group: Laser photocoagulation of subfoveal neovascular lesions in age-related macular degeneration: Results of a randomized clinical trial. *Arch Ophthalmol* 109:1220, 1991.

83. Macular Photocoagulation Study Group: Laser photocoagulation of subfoveal neovascular lesions of age-related macular degeneration. Updated findings from two clinical trials. *Arch Ophthalmol* 111:1200, 1993.

84. Macular Photocoagulation Study Group: Persistent and recurrent neovascularization after krypton laser photocoagulation for neovascular lesions of age-related macular degeneration. *Arch Ophthalmol* 108:825, 1990.

85. Macular Photocoagulation Study Group: Visual outcome after laser photocoagulation for subfoveal choroidal neovascularization secondary to age-related macular degeneration. The influence of initial lesion size and initial visual acuity. *Arch Ophthalmol* 112:480, 1994.

86. Macular Photocoagulation Study Group: Recurrent choroidal neovascularization after argon laser photocoagulation for neovascular maculopathy. *Arch Ophthalmol* 104:503, 1986.

87. Macular Photocoagulation Study Group: Persistent and recurrent neovascularization after argon laser photocoagulation for subfoveal choroidal neovascularization of age-related macular degeneration. *Arch Ophthalmol* 112:489, 1994.

88. Macular Photocoagulation Study Group: Laser photocoagulation of subfoveal recurrent neovascular lesions in age-related macular degeneration. Results of a randomized clinical trial. *Arch Ophthalmol* 109:1232, 1991.

89. McHugh JDA, Marshall J, ffytche TJ, et al: Initial clinical experience using a diode laser in the treatment of retinal vascular disease. *Eye* 3:516, 1989.

90. Balles MW, Puliafito CA, D'Amico DJ, et al: Semiconductor diode laser photocoagulation in retinal vascular disease. *Ophthalmology* 97:1553, 1990.

91. Balles MW, Puliafito CA: Semiconductor diode lasers: A new laser light source in ophthalmology. *Int Ophthalmol Clin* 30:77, 1990.

92. Ulbig MW, McHugh DA, Hamilton AMP: Photocoagulation of choroidal neovascular membranes with a diode laser. *Br J Ophthalmol* 77:218, 1993.

93. Sato Y, Berkowitz BA, Wilson CA, et al: Blood-retinal barrier breakdown caused by diode vs argon laser endophotocoagulation. *Arch Ophthalmol* 110:277, 1992.

94. Haller JA, Lim JI, Goldberg MF: Pilot trial of transscleral diode laser retinopexy in retinal detachment surgery. *Arch Ophthalmol* 111:952, 1993.

95. Charles S: Endophotocoagulation. *Retina* 1:117, 1981.

96. Thomas MA, Halperin LS: Subretinal endolaser treatment of a choroidal bleeding site. *Am J Ophthalmol* 109:742, 1990.

97. Jaffe GJ, Mieler WF, Burke JM, et al: Photoablation of ocular melanoma with a high-powered argon endolaser. *Arch Ophthalmol* 107:113, 1989.

98. Duker JS, Federman JL, Schubert H, et al: Semiconductor diode laser endophotocoagulation. *Ophthalmic Surg* 20:717, 1989.

99. Smiddy WE: Diode laser photocoagulation. *Arch Ophthalmol* 110:1172, 1992.

100. Diskin J, Maguire AM, Margherio RR: Choroidal folds induced with diode endolaser. *Arch Ophthalmol* 110:754, 1992.

101. Peyman GA, Lee KJ: Multifunction endolaser probe. *Am J Ophthalmol* 114:103, 1992.

102. Peyman GA, D'Amico DJ, Alturki WA: An endolaser probe with aspiration capability. *Arch Ophthalmol* 110:718, 1992.

103. Uram M: Ophthalmic laser microendoscope ciliary process ablation in the management of neovascular glaucoma. *Ophthalmology* 99:1823, 1992.

104. Uram M: Ophthalmic laser microendoscope endophotocoagulation. *Ophthalmology* 99:1829, 1992.

105. Mizuno K: Binocular indirect argon laser photocoagulator. *Br J Ophthalmol* 65:425, 1981.

106. Friberg TR: Principles of photocoagulation using binocular indirect ophthalmoscope laser delivery systems. *Int Ophthalmol Clin* 30:89, 1990.

107. Friberg TR: Clinical experience with a binocular indirect ophthalmoscope laser delivery system. *Retina* 7:28, 1987.

108. Friberg TR, Eller AW: Pneumatic repair of primary and secondary retinal detachments using a binocular indirect ophthalmoscope laser delivery system. *Ophthalmology* 95:187, 1988.

109. Augsburger JJ, Faulkner CB: Indirect ophthalmoscope argon laser treatment of retinoblastoma. *Ophthalmic Surg* 23:591, 1992.

110. Augsburger JJ, Mullen D, Kleineidam M: Planned combined I-125 plaque irradiation and indirect ophthalmoscope laser therapy for choroidal malignant melanoma. *Ophthalmic Surg* 24:76, 1993.

111. Irvine WD, Smiddy WE, Nicholson DH: Corneal and iris burns with the laser indirect ophthalmoscope. *Am J Ophthalmol* 110:311, 1990.

112. Morley MG, Frederick AR: Melted haptic as a complication of the indirect ophthalmic laser delivery system. *Am J Ophthalmol* 113:584, 1992.

113. Palmer EA, Flynn JT, Hardy RJ, et al: Incidence and early course of retinopathy of prematurity. *Ophthalmology* 98:1628, 1991.

114. Cryotherapy for Retinopathy of Prematurity Cooperative Group: Multicenter trial of cryotherapy for retinopathy of prematurity: One-year outcome—structure and function. *Arch Ophthalmol* 108:1408, 1990.

115. Cryotherapy for Retinopathy of Prematurity Cooperative Group: Multicenter trial of cryotherapy for retinopathy of prematurity: Preliminary results. *Arch Ophthalmol* 106:471, 1988.

116. Nagata M, Kanenari A, Fukuda T, et al: Photocoagulation for the treatment of retinopathy of prematurity. *Jpn J Clin Ophthalmol* 24:419, 1968.

117. Landers MB, Semple HC, Ruben JB, et al: Argon laser photocoagulation for advanced retinopathy of prematurity. *Am J Ophthalmol* 110:429, 1990.

118. Landers MB, Toth CA, Semple HC, et al: Treatment of retinopathy of prematurity with argon laser photocoagulation. *Arch Ophthalmol* 110:44, 1991.

119. Schechter RJ: Laser treatment of retinopathy of prematurity. *Arch Ophthalmol* 111:730, 1993.

120. Iverson DA, Trese MT, Orgel IK, et al: Laser photocoagulation for threshold retinopathy of prematurity. *Arch Ophthalmol* 109:1342, 1991.

121. McNamara JA, Tasman W, Brown GC, et al: Laser photocoagulation for stage 3+ retinopathy of prematurity. *Ophthalmology* 98:576, 1991.

122. Fleming TN, Runge PE, Charles ST: Diode laser photocoagulation for prethreshold, posterior retinopathy of prematurity. *Am J Ophthalmol* 114:589, 1992.

123. Capone A, Diaz-Rohena R, Sternberg P, et al: Diode-laser photocoagulation for zone 1 threshold retinopathy of prematurity. *Am J Ophthalmol* 116:444, 1993.

124. McNamara JA, Tasman W, Vander JF, et al: Diode laser photocoagulation for retinopathy of prematurity: Preliminary results. *Arch Ophthalmol* 110:1714, 1992.

125. Hunter DG, Repka MX: Diode laser photocoagulation for threshold retinopathy of prematurity: A randomized study. *Ophthalmology* 100:238, 1993.

126. Tasman W: Threshold retinopathy of prematurity revisited. *Arch Ophthalmol* 110:623, 1992.

127. Fleck BW: How large must an iridotomy be? *Br J Ophthalmol* 74:583, 1990.

128. Meyer-Schwickerath G: Erfahrungen mit der Lichtkoagulation der Netzhaut und der Iris. *Doc Ophthalmol* 10:119, 1956, p. 91.

129. Rodrigues MM, Streeten B, Spaeth GL, et al: Argon laser iridotomy on primary angle closure or pupillary block glaucoma. *Arch Ophthalmol* 96:2222, 1978.

130. Jacobson JJ, Schuman JS, El Koumy H, et al: Diode laser peripheral iridectomy. *Int Ophthalmol Clin* 30:120, 1990.

131. Schuman JS, Puliafito CA, Jacobson JJ: Semiconductor diode laser peripheral iridotomy. *Arch Ophthalmol* 108:1207, 1990.

132. Wise JB, Witter SL: Argon laser therapy for open angle glaucoma: A pilot study. *Arch Ophthalmol* 97:319, 1979.

133. Allf BE, Shields MB: Early intraocular pressure response to laser trabeculoplasty 180 degrees without apraclonidine versus 360 degrees with apraclonidine. *Ophthalmic Surg* 22:539, 1991.

134. Spurny RC, Lederer CM: Krypton laser trabeculoplasty: A clinical report. *Arch Ophthalmol* 102:1626, 1984.

135. Wise JB: Glaucoma treatment by trabecular tightening with the argon laser. *Int Ophthalmol Clin* 21:69, 1981.

136. Schwartz LW, Spaeth GL, Traverso C, et al: Variation of techniques on the results of argon laser trabeculoplasty. *Ophthalmology* 90:781, 1983.

137. Rouhiainen HJ, Teräsvirta ME, Tuovinen EJ: Laser power and postoperative intraocular pressure increase in argon laser trabeculoplasty. *Arch Ophthalmol* 105:1352, 1987.

138. Melamed S, Epstein DL: Alterations of aqueous humour outflow following argon laser trabeculoplasty in monkeys. *Br J Ophthalmol* 71:776, 1987.

139. Raviola G: Schwalbe line's cells: A new cell type in the trabecular meshwork of *Macaca mulatta*. *Invest Ophthalmol Vis Sci* 22:45, 1982.

140. Van Buskirk EM, Pond V, Rosenquist RC, et al: Argon laser trabeculoplasty: Studies of mechanism of action. *Ophthalmology* 91:1005, 1984.

141. Shingleton BJ, Richter CU, Dharma SK, et al: Long-term efficacy of argon laser trabeculoplasty. A 10-year follow-up study. *Ophthalmology* 100:1324, 1993.

142. Ticho U, Nesher R: Laser trabeculoplasty in glaucoma. Ten-year evaluation. *Arch Ophthalmol* 107:844, 1989.

143. The Glaucoma Laser Trial Research Group: The glaucoma laser trial (GLT): 2. Re-

sults of argon laser trabeculoplasty versus topical medicines. *Ophthalmology* 97:1403, 1990.

144. Lichter PR: Practice implications of the glaucoma laser trial (editorial). *Ophthalmology* 97:1401, 1990.

145. McHugh D, Marshall J, ffytche TJ, et al: Diode laser trabeculoplasty (DLT) for primary open-angle glaucoma and ocular hypertension. *Br J Ophthalmol* 74:743, 1990.

146. McHugh D, Marshall J, ffytche TJ, et al: Ultrastructural changes of human trabecular meshwork after photocoagulation with a diode laser. *Invest Ophthalmol Vis Sci* 33:2664, 1992.

147. Brancato R, Carassa R, Trabucchi G: Diode laser compared with argon laser for trabeculoplasty. *Am J Ophthalmol* 112:50, 1991.

148. Moriarty AP, McHugh JDA, ffytche TJ, et al: Long-term follow-up of diode laser trabeculoplasty for primary open-angle glaucoma and ocular hypertension. *Ophthalmology* 100:1614, 1989.

149. Traverso CE, Greenidge KC, Spaeth GL: Formation of peripheral anterior synechiae following argon laser trabeculoplasty. A prospective study to determine relationship to position of laser burns. *Arch Ophthalmol* 102:861, 1984.

150. Bellows AR, Grant WM: Cyclocryotherapy in advanced inadequately controlled glaucoma. *Am J Ophthalmol* 75:679, 1973.

151. Lee P-F, Pomerantzeff O: Transpupillary cyclophoto-coagulation of rabbit eyes: An experimental approach to glaucoma surgery. *Am J Ophthalmol* 71:911, 1971.

152. Weekers R, Lavergne G, Watillion M, et al: Effects of photocoagulation of ciliary body upon ocular tension. *Am J Ophthalmol* 52:156, 1961.

153. Beckman H, Kinoshita A, Rota AN, et al: Transscleral ruby laser irradiation of the ciliary body in the treatment of intractable glaucoma. *Trans Am Acad Ophthalmol Otolaryngol* 76:423, 1972.

154. Wilensky JT, Welch D, Mirolovich M: Transscleral cyclocoagulation using a neodymium:YAG laser. *Ophthalmic Surg* 16:95, 1985.

155. Schubert HD, Federman JL: A comparison of CW Nd:YAG contact transscleral cyclophotocoagulation with cyclocryopexy. *Invest Ophthalmol Vis Sci* 30:536, 1989.

156. Vogel A, Dlugos CH, Nuffer R, et al: Optical properties of human sclera, and their consequences of transscleral laser applications. *Lasers Surg Med* 11:331, 1991.

157. Barraquer RI, Kargacin M: Nd:YAG laser diascleral cyclophotocoagulation: Survival analysis after four years. *Dev Ophthalmol* 22:132, 1991.

158. Suzuki Y, Araie M, Yumita A, et al: Transscleral Nd:YAG cyclophotocoagulation versus cyclocryotherapy. *Graefe's Arch Clin Exp Ophthalmol* 229:33, 1991.

159. Schuman JS, Jacobson JJ, Puliafito CA, et al: Experimental use of semiconductor diode laser in contact transscleral cyclophotocoagulation in rabbits. *Arch Ophthalmol* 108:1152, 1990.

160. Schuman JS, Noecker RJ, Puliafito CA, et al: Energy levels and probe placement in contact transscleral semiconductor diode laser cyclophotocoagulation in human cadaver eyes. *Arch Ophthalmol* 109:1534, 1991.

161. Simmons RB, Prum BE, Shields SR, et al: Videographic and histologic comparison of Nd:YAG and diode laser contact transscleral cyclophotocoagulation. *Am J Ophthalmol* 117:337, 1994.

162. Gaasterland DE, Abrams DA, Belcher CD, et al: A multicenter study of contact diode laser transscleral cyclophotocoagulation in glaucoma patients. *Invest Ophthalmol Vis Sci* 33(Suppl):1019, 1992.

163. Berry J: Recurrent trichiasis: Treatment with laser photocoagulation. *Ophthalmic Surg* 10(7):36, 1979.

164. Campbell DC: Thermoablation treatment for trichiasis using the argon laser. *Aust NZ J Ophthalmol* 18:427, 1990.

165. Hornblass A, Coden DJ: Lasers in oculoplastic and orbital surgery. *Int Ophthalmol Clin* 29:265, 1989.

166. Noe JM, Barsky SH, Geer DE, et al: Port wine stains and the response to argon laser therapy: Successful treatment and the predictive role of color, age, and biopsy. *Plastic Reconstr Surg* 65:130, 1980.

167. Apfelberg DB, Maser MR, White DN, et al: Benefits of contact and noncontact YAG laser for periorbital hemangiomas. *Ann Plastic Surg* 24:397, 1990.

168. Bosniak SL, Ginsberg G: Laser eyelid surgery: Evaluating the therapeutic options. *Ophthalmol Clin N Am* 6:479, 1993.

169. Shields JA, Parsons H, Shields CL, et al: The role of photocoagulation in the management of retinoblastoma. *Arch Ophthalmol* 108:205, 1990.

170. Shields JA, Glazer LC, Mieler WF, et al: Comparison of xenon arc and argon laser photocoagulation in the treatment of choroidal melanomas. *Am J Ophthalmol* 109:647, 1990.

171. Vogel M, Schäfer FP, Theuring S, et al: Results of dye-laser photocoagulation of the rabbit fundus. In Gitter KA, Schatz H, Yannuzzi LA (eds): *Laser Photocoagulation of Retinal Disease*. Pacific Medical Press, San Francisco, 1988, pp. 37–40.

172. Brancato R, Menchini U, Pece A, et al: Dye laser photocoagulation of macular subretinal neovascularization in pathological myopia. *Int Ophthalmol* 11:235, 1988.

173. Anderson RR, Parrish JA: Selective photothermolysis: Precise microsurgery by selective absorption of pulsed radiation. *Science* 220:524, 1983.

174. Krauss JM, Puliafito CA, Lin WZ, et al: Interferometric technique for investigation of laser thermal retinal damage. *Invest Ophthalmol Vis Sci* 28:1290, 1987.

175. Roider J, Michaud NA, Flotte TJ, et al: Response of the retinal pigment epithelium to selective photocoagulation. *Arch Ophthalmol* 110:1786, 1992.

176. Tse DT, Dutton JJ, Weingeist TA, et al: Hematoporphyrin photoradiation therapy for intraocular and orbital malignant melanoma. *Arch Ophthalmol* 102:833, 1984.

177. Diamond I, Granelli SG, McDonagh AF, et al: Photodynamic therapy of malignant tumours. *Lancet* 2:1175, 1972.

178. Kramer M, Miller JW, Michaud N, et al: Photodynamic therapy (PDT) of experimental choroidal neovascularization (CNV) using liposomal benzoporphyrin derivative monoacid (BPD-MA): Refinement of dosimetry. *Invest Ophthalmol Vis Sci* 35(Suppl):1503, 1994.

179. Howard MA, Hu LK, Gonzalez VH, et al: Photodynamic therapy (PDT) of pigmented choroidal melanoma using a liposomal preparation of benzoporphyrin derivative (BPD). *Invest Ophthalmol Vis Sci* 35(Suppl):1722, 1994.

180. Destro M, Puliafito CA: Indocyanine green videoangiography of choroidal neovascularization. *Ophthalmology* 96:846, 1989.

181. Slakter JS, Yannuzzi LA, Sorenson JA, et al: A pilot study of indocyanine green angiography-guided laser photocoagulation of occult choroidal neovascularization in age-related macular degeneration. *Arch Ophthalmol* 112:465, 1994.

182. Reichel E, Duker JS, Puliafito CA: Indocyanine green dye enhanced diode laser photocoagulation of poorly defined subfoveal neovascularization. *Invest Ophthalmol Vis Sci* 35(Suppl):1563, 1994.

183. Puliafito CA, Guyer DR, Monés JM, et al: Indocyanine-green digital angiography and dye-enhanced diode laser photocoagulation of choroidal neovascularization. *Invest Ophthalmol Vis Sci* 32(Suppl):712, 1991.

184. Spikes JD: Phthalocyanines as photosensitizers in biological systems and for the photodynamic therapy of tumors. *Photochem Photobiol* 43:691, 1986.

185. Evensen JF, Moan J: A test of different photosensitizers for photodynamic treatment of cancer in a murine tumor model. *Photochem Photobiol* 46:859, 1987.

186. Bauman WC, Monés JM, Tritten J-J, et al: Transpupillary phthalocyanine photodynamic therapy of experimental posterior malignant melanoma. *Invest Ophthalmol Vis Sci* 32(Suppl):713, 1991.

187. Ozler SA, Nelson JS, Liggett PE, et al: Photodynamic therapy of experimental subchoroidal melanoma using chloroaluminum sulfonated phthalocyanine. *Arch Ophthalmol* 110:555, 1992.

188. Miller JW, Stinson WP, Gregory WA, et al: Phthalocyanine photodynamic therapy of experimental iris neovascularization. *Ophthalmology* 98:1711, 1991.

189. Naoumidis LP, Tsilimbaris MK, Georgiades A, et al: A diode laser mounted on a slit lamp for ophthalmic photodynamic applications of phthalocyanine. *Am J Ophthalmol* 115:111, 1993.

190. Firey PA, Rodgers MAJ: Photoproperties of a silicon naphthalocyanine: A potential photosensitizer for photodynamic therapy. *Photochem Photobiol* 45:535, 1987.

191. Garrett J, Reddy S, Ryan S, et al: Photodynamic therapy (PDT) of experimental ocular melanoma with silicon naphthalocyanine (SlNc) in rabbits. *Invest Ophthalmol Vis Sci* 35(Suppl):2120, 1994.

192. Zeimer RC, Khoobehi B, Niesman MR, et al: A potential method for local drug and dye delivery in the ocular vasculature. *Invest Ophthalmol Vis Sci* 29:1179, 1988.

193. Khoobehi B, Peyman GA, Niesman MR, et al: Measurement of retinal blood velocity and flow rate in primates using a liposome-dye system. *Ophthalmology* 96:905, 1989.

194. Khoobehi B, Char CA, Peyman GA: Assessment of laser-induced release of drugs from liposomes: An in vitro study. *Lasers Surg Med* 10:60, 1990.

195. Khoobehi B, Peyman GA, Vo K: Laser-triggered repetitive fluorescein angiography. *Ophthalmology* 99:72, 1992.

196. Ren Q, Simon G, Parel J-M, et al: Laser scleral buckling for retinal reattachment. *Am J Ophthalmol* 115:758, 1993.

197. Burstein NL, Williams JM, Nowicki MJ, et al: Corneal welding using hydrogen fluoride lasers. *Arch Ophthalmol* 110:12, 1992.

198. Khadem JJ, Truong TV, Ernest JT: Laser activated tissue glue. *Invest Ophthalmol Vis Sci* 34(Suppl):1247, 1993.

199. Wolf MD, Arrindell L, Han DP: Retinectomies treated by diode laser activated indocyanine green dye-enhanced fibrinogen glue. *Invest Ophthalmol Vis Sci* 33(Suppl): 1316, 1992.

200. McClung FJ, Hellwarth RW: Giant optical pulsating from ruby. *J Appl Phys* 33:828, 1967.

201. Krasnov M: Laser-puncture of the anterior chamber angle in glaucoma. *Vestn Oftalmol* 3:27, 1972.

202. Aron-Rosa D, Aron JJ, Greisemann J, et al: Use of the neodymium-YAG laser to open the posterior capsule after lens implant surgery: A preliminary report. *J Am Intraocul Implant Soc* 6:352, 1980.

203. Fankhauser F, Roussel P, Steffen J, et al: Clinical studies on the efficiency of a high power laser radiation upon some structures of the anterior segment of the eye. *Int Ophthalmol* 3:129, 1981.

204. Steinert RF, Puliafito CA, Trokel S: Plasma formation and shielding by three ophthalmic Nd-YAG lasers. *Am J Ophthalmol* 96:427, 1983.

205. Fradin DW, Bloembergen N, Letellier JP: Dependence of laser-induced breakdown field strength on pulse duration. *Appl Phys Lett* 22:635, 1973.

206. Ready JF: *Effects of High-Power Laser Radiation.* Academic Press, New York, 1971, pp. 133–143, 215–217.

207. Hu C-L, Barnes FS: The thermal-chemical damage in biological material under laser irradiation. *IEEE Trans Biomed Eng* 17:220, 1970.

208. Felix MP, Ellis AT: Laser-induced liquid breakdown—A step-by-step account. *Appl Phys Lett* 19:484, 1971.

209. Lauterborn W: High-speed photography of laser-induced breakdown in liquids. *Appl Phys Lett* 21:27, 1972.

210. Brewer RJ, Rieckhoff KE: Stimulated Brillouin scattering in liquids. *Phys Rev Lett* 13:334, 1964.

211. Cleary SF, Hamrick PE: Laser-induced acoustic transients in the mammalian eye. *J Acoust Soc Am* 46:1037, 1969.

212. Van der Zypen E, Fankhauser F, Bebie H, et al: Changes in the ultrastructure of the iris after irradiation with intense light. *Adv Ophthalmol* 39:59, 1979.

213. Fujimoto JG, Lin WZ, Ippen IP, et al: Time-resolved studies of Nd:YAG laser-induced breakdown: Plasma formation, acoustic wave generation, and cavitation. *Invest Ophthalmol Vis Sci* 26:1771, 1985.

214. Carome EF, Carreria EM, Prochaska CJ: Photographic studies of laser-induced pressure impulses in liquids. *Appl Phys Lett* 11:64, 1967.

215. Mainster MA, Sliney DH, Belcher CD, et al: Laser photodisruptors: Damage mechanisms, instrument design, and safety. *Ophthalmology* 90:973, 1983.

216. Taboada J: Interaction of short laser pulses with ocular tissues. In Trokel S (ed): *YAG Laser Ophthalmic Microsurgery*. Appleton-Century-Crofts, Norwalk, CT, 1983, pp. 15–38.

217. Smith WL, Liu P, Bloembergen N: Superbroadening in water and deuterium by self-focused picosecond pulses from a neodymium doped YAlG laser. *Phys Rev* (A) 15:2396, 1977.

218. Anthes JP, Bass M: Direct observation of the dynamics of picosecond-pulse optical breakdown. *Appl Phys Lett* 31:412, 1977.

219. Ashkinadze BM, Vladimirov VI, Likhachev VA, et al: Breakdown in dielectrics caused by intense laser radiation. *Sov Phys JETP* 23:788, 1966.

220. Loertscher H: Laser-induced breakdown for ophthalmic applications. In Trokel S (ed): *YAG Laser Ophthalmic Microsurgery*. Appleton-Century-Crofts, Norwalk, CT, 1983, p. 39.

221. Sinskey RM, Cain W: The posterior capsule and phacoemulsification. *J Am Intraocul Implant Soc* 4:206, 1978.

222. Steinert RF, Puliafito CA, Kumar SR, et al: Cystoid macular edema, retinal detachment and glaucoma after Nd:YAG laser posterior capsulotomy. *Am J Ophthalmol* 112:373, 1991.

223. Parker WT, Clorfeine GS, Stocklin RD: Marked intraocular pressure rise following Nd-YAG laser capsulotomy. *Ophthalmic Surg* 15:103, 1984.

224. Epstein DL, Jedziniak JA, Grant WM: Obstruction of aqueous outflow by lens particles and by heavy molecular-weight soluble lens proteins. *Invest Ophthalmol Vis Sci* 17:272, 1978.

225. Richter CU, Arzeno G, Pappas HR, et al: Prevention of intraocular pressure elevation following neodymium-YAG posterior capsulotomy. *Arch Ophthalmol* 103:912, 1985.

226. Silverstone DE, Novack GD, Kelley EP, et al: Prophylactic treatment of intraocular pressure elevations after neodymium-YAG laser capsulotomies and extracapsular cataract extraction with levobunolol. *Ophthalmology* 95:713, 1988.

227. Del Priore LV, Robin AL, Pollack IP: Neodymium-YAG and argon laser iridotomy. Long-term follow-up in a prospective, randomized clinical trial. *Ophthalmology* 95:1207, 1988.

228. Schwartz L: Laser iridectomy. In Schwartz L, Spaeth G, Brown G (eds): *Laser Therapy of the Anterior Segment: A Practical Approach*. Charles B Slack, Thorofare, NJ, 1984, pp. 29–58.

229. Panek WC, Lee DA, Christensen RE: The effects of Nd:YAG laser iridectomy on the corneal endothelium. *Am J Ophthalmol* 111:505, 1991.

230. Robin AL, Pollack IP: Q-switched neodymium-YAG laser iridotomy in patients in whom the argon laser fails. *Arch Ophthalmol* 104:531, 1986.

231. Zborwski-Gutman L, Rosner M, Blumenthal M, et al: Sequential use of argon and Nd–YAG lasers to produce an iridotomy—a pilot study. *Metab Pediatr Syst Ophthalmol* 11:58, 1988.

232. Puliafito CA, Wasson PJ, Steinert RF, et al: Nd–YAG laser surgery on experimental vitreous membranes. *Arch Ophthalmol* 102:843, 1984.

233. Krauss JM, Puliafito CA, Miglior S, et al: Vitreous changes after neodymium-YAG laser photodisruption. *Arch Ophthalmol* 104:592, 1986.

234. Pepose JS, Ubels JL: The cornea. In Hart WM (ed): *Adler's Physiology of the Eye*, 9th ed., Mosby Year Book, 1992, pp. 29–70.

235. Krauss JM, Puliafito CA, Steinert RF: Laser interactions with the cornea. *Surv Ophthalmol* 31:37, 1986.

236. Patel CKN: Interpretation of CO_2 optical laser experiments. *Phys Rev Lett* 12:588, 1964.

237. Beckman H, Fuller TA: Carbon dioxide laser scleral dissection and filtering procedure for glaucoma. *Am J Ophthalmol* 88:73, 1979.

238. Schachat A, Iliff WJ, Kashima HK: Carbon dioxide laser therapy of recurrent squamous papilloma of the conjunctiva. *Ophthalmic Surg* 13:916, 1982.

239. Meyers SM, Bonner RF, Rodrigues MM, et al: Phototransection of vitreal membranes with the carbon dioxide laser in rabbits. *Ophthalmology* 90:563, 1983.

240. Beckman H, Rota A, Barraco R: Limbectomies, keratectomies, and keratostomies performed with a rapid-pulsed carbon dioxide laser. *Am J Ophthalmol* 71:1277, 1971.

241. Keates RH, Pedrotti LS, Weichel H, et al: Carbon dioxide laser beam control for corneal surgery. *Ophthalmic Surg* 12:117, 1981.

242. Peyman GA, Larson B, Raichand M, et al: Modification of rabbit corneal curvature with use of carbon dioxide laser burns. *Ophthalmic Surg* 11:325, 1980.

243. Keates RH, Levy SN, Fried S, et al: Carbon dioxide laser use in wound sealing and epikeratophakia. *J Cataract Refract Surg* 13:290, 1987.

244. Bachem A: Ophthalmic ultraviolet action spectra. *Am J Ophthalmol* 41:969, 1956.

245. Slavin W: Stray light in ultraviolet, visible, and near-infrared spectral photometry. *Anal Chem* 35:561, 1963.

246. Beaven GH, Holiday ER: Ultraviolet absorption spectra of proteins and amino acids. *Adv Prot Chem* 7:319, 1952.

247. Wetlaufer DB: Ultraviolet spectra of proteins and amino acids. *Adv Prot Chem* 17:303, 1962.

248. Loofbourow JR, Gould BS, Sizer IW: Studies on the ultraviolet absorption spectra of collagen. *Arch Biochem* 22:406, 1949.

249. Smith KC: Ultraviolet radiation effects on molecules and cells. In Smith KC (ed): *The Science of Photobiology*. Plenum Press, New York, 1985, pp. 113–141.

250. O'Brien WJ: Measurement of corneal DNA content. *Invest Ophthalmol Vis Sci* 18:538, 1979.

251. Dougherty AM, Causley GC, Johnson WC: Flow dichroism evidence for tilting of the bases when DNA is in solution. *Proc Natl Acad Sci* (USA) 80:2194, 1983.

252. Stone AL: Optical rotary dispersion of mucopolysaccharides III: Ultraviolet circular dichroism and conformational specificity in amide groups. *Biopolymers* 10:739, 1971.

253. Kurzel RB: On the nature of the action spectrum for ultraviolet photokeratitis. *Ophthalmic Res* 10:312, 1978.

254. Buschke W, Friedenwald JS, Moses SG: Effects of ultraviolet irradiation on corneal epithelium: Mitosis, nuclear fragmentation, post-traumatic cell movements, loss of tissue cohesion. *J Cell Comp Physiol* 26:147, 1945.

255. Friedenwald JS, Buschke W, Crowell J, et al: Effects of ultraviolet irradiation on the corneal epithelium. *J Cell Comp Physiol* 32:161, 1948.

256. McCann J, Ames BN: Detection of carcinogens as mutagens in the salmonella/microsome test: Assay of 300 chemicals: Discussion. *Proc Natl Acad Sci* (USA) 73:950, 1976.

257. Norren DV, Vos JJ: Spectral transmission of the human ocular media. *Vision Res* 14:1237, 1974.

258. Ringvold A: Damage of the cornea epithelium caused by ultraviolet radiation: A scanning electron microscopic study in rabbit. *Acta Ophthalmol* 61:898, 1983.

259. Taboada J, Mikesell GW, Reed RD: Response of the corneal epithelium to KrF excimer laser pulses. *Health Phys* 40:677, 1981.

260. Zuclich JA: Ultraviolet induced damage in the primate cornea and retina. *Curr Eye Res* 3:27, 1984.

261. Tapaszto I, Vass Z: Alteration in mucopolysaccharide-compounds of tear and that of corneal epithelium caused by ultraviolet radiation. *Ophthalmologica* 58(Suppl):343, 1969.

262. Koliopoulos JX, Margaritis LH: Response of the cornea to far-ultraviolet light: An ultrastructural study. *Ann Ophthalmol* 11:765, 1979.

263. Regan JD, Trosko JE, Carrier WL: Evidence for excision of ultraviolet-induced pyrimidine dimers from the DNA of human cells *in vitro*. *Biophys J* 8:319, 1968.

264. Burlamacchi P: Laser sources. In Hillenkamp F, Pratesi R, Sacchi CA (eds): *Lasers in Biology and Medicine*. Plenum Press, New York, 1980, pp. 1–16.

265. Searles SK, Hart GA: Stimulated emission at 281.8 nm from XeBr. *Appl Phys Lett* 27:243, 1975.

266. McKee T, Nilson JA: Excimer applications. *Laser Focus* 18:51, 1982.

267. Deutsch TF, Geis MW: Self-developing UV photoresist using excimer laser exposure. *J Appl Phys* 54:7201, 1983.

268. Koren G, Yeh JT: Emission spectra, surface quality, and mechanism of excimer laser etching of polyimide films. *Appl Phys Lett* 44:1112, 1984.

269. Srinivasan R, Leigh WJ: Ablative photodecomposition on poly(ethylene terephthalate) films. *J Am Chem Soc* 104:6784, 1982.

270. Garrison BJ, Srinivasan R: Microscopic model for the ablative photodecomposition of polymers by far-ultraviolet radiation (193 nm). *Appl Phys Lett* 44:849, 1984.

271. Jellinek HH, Srinivasan R: Theory of etching of polymers by far-ultraviolet, high-intensity pulsed laser and long-term irradiation. *J Phys Chem* 88:3048, 1984.

272. Melcher RL: Thermal and acoustic techniques for monitoring pulsed laser processing. *Springer Ser Chem Phys* 39:418, 1984.

273. Srinivasan R: Kinetics of the ablative photodecomposition of organic polymers in the far-ultraviolet (193 nm). *J Vacuum Sci Technol* B 11:923, 1983.

274. Srinivasan R, Braren B: Ablative photodecomposition of polymer films by pulsed far-ultraviolet (193 nm) laser radiation: Dependence of etch depth on experimental conditions. *J Polym Sci Polym Chem Ed* 22:2601, 1984.

275. Gorodetsky G, Kazyaka TG, Melcher RL, et al: Calorimetric and acoustic study of ultraviolet laser ablation of polymers. *Appl Phys Lett* 46:828, 1985.

276. Parrish JA: Ultraviolet-laser ablation. *Arch Dermatol* 121:599, 1985.

277. Srinivasan R: Ultraviolet laser ablation of organic polymer films. *Springer Ser Chem Phys* 39:343, 1984.

278. Srinivasan R, Wynne JJ, Blum SE: Far-UV photoetching of organic material. *Laser Focus* 19:62, 1983.

279. Linsker R, Srinivasan R, Wynne JJ, et al: Far-ultraviolet laser ablation of atherosclerotic lesions. *Lasers Surg Med* 4:201, 1984.

280. Lane RJ, Linsker R, Wynne JJ, et al: Ultraviolet-laser ablation of the skin. *Arch Dermatol* 121:609, 1985.

281. Mohr FW, Grundfest WS, Litvack F, et al: Excimer laser angioplasty. In Ginsburg R, White JC (eds): *Primer on Laser Angioplasty*. Futura, 1989, Mount Kisco, NY, pp. 181–211.

282. Rowsey JJ, Balyeat HD, Rabinovitch B, et al: Predicting the results of radial keratotomy. *Ophthalmology* 90:642, 1983.

283. Villasenor RA, Salz J, Steel D, et al: Changes in corneal thickness during radial keratotomy. *Ophthalmic Surg* 12:341, 1981.

284. Hoffer KJ, Darrin JJ, Pettit TH, et al: The UCLA clinical trial of radial keratotomy: Preliminary report. *Ophthalmology* 88:729, 1981.

285. Fyodorov SN, Durnev VV: Operation of dosaged dissection of corneal circular ligament in cases of myopia of mild degree. *Ann Ophthalmol* 11:1885, 1979.

286. Gelender H, Flynn HW, Mandelbaum SH: Bacterial endophthalmitis resulting from radial keratotomy. *Am J Ophthalmol* 93:323, 1982.

287. Swinger CA, Barraquer JI: Keratophakia and keratomileusis—clinical results. *Ophthalmology* 88:709, 1981.

288. Trokel SL, Srinivasan R, Braren B: Excimer laser surgery of the cornea. *Am J Ophthalmol* 96:710, 1983.

289. Puliafito CA, Steinert RF, Deutsch TF, et al: Excimer laser ablation of the cornea and lens: Experimental studies. *Ophthalmology* 92:741, 1985.

290. Walsh JY, Deutsch TF, Flotte T, et al: Comparison of tissue ablation by pulsed CO_2 and excimer lasers. Paper TuL2, *Technical Digest*, Conference on Lasers and Electro-Optics. San Francisco, June 9–13, 1986.

291. Kerr-Muir MG, Trokel SL, Marshall J, et al: Ultrastructural comparison of conventional surgical and argon fluoride excimer laser keratectomy. *Am J Ophthalmol* 103:448, 1987.

292. Feld JR, Lin CP, Puliafito CA: Study of the mechanism of excimer laser corneal ablation using calorimetric measurements. *Invest Ophthalmol Vis Sci* 34(Suppl):802, 1993.

293. Krueger RR, Trokel SL: Quantitation of corneal ablation by ultraviolet laser light. *Arch Ophthalmol* 103:1741, 1985.

294. Krueger RR, Trokel SL, Schubert HD: Interaction of ultraviolet laser light with the cornea. *Invest Ophthalmol Vis Sci* 26:1455, 1985.

295. Boettner EA, Wolter JR: Transmission of the ocular media. *Invest Ophthalmol* 1:776, 1962.

296. Peyman GA, Kuszak JR, Weckstrom K, et al: Effects of XeCl excimer laser on the eyelid and anterior segment structures. *Arch Ophthalmol* 104:118, 1986.

297. Puliafito CA, Wong K, Steinert RF: Quantitative and ultrastructural studies of excimer laser ablation of the cornea at 193 and 248 nanometers. *Lasers Surg Med* 7:155, 1987.

298. Fantes FE, Waring GO: Effect of excimer laser radiant exposure on uniformity of ablated corneal surface. *Lasers Surg Med* 9:533, 1989.

299. Berns MW, Liaw L-H, Oliva A, et al: An acute light and electron microscopic study of ultraviolet 193-nm excimer laser corneal incisions. *Ophthalmology* 95:1422, 1988.

300. Campos M, Wang XW, Hertzog L, et al: Ablation rates and surface ultrastructure of 193 nm excimer laser keratectomies. *Invest Ophthalmol Vis Sci* 34:2493, 1993.

301. Baron WS, Munnerlyn C: Predicting visual performance following excimer photorefractive keratectomy. *Refract Corneal Surg* 8:355, 1992.

302. Dehm EJ, Puliafito CA, Adler CM, et al: Corneal endothelial injury in rabbits following excimer laser ablation at 193 and 248 nm. *Arch Ophthalmol* 104:1364, 1986.

303. Zabel R, Tuft S, Marshall J: Excimer laser photorefractive keratectomy: Endothelial morphology following area ablation of the cornea. *Invest Ophthalmol Vis Sci* 29(Suppl):390, 1988.

304. Nuss RC, Puliafito CA, Dehm EJ: Unscheduled DNA synthesis following excimer laser ablation of the cornea in vivo. *Invest Ophthalmol Vis Sci* 28:287, 1987.

305. Kochevar IE: Cytotoxicity and mutagenicity of excimer laser radiation. *Lasers Surg Med* 9:440, 1989.

306. Müller-Stolzenburg NW, Müller GJ, Buchwald HJ, et al: UV exposure of the lens during 193-nm excimer laser corneal surgery. *Arch Ophthalmol* 108:915, 1990.

307. Cotliar AM, Schubert HD, Mandel ER, et al: Excimer laser radial keratotomy. *Ophthalmology* 92:206, 1985.

308. Steinert RF, Puliafito CA: Corneal incisions with the excimer laser. In Sanders DR, Hofmann RF, Salz JJ (eds): *Refractive Corneal Surgery*. Charles Slack, Thorofare, NJ, 1986, pp. 401–410.

309. Marshall J, Trokel S, Rothery S, et al: Photoablative reprofiling of the cornea using an excimer laser. Photorefractive keratectomy. *Lasers Ophthalmol* 1:21, 1986.

310. Puliafito CA, Stern D, Krueger RR, et al: High-speed photography of excimer laser ablation of the cornea. *Arch Ophthalmol* 105:1255, 1987.

311. Missotten L, Boving R, François G, et al: Experimental excimer laser keratomileusis. *Bull Soc Belge Ophthalmol* 220:103, 1987.

312. Aron Rosa DS, Boerner CF, Gross M, et al: Wound healing following excimer laser radial keratotomy. *J Cataract Refract Surg* 14:173, 1988.

313. L'Esperance FA, Taylor DM, Del Pero RA, et al: Human excimer laser corneal surgery. *Trans Am Ophthalmol Soc* 86:208, 1988.

314. Hanna KD, Chastang JC, Pouliquen Y, et al: Excimer laser keratectomy for myopia with rotating-slit delivery system. *Arch Ophthalmol* 106:245, 1988.

315. Marshall J, Trokel SL, Rothery S, et al: Long-term healing of the central cornea after photorefractive keratectomy using an excimer laser. *Ophthalmology* 95:1411, 1988.

316. Munnerlyn CR, Koons SJ, Marshall J: Photorefractive keratectomy: A technique for laser refractive keratectomy. *J Cataract Refract Surg* 14:46, 1988.

317. Keates RH, Bloom RT, Ren Q, et al: Fibronectin on excimer laser and diamond knife incisions. *J Cataract Refract Surg* 15:404, 1989.

318. Hanna KD, Chastang JC, Asfar L, et al: Scanning slit delivery system. *J Cataract Refract Surg* 15:390, 1989.

319. Hanna KD, Pouliquen Y, Waring GO, et al: Corneal stromal wound healing in rabbits after 193-nm excimer laser surface ablation. *Arch Ophthalmol* 107:895, 1989.

320. Tuft SJ, Zabel RW, Marshall J: Corneal repair following keratectomy: A comparison between conventional surgery and laser photoablation. *Invest Ophthalmol Vis Sci* 30:1769, 1989.

321. Goodman GL, Trokel SL, Stark WJ, et al: Corneal healing following laser refractive keratectomy. *Arch Ophthalmol* 107:1799, 1989.

322. McDonald MB, Frantz JM, Klyce SD, et al: One-year refractive results of central photorefractive keratectomy for myopia in the nonhuman primate cornea. *Arch Ophthalmol* 108:40, 1990.

323. Del Pero RA, Gigstad JE, Roberts AD, et al: A refractive and histopathologic study of excimer laser keratectomy in primates. *Am J Ophthalmol* 109:419, 1990.

324. Fantes FE, Hanna KD, Waring GO, et al: Wound healing after excimer laser keratomileusis (photorefractive keratectomy) in monkeys. *Arch Ophthalmol* 108:665, 1990.

325. Hanna KD, Pouliquen YM, Savoldelli M, et al: Corneal wound healing in monkeys 18 months after excimer laser photorefractive keratectomy. *Refract Corneal Surg* 6:340, 1990.

326. Sher NA, Barak M, Daya S, et al: Excimer laser photorefractive keratectomy in high myopia. A multicenter study. *Arch Ophthalmol* 110:935, 1992.

327. Taylor DM, L'Esperance FA, Warner JW, et al: Experimental corneal studies with the excimer laser. *J Cataract Refract Surg* 15:384, 1989.

328. L'Esperance FA, Taylor DM, Warner JW: Human excimer laser keratectomy: Short-term histopathology. *J Refract Surg* 4:118, 1988.

329. McDonald MB, Frantz JM, Klyce SD, et al: Central photorefractive keratectomy for myopia. The blind eye study. *Arch Ophthalmol* 108:799, 1990.

330. Seiler T, Kahle G, Kriegerowski M: Excimer laser (193 nm) myopic keratomileusis in sighted and blind human eyes. *Refract Corneal Surg* 6:165, 1990.

331. Zabel RW, Sher NA, Ostrov CS, et al: Myopic excimer laser keratectomy: A preliminary report. *Refract Corneal Surg* 6:329, 1990.

332. Brancato R, Tavola A, Carones F, et al: Excimer laser photorefractive keratectomy for myopia: Results in 1165 eyes. *Refract Corneal Surg* 9:95, 1993.

333. Holschbach A, Derse M, Seiler T, et al: Preliminary two year results of photorefractive keratectomy. *Invest Ophthalmol Vis Sci* 34(Suppl):798, 1993.

334. Epstein D, Hamberg-Nyström H, Fagerholm P, et al: Three-year follow-up of excimer laser photorefractive keratectomy for myopia. *Invest Ophthalmol Vis Sci* 35(Suppl):1650, 1994.

335. Tengroth B, Epstein D, Fagerholm P, et al: Excimer laser photorefractive keratectomy for myopia. Clinical results in sighted eyes. *Ophthalmology* 100:739, 1993.

336. Gartry DS, Kerr Muir MG, Lohmann CP, et al: The effect of topical corticosteroids on refractive outcome and corneal haze after photorefractive keratectomy. *Arch Ophthalmol* 110:944, 1992.

337. Stevens JD, Steele AD, Ficker LA, et al: Prospective randomized study of two topical steroid regimes after excimer laser PRK. *Invest Ophthalmol Vis Sci* 35(Suppl):1651, 1994.

338. Liu JC, McDonald MB, Varnell R, et al: Myopic excimer laser photorefractive keratectomy: An analysis of clinical correlations. *Refract Corneal Surg* 6:321, 1990.

339. Campos M, Hertzog L, Garbus JJ, et al: Corneal sensitivity after photorefractive keratectomy. *Am J Ophthalmol* 114:51, 1992.

340. Ficker LA, Bates AK, Steele AD: Excimer laser photorefractive keratectomy for myopia: 12 month follow-up. *Eye* 7:617, 1993.

341. Piebenga LW, Matta CS, Deitz MR, et al: Excimer photorefractive keratectomy for myopia. *Ophthalmology* 100:1335, 1993.

342. Seiler T, Holschbach A, Derse M, et al: Complications of myopic photorefractive keratectomy with the excimer laser. *Ophthalmology* 101:153, 1994.

343. Heitzmann J, Binder PS, Kassar BS, et al: The correction of high myopia using the excimer laser. *Arch Ophthalmol* 111:1627, 1993.

344. Förster W: Time-delayed, two-step excimer laser photorefractive keratectomy to correct high myopia. *Refract Corneal Surg* 9:465, 1993.

345. Binder PS, Anderson JA, Lambert RW, et al: Endothelial cell loss associated with excimer laser. *Ophthalmology* 100(9A):107, 1993.

346. Krueger RR, Saedy NF, McDonnell PJ: Clinical analysis of topographic steep central islands following excimer laser photorefractive keratectomy (PRK). *Invest Ophthalmol Vis Sci* 35(Suppl):1740, 1994.

347. Kawesch GM, Maloney RK, Derse M, et al: Contour of the ablation zone after photorefractive keratectomy. *Invest Ophthalmol Vis Sci* 33(Suppl):1105, 1992.

348. Moreira H, Garbus JJ, Fasano A, et al: Multifocal corneal topographic changes with excimer laser photorefractive keratectomy. *Arch Ophthalmol* 110:994, 1992.

349. Seiler T, Bende T, Wollensak J, et al: Excimer laser keratectomy for correction of astigmatism. *Am J Ophthalmol* 105:117, 1988.

350. McDonnell PJ, Moreira H, Garbus J, et al: Photorefractive keratectomy to create toric ablations for correction of astigmatism. *Arch Ophthalmol* 109:710, 1991.

351. Campos M, Hertzog L, Garbus J, et al: Photorefractive keratectomy for severe post-keratoplasty astigmatism. *Am J Ophthalmol* 114:429, 1992.

352. Taylor HR, Guest CS, Kelly P, et al: Comparison of excimer laser treatment of astigmatism and myopia. *Arch Ophthalmol* 111:1621, 1993.

353. Lieurance RC, Patel AC, Wan WL, et al: Excimer laser cut lenticules for epikeratophakia. *Am J Ophthalmol* 103:475, 1987.

354. Serdarevic ON, Hanna K, Gribomont A-C, et al: Excimer laser trephination in penetrating keratoplasty: Morphologic features and wound healing. *Ophthalmology* 95:493, 1988.

355. Lang GK, Schroeder E, Koch JW, et al: Excimer laser keratoplasty part 2: Elliptical keratoplasty. *Ophthalmic Surg* 20:342, 1989.

356. Gabay S, Slomovic A, Jares T: Excimer laser-processed donor corneal lenticules for lamellar keratoplasty. *Am J Ophthalmol* 107:47, 1989.

357. Serdarevic O, Darrell RW, Krueger RR, et al: Excimer laser therapy for experimental *Candida* keratitis. *Am J Ophthalmol* 99:534, 1985.

358. Dausch D, Schröder E: Die behandlung von hornhaut- und skleraerkrankungen mit dem excimerlaser: Ein vorläufiger erfahrungsbericht. *Fortschr Ophthalmol* 87:115, 1990.

359. Steinert RF, Puliafito CA: Excimer laser phototherapeutic keratectomy for a corneal nodule. *Refract Corneal Surg* 6:352, 1990.

360. Sher NA, Bowers RA, Zabel RW, et al: Clinical use of the 193-nm excimer laser in the treatment of corneal scars. *Arch Ophthalmol* 109:491, 1991.

361. Chamon W, Azar DT, Stark WJ, et al: Phototherapeutic keratectomy. *Ophthalmol Clin N Am* 6:399, 1993.

362. Bowers RA, Sher NA, Gothard TW, et al: The clinical use of 193 nm excimer laser in the treatment of corneal scars. *Invest Ophthalmol Vis Sci* 32(Suppl):720, 1991.

363. Kornmehl EW, Steinert RF, Puliafito CA: A comparative study of masking fluids for excimer laser phototherapeutic keratectomy. *Arch Ophthalmol* 109:860, 1991.

364. Klyce SD, Wilson SE, McDonald MB, et al: Corneal topography after excimer laser keratectomy. *Invest Ophthalmol Vis Sci* 32(Suppl):721, 1991.

365. Loertscher H, Mandelbaum S, Parrish RK, et al: Preliminary report on corneal incisions created by a hydrogen fluoride laser. *Am J Ophthalmol* 102:217, 1986.

366. Thompson KP, Barraquer E, Parel J-M, et al: Potential use of lasers for penetrating keratoplasty. *J Cataract Refract Surg* 15:397, 1989.

367. Peyman GA, Badaro RM, Khoobehi B: Corneal ablation in rabbits using an infrared (2.9-μm) erbium:YAG laser. *Ophthalmology* 96:1160, 1989.

368. Tsubota K: Application of erbium:YAG laser in ocular ablation. *Ophthalmologica* 200:117, 1990.

369. Seiler T, Berlin M, Genth U: Fundamental mode photoablation (FMP) using an electro-optically Q-switched Er:YAG laser. *Invest Ophthalmol Vis Sci* 35(Suppl):2017, 1994.

370. Stern D, Puliafito CA, Dobi ET, et al: Infrared laser surgery of the cornea: Studies with a Raman-shifted neodymium:YAG laser at 2.80 and 2.92 μm. *Ophthalmology* 95:1434, 1988.

371. Stern D, Schoenlein RW, Puliafito CA, et al: Corneal ablation by nanosecond, picosecond, and femtosecond lasers at 532 and 625 nm. *Arch Ophthalmol* 107:587, 1989.

372. Gasset AR, Kaufman H: Thermokeratoplasty in the treatment of keratoconus. *Am J Ophthalmol* 79:226, 1975.

373. Seiler T, Matallana M, Bende T: Laser thermokeratoplasty by means of a pulsed holmium:YAG laser for hyperopic correction. *Refract Corneal Surg* 6:335, 1990.

374. Seiler T: Ho:YAG laser thermokeratoplasty for hyperopia. *Ophthalmol Clin N Am* 5:773, 1992.

375. Moreira H, Campos M, Sawusch MR, et al: Holmium laser thermokeratoplasty. *Ophthalmology* 100:752, 1993.

376. Chow DR, Chen JC, Saheb NA, et al: Holmium laser scleroplasty—A new idea in refractive surgery. *Invest Ophthalmol Vis Sci* 35(Suppl):2021, 1994.

377. Simon G, Ren Q: Laser refractive scleroplasty (LRS) for astigmatic correction. *Invest Ophthalmol Vis Sci* 35(Suppl):2021, 1994.

378. Wapner F, Eaton A, Schubert H, et al: Micropulsed diode laser dye-enhanced thermokeratoplasty. *Invest Ophthalmol Vis Sci* 33(Suppl):769, 1992.

379. Korn EL: Use of the carbon dioxide laser for removal of lesions adjacent to the punctum. *Ann Ophthalmol* 22:230, 1990.

380. Korn E: Tarsorrhaphy: A laser-assisted approach. *Ann Ophthalmol* 22:154, 1990.

381. Kennerdell JS, Maroon JC: Use of the carbon dioxide laser in the management of orbital plexiform neurofibromas. *Ophthalmic Surg* 21:138, 1990.

382. Bosniak SL, Novick NL, Sachs ME: Treatment of recurrent squamous papillomata of the conjunctiva by carbon dioxide laser vaporization. *Ophthalmology* 93:1078, 1986.

383. Kennerdell JS, Maroon JC, Garrity JA, et al: Surgical management of orbital lymphangioma with the carbon dioxide laser. *Am J Ophthalmol* 102:308, 1986.

384. Korn EL, Glotzbach RK: Carbon dioxide laser repair of medial ectropion. *Ophthalmic Surg* 19:653, 1988.

385. Gonnering RS, Lyon DB, Fisher JC: Endoscopic laser-assisted lacrimal surgery. *Am J Ophthalmol* 111:152, 1991.

386. Mittelman H, Apfelberg DB: Carbon dioxide laser blepharoplasty—Advantages and disadvantages. *Ann Plast Surg* 24:1, 1990.

387. Woog JJ, Metson R, Puliafito CA: Holmium:YAG endonasal laser dacryocystorhinostomy. *Am J Ophthalmol* 116:1, 1993.

388. Troutman RC, Véronneau-Troutman S, Jakobiec FA, et al: A new laser for colla-

gen wounding in corneal and strabismus surgery: A preliminary report. *Trans Am Ophthalmol Soc* 84:117, 1986.

389. Nanevicz TM, Prince MR, Gawande AA, et al: Excimer laser ablation of the lens. *Arch Ophthalmol* 104:1825, 1986.

390. Bath PE, Mueller G, Apple DJ, et al: Excimer laser lens ablation. *Arch Ophthalmol* 105:1164, 1987.

391. Keates RH, Bloom RT, Schneider RT, et al: Absorption of 308-nm excimer laser radiation by balanced salt solution, sodium hyaluronate, and human cadaver eyes. *Arch Ophthalmol* 108:1611, 1990.

392. L'Esperance FA, Mittl RN, James WA: Carbon dioxide laser trabeculostomy for the treatment of neovascular glaucoma. *Ophthalmology* 90:821, 1983.

393. Jaffe GJ, Williams GA, Mieler WF, et al: Ab interno sclerostomy with a high-powered argon endolaser. *Am J Ophthalmol* 106:391, 1988.

394. Berlin MS, Rajacich G, Duffy M, et al: Excimer laser photoablation in glaucoma filtering surgery. *Am J Ophthalmol* 103:713, 1987.

395. Seiler T, Kriegerowski M, Bende T, et al: Partial trabeculectomy with the excimer laser (193 nm). *Invest Ophthalmol Vis Sci* 29(Suppl):239, 1988.

396. Allan BDS, van Saarloos PP, Russo AV, et al: Excimer laser sclerostomy: The in vitro development of a modified open mask delivery system. *Eye* 7:47, 1993.

397. Margolis TI, Farnath DA, Puliafito CA: Mid infrared laser sclerostomy. *Invest Ophthalmol Vis Sci* 29(Suppl):366, 1988.

398. Eaton AM, Odrich SA, Schubert HD, et al: Holmium and thulium laser sclerostomy: A comparative study. *Invest Ophthalmol Vis Sci* 32(Suppl):860, 1991.

399. Iwach AG, Hoskins HD, Drake MV, et al: Update of the subconjunctival THC:YAG (holmium) laser sclerostomy ab externo clinical trial: 30-month report. *Ophthalmic Surg* 25:13, 1994.

400. McAllister JA, Watts PO: Holmium laser sclerostomy: A clinical study. *Eye* 7:656, 1993.

401. Namazi N, Schuman JS, Wang N, et al: Acute and long-term effects of THC:YAG sclerostomy with adjunctive antimetabolite therapy in rabbits. *Invest Ophthalmol Vis Sci* 33(Suppl):1266, 1992.

402. Engel JM, Blair NP, Harris D, et al: Use of the carbon dioxide laser in the drainage of subretinal fluid. *Arch Ophthalmol* 107:731, 1989.

403. Pellin MJ, Williams GA, Young CE, et al: Endoexcimer laser intraocular ablative photodecomposition. *Am J Ophthalmol* 99:483, 1985.

404. Margolis TI, Farnath DA, Destro M: Erbium-YAG laser surgery on experimental vitreous membranes. *Arch Ophthalmol* 107:424, 1989.

405. Lin CP, Stern D, Puliafito CA: High-speed photography of Er:YAG laser ablation in fluid. *Invest Ophthalmol Vis Sci* 31:2546, 1990.

406. Borirakchanyavat S, Puliafito CA, Kliman GH, et al: Holmium-YAG laser surgery on experimental vitreous membranes. *Arch Ophthalmol* 109:1605, 1991.

407. Huie TY, Chang CJ, Tso MOM: Localized surgical debridement of RPE by Q-switched neodymium:YAG laser. *Invest Ophthalmol Vis Sci* 34(Suppl):959, 1993.

408. Chong LP, Kohen L: A retinal laser which damages only the RPE: Ultrastructural study. *Invest Ophthalmol Vis Sci* 34(Suppl):960, 1993.

409. Roider J, Michaud N, Flotte T, et al: Selective RPE photocoagulation by 1 μsec laser pulses. *Invest Ophthalmol Vis Sci* 34(Suppl):960, 1993.

410. Lewis A, Palanker D, Hemo I, et al: Microsurgery of the retina with a needle-guided 193-nm excimer laser. *Invest Ophthalmol Vis Sci* 33:2377, 1992.

411. Webb RH, Hughes GW, Pomerantzeff O: Flying spot TV ophthalmoscope. *Appl Optics* 19:2991, 1980.

412. Mainster MA, Timberlake GT, Webb RH, et al: Scanning laser ophthalmoscopy: Clinical applications. *Ophthalmology* 89:852, 1982.

413. Dreher AW, Weinreb RN: Accuracy of topographic measurements in a model eye with the laser tomographic scanner. *Invest Ophthalmol Vis Sci* 32:2992, 1991.

414. Weinreb RN, Dreher AW, Bille JF: Quantitative assessment of the optic nerve head with the laser tomographic scanner. *Int Ophthalmol* 13:25, 1989.

415. Bartsch D-U, Intaglietta M, Bille JF, et al: Confocal laser tomographic analysis of the retina in eyes with macular hole formation and other focal macular diseases. *Am J Ophthalmol* 108:277, 1989.

416. Sjaarda RN, Frank DA, Glaser BM, et al: Assessment of vision in idiopathic macular holes with macular microperimetry using the scanning laser ophthalmoscope. *Ophthalmology* 100:1513, 1993.

417. Woon WH, Fitzke FW, Bird AC, et al: Confocal imaging of the fundus using a scanning laser ophthalmoscope. *Br J Ophthalmol* 76:470, 1992.

418. Weinreb RN, Dreher AW, Coleman A, et al: Histopathologic validation of Fourier-ellipsometry measurements of retinal nerve fiber layer thickness. *Arch Ophthalmol* 108:557, 1990.

419. Dreher AW, Bille JF, Weinreb RN: Active optical depth resolution improvement of the laser tomographic scanner. *Appl Optics* 28:804, 1989.

420. Cioffi GA, Robin AL, Eastman RD, et al: Confocal laser scanning ophthalmoscope: Reproducibility of optic nerve head topographic measurements with the confocal laser scanning ophthalmoscope. *Ophthalmology* 100:57, 1993.

421. Katsumi O, Timberlake GT, Hirose T, et al: Recording pattern reversal visual evoked response with the scanning laser ophthalmoscope. *Acta Ophthalmol* 67:243, 1989.

422. Timberlake GT, Van de Velde FJ, Jalkh AE: Clinical use of scanning laser ophthalmoscope retinal function maps in macular disease. *Lasers Light Ophthalmol* 2:211, 1989.

423. Wolf S, Arend O, Sponsel WE, et al: Retinal hemodynamics using scanning laser ophthalmoscopy and hemorheology in chronic open-angle glaucoma. *Ophthalmology* 100:1561, 1993.

424. Frambach DA, Dacey MP, Sadun A: Stereoscopic photography with a scanning laser ophthalmoscope. *Am J Ophthalmol* 116:484, 1993.

425. Scheider A, Kaboth A, Neuhauser L: Detection of subretinal neovascular membranes with indocyanine green and an infrared scanning laser ophthalmoscope. *Am J Ophthalmol* 113:45, 1992.

426. Kuck H, Inhoffen W, Schneider U, et al: Diagnosis of occult subretinal neovascularization in age-related macular degeneration by infrared scanning laser videoangiography. *Retina* 13:36, 1993.

427. Kiryu J, Ogura Y, Shahidi M, et al: Enhanced visualization of vitreoretinal interface by laser biomicroscopy. *Ophthalmology* 100:1040, 1993.

428. Sawa M, Sakanishi Y, Shimizu H: Fluorophotometric study of anterior segment barrier functions after extracapsular cataract extraction and posterior intraocular lens implantation. *Am J Ophthalmol* 97:197, 1984.

429. Shah SM, Spalton DJ, Allen RJ, et al: A comparison of the laser flare cell meter and fluorophotometry in assessment of the blood-aqueous barrier. *Invest Ophthalmol Vis Sci* 34:3124, 1993.

430. Sawa M, Tsurimaki Y, Tsuru T, et al: New quantitative method to determine protein concentration and cell number in aqueous in vivo. *Jpn J Ophthalmol* 32:132, 1988.

431. Shah SM, Spalton DJ, Smith SE: Measurement of aqueous cells and flare in normal eyes. *Br J Ophthalmol* 75:348, 1991.

432. Sawa M: Clinical application of laser flare-cell meter. *Jpn J Ophthalmol* 34:346, 1990.

433. Oshika T, Araie M, Masuda K: Diurnal variation of aqueous flare in normal human eyes measured with laser flare-cell meter. *Jpn J Ophthalmol* 32:143, 1988.

434. Oshika T, Kato S, Sawa M, et al: Aqueous flare intensity and age. *Jpn J Ophthalmol* 33:237, 1989.

435. Guex-Crosier Y, Pittet N, Herbort CP: Evaluation of laser flare-cell photometry in the appraisal and management of intraocular inflammation in uveitis. *Ophthalmology* 101:728, 1994.

436. Oshika T: Aqueous protein concentration in rhegmatogenous retinal detachment. *Jpn J Ophthalmol* 34:63, 1990.

437. Oshika T, Yoshimura K, Miyata N: Postsurgical inflammation after phacoemulsification and extracapsular extraction with soft or conventional intraocular lens implantation. *J Cataract Refract Surg* 18:356, 1992.

438. Altamirano D, Mermoud A, Pittet N, et al: Aqueous humor analysis after Nd:YAG laser capsulotomy with the laser flare-cell meter. *J Cataract Refract Surg* 18:554, 1992.

439. Mermoud A, Pittet N, Herbort CP: Inflammation patterns after laser trabeculoplasty measured with the laser flare meter. *Arch Ophthalmol* 110:368, 1992.

440. Nussenblatt RB, de Smet M, Podgor M, et al: The use of flarephotometry in the detection of cytomegalic virus retinitis in AIDS patients. *AIDS* 8:135, 1994.

441. Yoshitomi T, Wong AS, Daher E, et al: Aqueous flare measurement with laser flare-cell meter. *Jpn J Ophthalmol* 34:57, 1990.

442. Green DG: Testing the vision of cataract patients by means of laser-generated interference fringes. *Science* 168:1240, 1970.

443. Faulkner W: Laser interferometric prediction of postoperative visual acuity in patients with cataracts. *Am J Ophthalmol* 95:626, 1983.

444. Selenow A, Ciuffreda KJ, Mozlin R, et al: Prognostic value of laser interferometric visual acuity in amblyopia therapy. *Invest Ophthalmol Vis Sci* 27:273, 1986.

445. Palestine AG, Alter GJ, Chan CC, et al: Laser interferometry and visual prognosis in uveitis. *Ophthalmology* 92:1567, 1985.

446. Spurny RC, Zaldivar R, Belcher CD, et al: Instruments for predicting visual acuity. A clinical comparison. *Arch Ophthalmol* 104:196, 1986.

447. Fish GE, Birch DG, Fuller DG, et al: A comparison of visual function tests in eyes with maculopathy. *Ophthalmology* 93:1177, 1986.

448. Klett Z, Morris M, Gieser SC, et al: Assessment of contrast sensitivity. Part II: The relationship between objective lens opacity and laser interferometric contrast sensitivity in the cataract patient. *J Cataract Refract Surg* 17:45, 1991.

449. Hitzenberger CK: Optical measurement of the axial eye length by laser Doppler interferometry. *Invest Ophthalmol Vis Sci* 32:616, 1991.

450. Hitzenberger CK, Drexler W, Fercher AF: Measurement of corneal thickness by laser Doppler interferometry. *Invest Ophthalmol Vis Sci* 33:98, 1992.

451. Huang D, Wang J, Lin CP, et al: Micron-resolution ranging of cornea anterior chamber by optical reflectometry. *Lasers Surg Med* 11:419, 1991.

452. Huang D, Swanson EA, Lin CP, et al: Optical coherence tomography. *Science* 254:1178, 1991.

453. Swanson EA, Huang D, Hee MR, et al: High-speed optical coherence domain reflectometry. *Opt Lett* 17:151, 1992.

454. Izatt JA, Hee MR, Huang D, et al: High speed in vivo retinal imaging with optical coherence tomography. *Invest Ophthalmol Vis Sci* 35(Suppl):1729, 1994.

455. Prydal JI, Campbell FW: Study of precorneal tear film thickness and structure by interferometry and confocal microscopy. *Invest Ophthalmol Vis Sci* 33:1996, 1992.

456. Hitzenberger CK, Drexler W, Dolezal C, et al: Measurement of the axial length of cataract eyes by laser Doppler interferometry. *Invest Ophthalmol Vis Sci* 34:1886, 1993.

457. Riva CE, Ross B, Benedek GB: Laser Doppler measurements of blood flow in capillary tubes and retinal arteries. *Invest Ophthalmol* 11:936, 1972.

458. Riva CE, Feke GT, Eberli B, et al: Bidirectional LDV system for absolute measurement of retinal blood speed. *Appl Optics* 18:2302, 1979.

459. Milbocker MT, Feke GT: Intensified charge-coupled-device-based eyetracker and image stabilizer. *Appl Optics* 31:3719, 1992.

460. Grunwald JE: Effect of topical timolol on the human retinal circulation. *Invest Ophthalmol Vis Sci* 27:1713, 1986.

461. Sullivan PM, Davies GE, Caldwell G, et al: Retinal blood flow during hyperglycemia. A laser Doppler velocimetry study. *Invest Ophthalmol Vis Sci* 31:2041, 1990.

462. Riva CE, Petrig BL, Grunwald JE: Near infrared retinal laser Doppler velocimetry. *Lasers Ophthalmol* 1:211, 1987.

463. Grunwald JE, Riva CE, Martin DB, et al: Effect of an insulin induced decrease in blood glucose on the human diabetic retinal circulation. *Ophthalmology* 94:1614, 1987.

464. Sebag J, Delori FC, Feke GT, et al: Effects of optic atrophy on retinal blood flow and oxygen saturation in humans. *Arch Ophthalmol* 107:222, 1989.

465. Grunwald JE, Brucker AJ, Grunwald SE, et al: Retinal hemodynamics in proliferative diabetic retinopathy: A laser Doppler velocimetry study. *Invest Ophthalmol Vis Sci* 34:66, 1993.

466. Pratesi R, Brancato R, Trabucchi G: Miniature laser for retinal photocoagulation: The self-doubling 532 nm neodymium-yttrium aluminum borate (NYAB) microlaser. *Invest Ophthalmol Vis Sci* 33(Suppl):1317, 1992.

467. Inderfurth JHC, Ferguson RD, Puliafito CA, et al: Reflectance monitoring during retinal photocoagulation in humans—steps toward the development of an automated feedback-controlled photocoagulator. *Invest Ophthalmol Vis Sci* 35(Suppl):1374, 1994.

468. Seiler T, Genth U, Holschbach A, et al: Aspheric photorefractive keratectomy with excimer laser. *Refract Corneal Surg* 9:166, 1993.

469. Solomon KD, Chamon W, Green WR, et al: Superficial concentric excimer laser keratectomy in monkey eyes: A histopathologic analysis. *Invest Ophthalmol Vis Sci* 34(Suppl):797, 1993.

470. Maloney RK, Friedman M, Harmon T, et al: A prototype erodible mask delivery system for the excimer laser. *Ophthalmology* 100:542, 1993.

471. Talamo JH, Gollamudi S, Green WR, et al: Modulation of corneal wound healing after excimer laser keratomileusis using topical mitomycin C and steroids. *Arch Ophthalmol* 109:1141, 1991.

472. Liu JC, Steinemann TL, McDonald MB, et al: Effects of corticosteroids and mitomycin C on corneal remodeling after excimer laser photorefractive keratectomy. *Invest Ophthalmol Vis Sci* 32(Suppl):1248, 1991.

473. Lohmann CP, O'Brart D, Patmore A, et al: Plasmin in the tear fluid: A new therapeutic concept to reduce postoperative myopic regression and corneal haze after excimer laser photorefractive keratectomy. *Lasers Light Ophthalmol* 5:205, 1993.

474. Morlet N, Gillies MC, Crouch R, et al: Effect of topical interferon-alpha 2b on corneal haze after excimer laser photorefractive keratectomy in rabbits. *Refract Corneal Surg* 9:443, 1993.

475. Jain S, Hahn TW, Chen W, et al: Modulation of corneal wound healing following 193-nm excimer laser keratectomy using free radical scavengers. *Invest Ophthalmol Vis Sci* 35(Suppl):2015, 1994.

476. Lohmann CP, MacRobert I, Patmore A, et al: A histopathological study of surgical human specimen of tissue responsible for haze and regression after excimer laser photorefractive keratectomy. *Invest Ophthalmol Vis Sci* 35(Suppl):1723, 1994.

477. Sher NA, Frantz JM, Talley A, et al: Topical diclofenac in the treatment of ocular pain after excimer photorefractive keratectomy. *Refract Corneal Surg* 9:425, 1993.

478. Derse M, Seiler T: Repeated excimer laser treatment after photorefractive keratectomy. *Invest Ophthalmol Vis Sci* 33(Suppl):761, 1992.

479. Tengroth B, Fagerholm P, Epstein D, et al: Retreatment of regression after photorefractive keratectomy for myopia. *Invest Ophthalmol Vis Sci* 35(Suppl):1724, 1994.

480. Seiler T, Jean B: Photorefractive keratectomy as a second attempt to correct myopia after radial keratotomy. *Refract Corneal Surg* 8:211, 1992.

481. Thompson KP, Hanna K, Waring GO: Emerging technologies for refractive surgery: Laser adjustable synthetic epikeratoplasty. *Refract Corneal Surg* 5:46, 1989.

482. Pallikaris IG, Papatzanaki ME, Stathi EZ, et al: Laser in situ keratomileusis. *Lasers Surg Med* 10:463, 1990.

483. Shimmick JK, Koons SJ, Telfair WB, et al: Analysis of an offset slit excimer laser delivery system for the treatment of hyperopia. *Invest Ophthalmol Vis Sci* 35(Suppl):1487, 1994.

484. McDonald MB, Telfair WB, Nesburn AB, et al: Excimer laser hyperopia PRK phase I: The blind eye study. *Invest Ophthalmol Vis Sci* 35(Suppl):1488, 1994.

485. Hogan C, McDonald M, Byrd T, et al: Effect of excimer laser photorefractive keratectomy on contrast sensitivity. *Invest Ophthalmol Vis Sci* 32(Suppl):721, 1991.

486. Gobbi PG, Carena M, Fortini A, et al: Automatic eye tracker for excimer laser photorefractive keratectomy. *Invest Ophthalmol Vis Sci* 35(Suppl):2017, 1994.

487. Stern D, Lin W-Z, Puliafito CA, et al: Femtosecond optical ranging of corneal incision depth. *Invest Ophthalmol Vis Sci* 30:99, 1989.

488. Ren Q, Gailitis RP, Thompson K, et al: Corneal refractive surgery using an ultraviolet (213 nm) solid state laser. *Proc Ophthalmic Technologies*, SPIE 1423:129, 1991.

489. Gailitis RP, Ren Q, Thompson KP, et al: Solid state UV laser ablation (213 nm) of the cornea and synthetic epikeratoplasty material. *Invest Ophthalmol Vis Sci* 32(Suppl):996, 1991.

490. Manns F, Ren Q, Parel J-M, et al: Investigation of an algorithm for photo-refractive keratectomy (PRK) using a scanning beam delivery system. *Invest Ophthalmol Vis Sci* 34(Suppl):800, 1993.

491. Ren Q, Simon G, Parel J-M, et al: Ultraviolet solid-state laser (213-nm) photorefractive keratectomy. In vitro study. *Ophthalmology* 100:1828, 1993.

492. Ren Q, Simon G, Legeais J-M, et al: Ultraviolet solid-state laser (213-nm) photorefractive keratectomy. In vivo study. *Ophthalmology* 101:883, 1994.

493. Lin JT: A multiwavelength solid state laser for ophthalmic applications. *Proc Ophthalmic Technologies* II, SPIE 1644:266, 1992.

494. Feld JR, Lin CP, Woods WJ, et al: Cornea ablation studies at wavelengths between 205 and 225 nm using a tunable solid state laser. *Invest Ophthalmol Vis Sci* 33(Suppl):1105, 1992.

495. Bende T, Jean B, Matallana M, et al: Photoablation with the free electron laser between 2.8 and 6.2 microns wavelength. *Invest Ophthalmol Vis Sci* 34(Suppl):1246, 1993.

496. Logan RA, O'Day DM, Haglund RF, et al: Preliminary observations on the effects of the free electron laser on corneal tissue. *Invest Ophthalmol Vis Sci* 34(Suppl):1246, 1993.

497. Fowler WC, Wehrly SR, Imami NR, et al: Keratorefractive change induced by circumferential infrared wavelength ablations utilizing the free electron laser in porcine corneas. *Invest Ophthalmol Vis Sci* 35(Suppl):2027, 1994.

498. Zysset B, Fujimoto JG, Puliafito CA, et al: Picosecond optical breakdown: Tissue effects and reduction of collateral damage. *Lasers Surg Med* 9:193, 1989.

499. Rowsey JJ, Bowyer BL, Margo CE, et al: Intrastromal ablation of corneal tissue using a frequency doubled Nd:YAG laser. *Invest Ophthalmol Vis Sci* 34(Suppl):1247, 1993.

500. Rowsey JJ, Bowyer BL, Johnson DE, et al: Phoenix laser: Refractive changes in the cat corneal model. *Invest Ophthalmol Vis Sci* 35(Suppl):2027, 1994.

501. Bado P, Bouvier M, Coe JS: Nd:YLF mode-locked oscillator and regenerative amplifier. *Opt Lett* 12:319, 1987.

502. Frueh BE, Bille JF, Brown SI: Intrastromal relaxing incisions with a picosecond infrared laser. *Lasers Light Ophthalmol* 4:165, 1992.

503. Remmel RM, Dardenne CM, Bille JF: Intrastromal tissue removal using an infrared picosecond Nd:YLF ophthalmic laser operating at 1053 nm. *Lasers Light Ophthalmol* 4:169, 1992.

504. Niemz MH, Hoppeler TP, Juhasz T, et al: Intrastromal ablation for refractive corneal surgery using picosecond infrared laser pulses. *Lasers Light Ophthalmol* 5:149, 1993.

505. Nissen M, Speaker MG, Davidian ME, et al: Acute effects of intrastromal ablation with the Nd:YLF picosecond laser on the endothelium of rabbit eyes. *Invest Ophthalmol Vis Sci* 34(Suppl):1246, 1993.

506. Brown DB, O'Brien WJ, Schultz RO: Nd:YLF picosecond laser capabilities and ultrastructure effects in corneal ablations. *Invest Ophthalmol Vis Sci* 34(Suppl):1246, 1993.

507. Cooper HM, Schuman JS, Puliafito CA, et al: Picosecond neodymium:yttrium lithium fluoride laser sclerectomy. *Am J Ophthalmol* 115:221, 1993.

508. Park SB, Kim JC, Aquavella JV: Nd:YLF laser sclerostomy. *Ophthalmic Surg* 24:118, 1993.

509. Frangie JP, Park SB, Aquavella JV: Peripheral iridotomy using Nd:YLF laser. *Ophthalmic Surg* 23:220, 1992.

510. Taboada J, Poirier RH, Yee RW, et al: Intrastromal photorefractive keratectomy with a new optically coupled laser probe. *Refract Corneal Surg* 8:399, 1992.

511. Zelman J: Photophaco fragmentation. *J Cataract Refract Surg* 13:287, 1987.

512. Chambless WS: Neodymium:YAG laser phacofracture: An aid to phacoemulsification. *J Cataract Refract Surg* 14:180, 1988.

513. Maguen E, Martinez M, Grundfest W, et al: Excimer laser ablation of the human lens at 308 nm with a fiber delivery system. *J Cataract Refract Surg* 15:409, 1989.

514. Gailitis RP, Patterson SW, Samuels MA, et al: Comparison of laser phacovaporization using the Er-YAG and the Er-YSGG laser. *Arch Ophthalmol* 111:697, 1993.

515. Haefliger E, Parel J-M, Fantes F, et al: Accommodation of an endocapsular silicone lens (phaco-ersatz) in the nonhuman primate. *Ophthalmology* 94:471, 1987.

516. Netter FH: *Atlas of Human Anatomy.* Summit, NJ, Ciba-Geigy, 1989.

517. Warwick R: *Wolffs Anatomy of the Eye and Orbit*, 7th ed., WB Saunders, Philadelphia, 1976.

518. L'Esperance FA: *Ophthalmic Lasers*, 3rd ed. CV Mosby, St. Louis, 1989.

CHAPTER SEVEN

Advanced Technologies in Coronary Angioplasty: Lasers and Stents

ALEXANDER A. STRATIENKO, M.D., *Chattanooga Heart Institute, Chattanooga, TN*
WILLIAM P. SANTAMORE, Ph.D., *Thoracic and Cardiovascular Surgery Division, University of Louisville, Louisville, KY*

7.1 INTRODUCTION

Heart disease remains the primary cause of mortality and morbidity in the United States. The major cause of heart disease is atherosclerotic obstruction in the large arteries (coronary arteries) supplying blood to the heart. The heart is continuously working to pump blood through the body; as such, it has high energy and oxygen demands. Restrictions in this blood supply can cause chest pain (angina pectoris), heart attacks (myocardial infarctions), and electrical instability (ventricular arrhythmias). These small obstructions in the coronary arteries are the cause of the restrictions in blood flow to the heart muscle.

Accordingly, considerable effort has been directed to treat these obstructions or coronary artery disease. Changes in lifestyle (e.g., smoking, exercise) and medical therapy help to retard the development of these obstructions and to treat the symptoms. However, once the obstructions start to affect blood flow to the heart under resting or low-stress conditions, the primary mode of therapy is mechanical. This may involve placing a vessel between the main outflow vessel

New Frontiers in Medical Device Technology, Edited by Arye Rosen and Harel Rosen
ISBN 0-471-59189-0

from the heart (aorta) and the obstruction in the coronary artery, thus bypassing the obstruction. Coronary artery bypass surgery is effective in treating severe coronary disease, but this is a complicated procedure. The procedure requires the chest to be opened and the heart mechanically supported during the surgery. Further, obstructions can develop in vessels used to bypass the originally diseased vessels.

In 1979, Grüentzig[1] introduced a simpler procedure to treat coronary artery disease, called *angioplasty*. Angioplasty, in its original form, involved placement of a small balloon catheter within the obstruction in the coronary artery. Inflation of the balloon results in an expansion of the obstructed coronary artery. This expansion enabled normal levels of blood flow to pass through the coronary artery, thereby treating the disease.

Because of its effectiveness and relative ease of use, balloon angioplasty has grown rapidly. About 10,000 patients were treated by balloon angioplasty in 1980, a figure estimated at about 250,000 patients in 1988 and 500,000 in 1992.[2]

7.1.1 Limitations of Balloon Angioplasty

Despite the efficacy and safety of percutaneous transluminal coronary angioplasty (PTCA) as an intracoronary intervention, PTCA may have limited ability to treat diffuse disease, calcified lesions, and complete occlusions. Restenosis associated with PTCA is considered related to the magnitude of vascular injury induced by the intervention and to the residual obstruction present at the procedure's conclusion. About 30–40% of patients whose initial PTCA was successful develop restenosis within 6 months. This high restenosis rate has been called the "Achilles heel" of coronary angioplasty. Furthermore, in approximately 7% of cases of balloon angioplasty, acute occlusion of the vessel occurs. This results from arterial dissection, thrombus formation, or a combination of the two. In approximately half of these cases of acute occlusion, establishing good coronary blood flow is not possible with catheter techniques and therefore, emergency coronary artery bypass graft surgery is necessary.

The hope of reducing the restenosis rate and the incidence of abrupt vessel closure has been the driving forces behind experimentation with other devices such as the laser and intravascular stents. The laser improves luminal diameter by vaporizing and ablating atheromatous tissue. Ideally, by removing atheromatous tissue, laser treatment may reduce the frequency of restenosis. The stent, by nature, enlarges the vessel and prevents elastic recoil, thus decreasing the potential for clinically significant restenosis.

7.2 LASER SYSTEM

The basic integrated laser angioplasty system consists of the laser generator, fiberoptic laser delivery catheter, and a means for guiding the catheter to its target site. Various lasers are used. Our institution is currently using an excited dimer or excimer laser, which uses xenon chloride excitation to produce the laser light (Advanced Interventional Systems, Irvine, CA). The short UV wavelength (308 nm) has much higher energy than that generated by other medical lasers, such as the argon, Nd:YAG, and CO_2 types. In addition, the short wavelength has less tissue penetration. These characteristics make the excimer laser especially attractive for vascular applications. Limited tissue penetration reduces diffusion of heat energy through tissue, reducing thermal injury. The high energy per pulse permits rapid vaporization and destruction of material that might not be vaporized by lower-energy lasers.

The engineering of the laser energy source is not unique. However, ablating obstructions in the small, tortuous coronary arteries requires a highly refined fiberoptic catheter capable of enduring several bends within a guiding catheter, even before it reaches the coronary system. Thus, the main engineering problem in the clinical use of lasers is how to deliver this laser energy into the coronary arteries. Currently, laser delivery catheters used clinically incorporate multiple silica fibers. Each fiber is composed of a central core that transmits the laser energy. However, silica fiber has limitations; for instance, it can transmit wavelengths ranging only from 300 to 2100 nm. Moreover, the capacity of the fiber to transmit laser energy varies with the fiber diameter. Large fiber transmits more laser energy but is stiffer and less flexible; small-diameter fiber transmits less energy but is more flexible. Thus, to manipulate a fiberoptic catheter in the coronary artery, fibers having a diameter $\leq 200\ \mu m$ are used; some power transmission is compromised in order to achieve the flexibility and maneuverability to torque.

Positioning the laser catheter inside the coronary artery is another problem. Presently, the most practical way to guide the laser catheter into the diseased target site is with the aid of fluoroscopic x-ray imaging with the injection of contrast media. This provides an overview of the vascular tree and shows the location of obstructive lesions. The application of digital imaging and "roadmapping" (computer subtracted enhanced pathways superimposed over live fluoroscopic images) has provided easier and more precise catheter placement, especially in the peripheral arteries. Typically, the laser catheter has a hollow core (lumen) and "tracks" over a flexible guidewire (0.014 in. in diameter) used to lead the laser catheter to the target site and through the diseased obstruction.

7.2.1 Laser–Tissue Interaction

Laser energy is carried by silica fiber and is directed coaxially along the central axis in the atherosclerotic arteries. The light vaporizes plaque adjacent to the stenotic lumen and the channel diameter is widened.[2] In completely obstructed arteries, the beam can be directed to clear a new passageway through the plaque obstruction. When laser energy is absorbed by atherosclerotic plaque tissue, the light energy is transformed into thermal energy, and the area of solid plaque becomes a vaporized crater. The crater depth depends on the physical properties of the laser beam; a deeper crater results from higher-power intensity, longer-duration exposure, and greater focusing of the beam. Lasing converts solids and liquids to gaseous products, including water, carbon dioxide, nitrogen, hydrogen, and light hydrocarbons.[3] Rapid formation of these gases can create injury to vascular structures. The vaporized crater is larger in proportion to the length of the laser exposure and diameter of lasing catheter. Adjacent to and beyond the new lumen may be an area of acoustical injury, where cells and noncellular material may have been disrupted.

7.2.2 Risks and Complications

The exciting potential of lasers to vaporize atherosclerotic deposits is tempered by the observation in early tissue experiments and live-animal studies of certain dangers inherent to this procedure.[4] With any current laser or laser delivery system, there is always at least some risk of vascular perforation. Use of high energy for long duration to vaporize plaque tends to increase this risk. Perforations in the coronary artery wall can lead to cardiac tamponade, hemodynamic and electrical instability, and patient death. Delivery of "pulsed" energy, as with the excimer laser, reduces the risk of perforation.

Thrombogenic and embolic complications are another potential problem. The danger of distal embolization of clot and debris is of concern. Clinical experience suggests that calcified lesions do not respond well to lasing.

Furthermore, the transmission of laser energy through blood exposes the blood to intense heat with the possibility of more gas formation. Accordingly, using a "bloodless field" is now being studied, where laser energy is delivered during the injection of saline or contrast medium around the laser catheter.

7.2.3 Laser Ablation of Obstructed Coronary Arteries

Lasers were initially used in patients with coronary disease during bypass surgery. Choy and associates[5] utilized a fiberoptic catheter to deliver direct continuous-wave (CW) argon energy to the atherosclerotic lesion proximal to the native coronary bypass graft site in eight patients. Recanalization was achieved, but one patient had mechanical or laser perforation and another had severe thermal damage to the vessel wall. By 1 month, all except one of the lased native vessels had occluded. It is possible that the high energy or the sharp optical fiber injured the intraluminal wall, leading to restenosis and occlusion. Livesay and colleagues[6] performed intraoperative laser endarterectomy on six coronary arteries in three patients. At restudy, five of the six arteries and all bypass grafts remained patent.

The excimer laser has been evaluated for PTCA. Litvack and coworkers[8] used a 5F (size 5 French) catheter comprising twelve 200-μm silica fibers concentrically arranged around a guidewire channel. The excimer was used to treat 53 lesions (44 stenoses and 9 occlusions) in 41 patients (35 native arteries and 6 vein grafts). The laser improved coronary luminal diameter in 39 of the 53 lesions (74%) in 31 of 41 patients (76%). Mean percent stenosis improved from 89 to 50% after lasing. The mean lumen created was approximately 1.6 mm; except for six patients, all had balloon dilation following laser treatment. Complications included one occlusion requiring emergent bypass surgery, two dissections, four coronary spasm cases (10%), and one myocardial infarction. No vascular perforations were reported.

Karsch and associates[9] employed an excimer laser using a different, smaller-diameter 4F catheter comprising 20 silica fibers. The procedure was performed successfully in 25 of 30 patients (83%); mean stenosis was reduced from 96% pretreatment to 64% after laser ablation.

Despite favorable initial results, late restenosis continues to be a nagging problem. Recent clinical studies indicate that the combined use of laser and standard balloon angioplasty might increase the rate of restenosis,[10] but that larger laser catheters followed by balloon angioplasty result in a greater opening of the vessel (which may possibly reduce restenosis rates).[11] The authors believe that the phase change from a solid to a gaseous material results in a tremendous volume expansion, which occurs almost instantaneously. This rapid expansion leads to extensive tissue damage. Additionally, the postprocedure lumen diameter obtained after lasing and adjunctive balloon angioplasty (which is required in approximately 90% of laser applications) is not significantly larger than the postprocedure lumen obtained by balloon angioplasty alone. Nonetheless, laser angioplasty appears to be a clinically safe procedure, although indentification of lesions and/or patient subsets that demonstrate a restenosis benefit remains elusive.

7.3 STENTS

The concept of an intravascular stent was introduced by Dotter and Judkins[12] in 1964 when they described a silastic tube delivered percutaneously to maintain vessel patency following angioplasty. Dotter later tested such a device in canine femoral and popliteal arteries, but lacking expansive ability, the tubes migrated. He further tested stainless-steel wire coil but encountered acute thrombosis and restenosis.[13] Nonetheless, his work underscored the importance of fixed deployment of a prosthesis and the importance of prompt endothelialization of stents for maintenance of vascular patency.

The potential of Dotter's early work with stents was not appreciated until the development of coronary angioplasty as a clinically useful technique in the early 1980s and the emergence of abrupt vessel closure and restenosis as major limitations of this technique. With work in peripheral arteries[14–18] providing the technical and scientific background, the concepts of peripheral stenting were applied to coronary arteries.

From an engineering point of view, several issues of stent design need to be considered. The stent has to be expandable after positioning in the coronary artery. A system to deploy the stent in the coronary artery must be developed.

Stent thromboresistance is a concern. Finally, in these days of financial concern, cost-effectiveness must be considered.

7.3.1 Stent Designs

Coronary stents can be grouped into three broad categories based on design and method of deployment: (1) stents that are delivered on a conventional balloon catheter to the arterial segment and expanded by balloon inflation, thus deforming stent material beyond its elastic limits (balloon-expandable stents); (2) spring models that are delivered in a constrained form and, because of inherent spring tension, are permitted to expand in the arterial segment when positioned and released from the delivery device (self-expanding stents); and (3) stents that exploit the peculiar metallurgic properties of nitinol (a nickel–titanium alloy) that can be shaped for deployment while cool and then revert to a predetermined dimension on rewarming to body temperature (thermal-memory stents).

7.3.1.1 Balloon-Expandable Designs

There are currently many different designs of balloon-expandable stents in various phases of development, but only three designs have entered clinical testing in the United States: the Palmaz–Schatz model, the Gianturco–Roubin model, and the Wiktor design. All three designs rely on deformation of stent metal beyond its elastic limit. Idiosyncracies of design pose both theoretical and practical advantages and disadvantages of each design.

The Palmaz–Schatz design is a rigid, monoconstructed, slotted stainless-steel tube made of 0.002 in. stainless steel, 15 mm in total length consisting of two 7-mm segments bridged by a 1-mm articulation strut (Fig. 7.1). Staggered rectangles are expanded to form a diamond configuration with a resultant low proportion of metal to open surface area (approximately 10%, depending on final diameter), thus minimizing thrombus formation. The expansion ratio of the design is high, approximately 4 : 1, with some shortening in length noted at larger final diameters. The design allows for rapid endothelialization with multiple foci of endothelium within the open spaces. The stent is radially noncompliant with little or no elastic recoil after deployment. Its rigid design and sharp leading edges make the use of a delivery sheath necessary for deployment in tortuous arteries or distal arterial segments. Once deployed, the diamond-shaped struts cross side branch ostia without early or late loss of flow in these branches.[19] However, the struts make future access to crossed side branches with guidewires impossible.

The Gianturco–Roubin design consists of 0.006 in. monofilamentous stainless steel formed into an interdigitating coil structure wrapped around a standard PTCA balloon catheter (Fig. 7.2). One advantage of the continuous-coil design over other designs is its ability to span ostia to side branches without inhibiting access to these branches with guidewires for future angioplasty. The device is smooth-surfaced and tightly wrapped enough to permit deployment without the use of a delivery sheath, although snagging and failed delivery occurs in approximately 11% of attempted deliveries for actual or threatened vessel

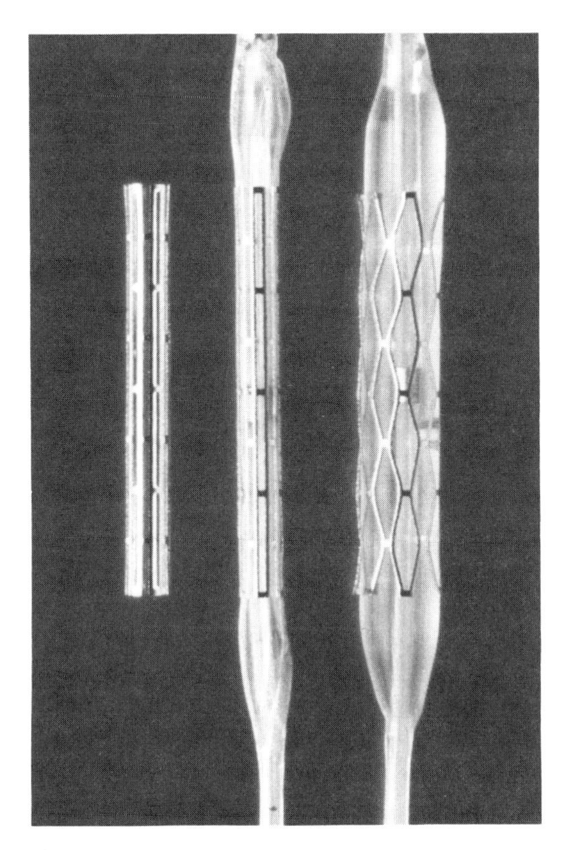

FIGURE 7.1 The Palmaz–Schatz coronary stent. Nonarticulated version first used.

closure.[20] It remains radially compliant after deployment and must be overdilated to accommodate for elastic recoil.[21,22] Perhaps because of its increased flexibility, the larger wire diameter, or patient selection, both thrombotic occlusion and restenosis appear to be more common with this stent compared to others on the basis of early clinical results.[23–28]

The Wiktor design is also a continuous-coil design, which has an advantage in the radio-opacity of the 0.005 in. tantalum wire from which it is formed (Fig. 7.3). While fluoroscopic visibility is a major asset of this stent, the fracture mechanics of the metal is uncertain. The early European clinical experience with this stent suggests the potential for thrombogenicity,[29] which may be due to, in part, inconsistent anticoagulation protocols.

7.3.1.2 Self-Expanding Design

Sigwart et al.[30] were the first to describe a self-expanding stent design for coronary deployment (Fig. 7.4). Unlike the balloon-expandable designs, this wire mesh stent is made of 0.003 in. stainless steel wire filament. The device is delivered to its target by a constraining catheter. When ready for deployment,

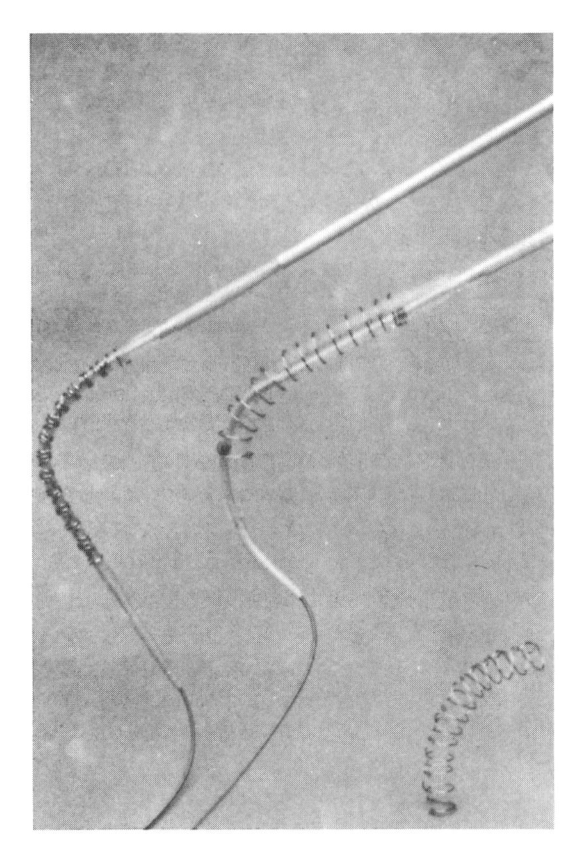

FIGURE 7.2 The Gianturco–Roubin continuous-coil stent.

the constraining catheter is withdrawn and spring tension forces the stent into contact with vessel intima until an equilibrium is reached between circumferential elastic resistance of the arterial wall and the dilating force of the prosthesis. Needless to say, the stent remains radially compliant, which, along with the high concentration of metal in the deployed state, may explain the high incidence of thrombotic closure reported to date.[31–34]

7.3.1.3 Other Designs

Other designs of coronary stents that have not yet entered clinical trials include those made of bioabsorbable material (particularly L-polylactide),[35] and thermal-memory stents made of nitinol.[36] The preclinical experience with these devices is too limited to warrant reporting at this time.

7.3.2 Stent Deployment

From an engineering prospective, a catheter system that enables placement of the stent in the small coronary arteries for deployment needs to be devised. Some

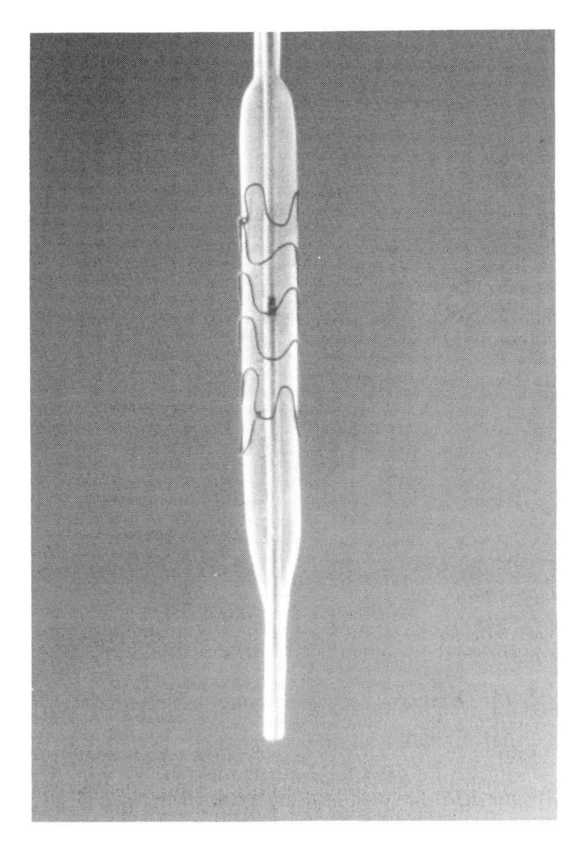

FIGURE 7.3 The Wiktor tantalum stent.

of the parameters in this design are apparent; for instance, the catheter needs to have radioopaque elements, as well as a balloon material that will not rupture during stent deployment. Stent and stent delivery system design are in progress, and different manufacturers have differing approaches to these problems. Thus, to illustrate the engineering issues, we will focus on the designs of the Palmaz–Schatz stent of Johnson & Johnson Interventional Systems Company.

7.3.2.1 Guiding Catheter

The guiding catheter for stent delivery is of critical importance. Two factors in catheter design prevail. First, the inside diameter of the catheter must permit easy passage of the stent delivery sheath without impeding flow of contrast medium. Repeated injections of radiodense contrast material are used in positioning the guide catheter by the coronary artery ostium. Second, the guiding catheter must engage the coronary artery ostium, without occluding the coronary artery. The optimal catheter tip design will fit the coronary artery coaxially, will allow the injection of contrast material, and will allow the advancement of the stent delivery

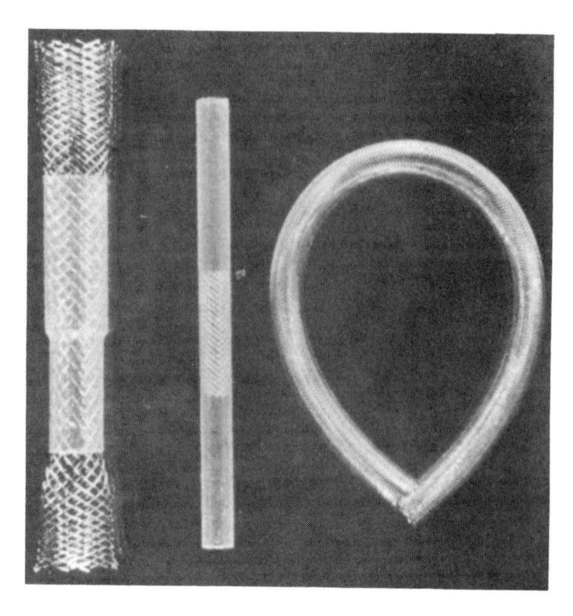

FIGURE 7.4 The Medinvent self-expanding stent.

system once engaged in the artery. Thus, the inside diameter must have a minimal size, while the external diameter should be as small as possible. This places large demands on the design of the catheter wall; it must be as thin as possible, and yet be radiodense and of sufficient strength to withstand the torque used in positioning the catheter by the coronary ostium.

7.3.2.2 Predilation

Predilation enlarges the stenotic portion of the coronary artery to enable passage of the stent delivery system. Therefore, predilation balloon size selection is based on undersizing the diameter and thereby reducing the possibility of arterial dissection prior to stenting. Since the smallest (expanded) stent delivery system currently available is 3.0 mm, the diameter of the balloon used for predilation is almost always 2.5 mm with a length of 20 mm. Seldom is it necessary to predilate with larger balloons for larger arteries.

7.3.2.3 Stent Delivery System

Each stent is premounted on an Interventional Systems balloon catheter that has a conventional coaxial design. The balloon tip is unique in that two radio-opaque markers straddle the stent crimped onto it. The balloon catheter and stent assembly are housed within a 4.9F delivery sheath. Collectively, the delivery sheath, the balloon, and the stent constitute the stent delivery system (Fig. 7.5).

The diameter of the stent–balloon assembly is selected to match the normal coronary artery size. The final diameter of the deployed stent, however, is determined by the size of the balloon used for final expansion.

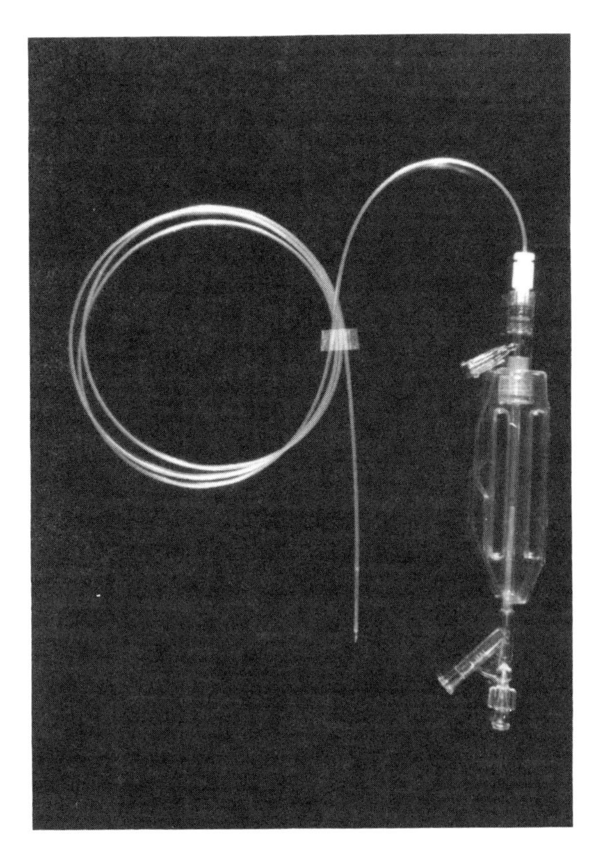

FIGURE 7.5 The Johnson and Johnson Interventional Systems Stent Delivery System. Removable plastic sleeve (arrows) at hub maintains fixed relationship between balloon catheter with stent and delivery sheath. (See text.)

As a unit, the stent delivery system is advanced to the coronary stenosis. Once correct stent position is assured, the balloon catheter is negatively aspirated (prepared) and the stent is deployed with one inflation of 6 atm pressure for 10 s. Following balloon deflation, the catheter is easily withdrawn under fluoroscopic guidance into the guiding catheter.

7.3.3 Thrombosis

Thrombosis or clotting by the stent remains the primary problem in using the stents, and the area for the most improvement. As outlined below, systemic agents are administered to prevent clotting. However, these agents can lead to bleeding complications.

7.3.3.1 Procedure Regimen

All patients in whom a coronary stent is implanted electively are pretreated with dipyridamole, aspirin, and low-molecular-weight dextran. This regimen is intended to minimize platelet adhesion to the newly deployed stent. During deployment, heparin is administered, and typically, no additional heparin is administered immediately after stent deployment. The arterial sheath (the tube in the femoral artery through which the guiding catheter was placed in the aorta) is removed when activated clotting time falls to $\leq$ 170 s. Anticipating additional systemic anticoagulation, great care is exercised in sheath removal. Hemostasis is achieved by manual direct pressure. Later, intravenous heparin and warfarin are given. After discharge from the hospital, daily aspirin and dipyridamole are continued for 3 months postimplantation.

7.3.3.2 Followup

Although there have been no documented cases of acute thrombosis with the Palmaz–Schatz stent (within 24 h of deployment), nor any cases of clinically apparent thrombotic occlusion more than 14 days following implantation, subacute thrombosis remains a low-frequency but predictable event (approximately 3%)[20,22,23] with peak occurrence between days 5 and 10 postimplantation. Since more than half of the cases of subacute thrombosis can be linked to inadequate anticoagulation,[37] careful monitoring of anticoagulation status is critical. Accordingly, prothrombin time is checked during the vulnerable period and patients at high risk of thrombosis are advised to remain within reasonable contact of medical care. These high-risk patients include those with implantation of stent after transient total occlusion of vessel during balloon angioplasty, those receiving the stent into a dissected vessel, those with suboptimal blood runoff and inflow into the stent, those with evidence of thrombus at the stent site at time of deployment or patients with significant "muscle at risk" in the distribution of the stented artery.

7.3.4 Restenosis

As mentioned at the beginning of this chapter, the hope of improving on the $\geq 30-40\%$ restenosis rate that occurs within 6 months after balloon angioplasty has been one of the driving forces to develop new devices. In this context, the early reports on stents look promising. Reporting the multicenter experience with the Palmaz–Schatz stent in the United States, Ellis and others[16] described the angiographic followup on 103 consecutive single stents. Of these stented lesions, 70% had prior restenosis, 36% were deployed in the left anterior descending artery, and 28% had lesion length $\geq$ 10 mm. Repeat angiography was performed $\geq$ 4 months after implantation with digital quantitation performed at a core laboratory. Using the $\geq 50\%$ luminal narrowing definition of restenosis, restenosis was seen in 20% of stents.

Marco and colleagues[38] reported their institution's experience in 91 patients with single and multiple stents in native vessels and saphenous vein bypass grafts. The indications for stenting were prior restenosis in 31%, acute dissection in 23%, and judged high risk of restenosis in 46%. Collectively, $>50\%$ luminal narrowing was present in 19 of 91 patients (21% by patient) at angiography 4–6 months following implantation.

7.3.4.1 Mechanisms of Prevention of Restenosis

Although the mechanisms by which a rigid stent may prevent restenosis are not proven, experimental and clinical observations suggest that two factors may play a role: attenuation of arterial wall stress and improvement in initial luminal profile.

In Ellis's study mentioned above, using multiple-regression analysis, they found that poststent percentage stenosis was independently and positively associated with the occurrence of restenosis at angiographic followup. Therefore, they concluded that improvement in the initial luminal diameter, and not lessened neointimal growth, was responsible for the decreased restenosis rate seen in coronary arteries stented with the Palmaz–Schatz stent.

A similar conclusion was reached by Kuntz et al.[39] when they compared the initial angiographic improvement in luminal diameter achieved by stenting, laser–balloon angioplasty, and atherectomy. The loss in luminal diameter at followup angiography was equivalent among the three devices and was normally distributed. Since the initial improvement was greatest with stenting, the followup loss in luminal diameter was lowest with stenting. Therefore, the restenosis rate observed was lowest with the stent. This is consistent with findings of Ellis, which suggest that restenosis occurs less frequently when the initial angiographic result is optimal.

7.3.5 Conclusions and Future Directions

Coronary stenting represents a new and exciting technology that addresses the major limitations of balloon angioplasty: abrupt vessel closure and restenosis. Early experience with several designs demonstrates the efficacy and safety of stent deployment. Prospective, randomized trials of coronary stenting as an alternative to balloon angioplasty are under way to conclusively determine a potential benefit of stenting on restenosis. Retrospective data from the Palmaz–Schatz stent suggest that in certain patient populations, stenting results in a lower rate of restenosis than balloon angioplasty. This benefit may be especially prominent in patients undergoing de novo stenting of short lesions. In consideration of this benefit, a small risk of subacute thrombosis lingers.

Perhaps the basic lesson to be derived from the early experience with coronary stenting is that deployment of an intravascular prosthesis is not the same procedure as balloon angioplasty. Potential complications and risks and benefits distinguish stenting from balloon angioplasty. As experience with the devices and the accompanying technologies evolve, this procedure will almost certainly become commonplace.

While clinical investigation continues, experimental work also continues with stents that have polymer coating to attenuate vascular spasm,[40] stents with pharmacologic activity to inhibit thrombus formation,[41,42] and seeding of stents with genetically engineered endothelial cells that are hypersecretors of plasminogen activator.[43,44] In this context, coronary stenting represents a new threshold of technology in our battle against the mechanical and biochemical elements of coronary artery disease.

REFERENCES

1. Gruentzig AR, Senning A, Siegenthaler WE: Nonoperative dilatation of coronary artery stenoses. Percutaneous transluminal angioplasty. *New Engl J Med* 301:61, 1979.
2. Lee G, Mason DT: Laser angioplasty: Current investigations and future prospects. *Primary Cardiol* 16:15, 1990.
3. Kaminow IP, Wiesenfeld JM, Choy DSJ: Argon laser disintegration of thrombus and atherosclerotic plaque. *Appl Optics* 23:1301, 1984.
4. Lee G, Ikeda RM, Theis JH: Acute and chronic complications of laser angioplasty. Vacular wall damage and formation of aneurysms in atherosclerotic rabbit. *Am J Cardiol* 53:290, 1984.
5. Choy DSJ, Stertzer SH, Myler RK: Human coronary laser recanalization. *Clin Cardiol* 7:377, 1984.
6. Livesay JJ, Leachman DR, Hogan PJ: Preliminary report on laser coronary endarterectomy in patient. *Circulation* 75:III-302, 1985.
7. Sanborn TA, Faxon DP, Kellett MA: Percutaneous coronary laser thermal angioplasty. *J Am Coll Cardiol* 8:1437, 1986.
8. Litvack F, Grundfest WS, Goldenberg T, et al: Percutaneous excimer laser angioplasty of aortocoronary saphenous vein grafts. *J Am Coll Cardiol* 14:803, 1989.
9. Karsch KR, Hasse KK, Mauser M, et al: Percutaneous coronary excimer laser angioplasty: Initial clinical results. *Lancet* 2:647, 1989.
10. Hasse KK, Mauser M, Baumbach A, et al: Restenosis after excimer laser coronary atherectomy. *Circulation* 82:III-672, 1990.
11. Torre SR, Sanborn TA, Sharma SK, et al: Percutaneous coronary excimer laser angioplasty: Quantitative angiographic analysis demonstrates improved angioplasty results with larger laser catheters. *Circulation* 82:671, 1990.
12. Dotter CT, Judkins MP: Transluminal treatment of arteriosclerotic obstruction. *Circulation* 30:654, 1964.
13. Dotter CT: Transluminally placed coil-spring endarterial tube grafts, long term patency in canine popliteal artery. *Invest Radiol* 4:329, 1969.
14. Dotter CT, Buschmann RW, McKinney MK, et al: Transluminal expandable nitinol coil stent grafting: preliminary report. *Radiology* 147:259, 1983.
15. Cragg A, Lung G, Rysavy J, et al: Nonsurgical placement of arterial endoprostheses: A new technique using nitinol wires. *Radiology* 147:261, 1983.
16. Palmaz JC, Sibbitt RR, Reuter SR, et al: Expandable intraluminal graft: A preliminary study. *Radiology* 156:73, 1985.

17. Maass D, Zollikofer CL, Largiader F, et al: Radiological follow-up of transluminally inserted vascular endoprostheses: An experimental study using expanding spirals. *Radiology* 150:659, 1984.

18. Rousseau H, Puel J, Joffre F, et al: Self-expanding endovascular prosthesis: An experimental study. *Radiology* 164:709, 1987.

19. Fischman D, Savage M, Cleman M, et al: Fate of lesion-related side branches following coronary artery stenting. *J Am Coll Cardiol* 17:280A, 1991.

20. Roubin GS, Hearn JA, Carlin SF, et al: Angiographic and clinical follow-up in patients receiving a balloon expandable, stainless steel stent (Cook Inc.) for prevention or treatment of acute closure after PTCA. *Circulation* 82:III-190, 1990.

21. Roubin GS, Robinson KA, King, SB, et al: Early and late results of intracoronary arterial stenting after coronary angioplasty in dogs. *Circulation* 76:891, 1987.

22. Leung DY, Glagov S, Mathews MB: Cyclic stretching stimulates synthesis of matrix components by arterial smooth muscle cells in vitro. *Science* 191:475, 1976.

23. Roubin GS, Cannon AD, Agrawal SK, et al: Intracoronary stenting for acute and threatened closure: Complicating percutaneous transluminal coronary angioplasty. *Circulation* 85:916, 1992.

24. Schatz RA, Goldberg S, Leon MB, et al: Coronary stenting following "suboptimal" coronary angioplasty results. *Circulation* 82:III-540, 1990.

25. Levine MJ, Leonard BM, Burke JA, et al: Clinical and angiographic results of balloon-expandable intracoronary stents in right coronary artery stenoses. *J Am Coll Cardiol* 16:332, 1990.

26. Schatz RA, Leon M, Baim D, et al: Short-term clinical results and complications with the Palmaz–Schatz coronary stent. *J Am Coll Cardiol* 15:117A, 1990.

27. Schatz RA, Leon MB, Baim DS, et al: Balloon expandable intracoronary stents: Initial results of a multicenter study. *Circulation* 80:II-174, 1989.

28. Schatz RA, Baim DS, Leon M, et al: Clinical experience with the Palmaz–Schatz coronary stent: Initial results of a multicenter study. *Circulation* 83:148, 1991.

29. Bertrand M, Kober G, Scheerder Y, et al: Initial multi-center human clinical experience with the Medtronic Wiktor coronary stent. *Circulation* 82:III-541, 1990.

30. Sigwart U, Puel J, Mirkowitch V, et al: Intravascular stents to prevent occlusion and restenosis after transluminal angioplasty. *New Engl J Med* 316:701, 1987.

31. Puel J, Rosseau H, Joffre F, et al: Intravascular stents to prevent restenosis after transluminal coronary angioplasty. *Circulation* 76:27, 1987.

32. Bertrand ME, Rickards AF, Serruys PW: Coronary stenting implanatation for primary and secondary prevention of restenosis after PTCA: Results of a pilot multicenter trial (1986–1987) (CASIS trial). *Eur Heart J* 9:55, 1988.

33. Puel J, Haddad J, Courtault A, et al: Angiographic follow-up of percutaneous coronary stenting. *Circulation* 78:II-408, 1988.

34. Serruys PW, Strauss BH, Beatt KJ, et al: Angiographic follow-up after placement of a self-expanding coronary-artery stent. *New Engl J Med* 324:13, 1991.

35. Chapman GD, Gammon RS, Bauman RP, et al: A bioabsorbable stent: Initial experimental results. *Circulation* 82:IV-72, 1990.

36. Cragg AH, Lund G, Rysavy JA, et al: Percutaneous arterial grafting. *Radiology* 150:45, 1984.

37. Cleman MW, Cabin HS, Leon M, et al: Major complications associated with intracoronary delivery with the Palmaz–Schatz stent. *J Am Coll Cardiol* 17:280A, 1991.

38. Marco J, Fajadet JC, Cassagneau BG, et al: Balloon expandable intracoronary stents: Immediate and mean term results in a series of 160 consecutive patients. *Circulation* 82:III-658, 1990.

39. Kuntz RE, Safian RD, Schmidt DA, et al: Restenosis following new coronary devices: The influence of post procedure luminal diameter. *J Am Coll Cardiol* 17:2A, 1991.

40. Bailey SR, Guy DM, Garcia OJ, et al: Polymer coating of Palmaz–Schatz stent attenuates vascular spasm after stent placement. *Circulation* 82:III-541, 1990.

41. Cavender JB, Anderson P, Roubin GS: The effects of heparin bounded tantalum stents on thrombosis and neointimal proliferation. *Circulation* 82:541, 1990.

42. Stratienko AA, Zhu D, Lambert CR, et al: Improved thromboresistance of heparin coated Palmaz-Schatz™ coronary stents in an animal model. *Circulation* 88:I-596, 1993.

43. Van der Geissen WJ, Serruys PW, Visser WJ, et al: Endothelialization of intravascular stents. *J Intervent Cardiol* 1:109, 1988.

44. Dichek DA, Neville RF, Zweibel JA, et al: Seeding of intravascular stents with genetically engineered endothelial cells. *Circulation* 80:1347, 1989.

CHAPTER EIGHT

Advances in Magnetic Resonance Imaging

REUBEN S. MEZRICH, M.D., Ph.D., *Department of Radiology, Robert Wood Johnson Medical School, New Brunswick, NJ*

8.1 INTRODUCTION

8.1.1 History

In 1946 Bloch and Purcell, in separate papers, described the phenomenon of nuclear magnetic resonance. In the early 1970s, Damadian and Lauterbur, again in separate works, described the application of magnetic resonance to medicine. Magnetic resonance imaging (MRI) was introduced into clinical practice by the early 1980s, and in less than 10 years has become a primary diagnostic tool in several clinical areas, especially neurology and orthopedics.

The clinically important characteristics of MRI are high resolution, intrinsic high soft-tissue contrast, direct visualization and measurement of blood and cerebrospinal fluid (CSF) flow, and the lack of ionizing radiation.

8.1.1.1 Increasing Clinical Utility

Magnetic resonance imaging is often the primary tool in the workup of CNS disease (Figs. 8.1–8.6) and has supplanted arthrographic studies in the study of joint disease, especially for the hip, knee, shoulder, and temporomandibular joint (Figs. 8.7–8.9). It is being increasingly used in the study of musculoskeletal,

New Frontiers in Medical Device Technology, Edited by Arye Rosen and Harel Rosen
ISBN 0-471-59189-0

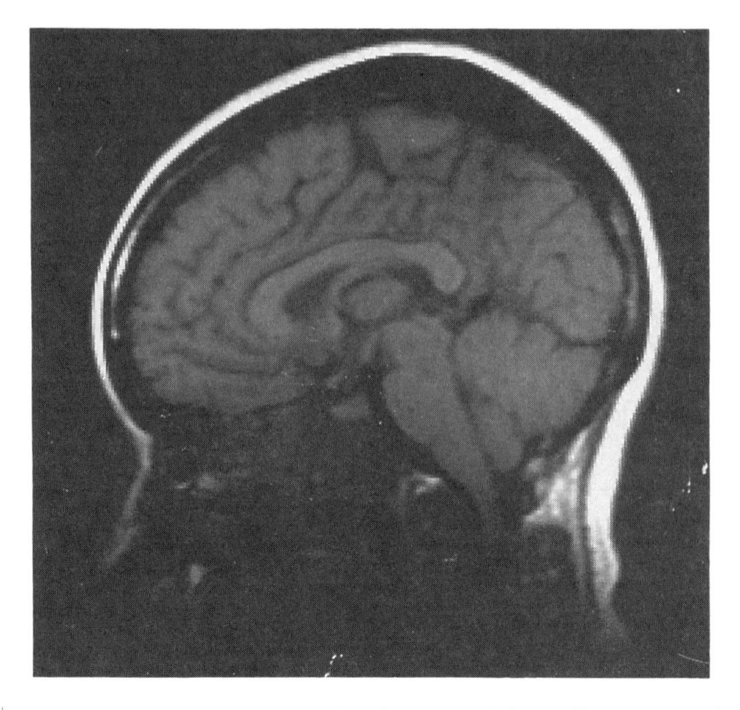

FIGURE 8.1 Sagittal view of the brain showing some of the midline structures including the corpus callosum, midbrain, pons, and pituitary gland. No abnormalities are seen on this view.

gynecologic, and prostatic disease and is an important adjunct in the study of hepatic disease (Fig. 8.10).

8.1.1.2 New Capabilities

Continued improvement in MRI technology, especially improved control of magnetic and radiofrequency field (RF), and in computer technology have led to continued growth in the clinical capabilities of the technique. Several important conceptual advances, especially in methods for high-speed scanning and flow visualization, will increase the clinical capabilities of MRI and add to the use of this modality.

8.1.2 Background

8.1.2.1 NMR

Magnetic resonance is based on the observation that magnetic moments placed in a magnetic field tend to align with the field direction. On an atomic scale such moments (or spins) are constrained in their ability to align by the allowable values of their angular momentum (e.g. hydrogen with allowable angular momentums of $\pm\frac{1}{2}$ will have nearly as many antiparallel as parallel to the magnetic

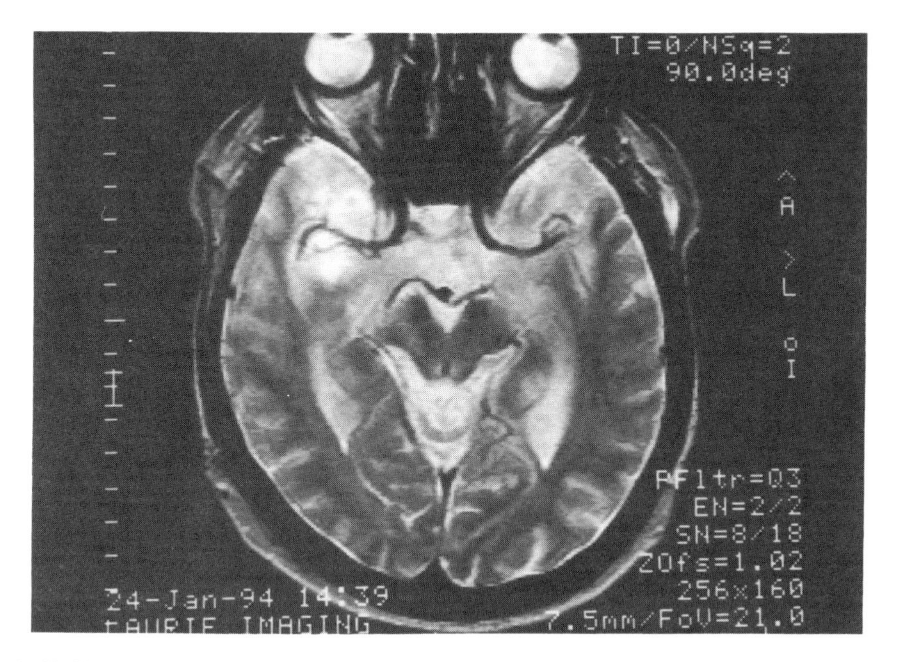

FIGURE 8.2 Axial view of the brain using T_2 technique. The images are at the level of the midbrain. Notice the high signal intensity of the fluid in the eyes, and the dark signal due to flowing blood in the middle cerebral arteries.

field direction), but in a sufficiently large magnetic field there will be a net magnetic moment directed along the field direction. This net magnetization can be perturbed, or excited by application of electromagnetic radiation of the proper frequency (ω), which is given by

$$\omega = \gamma B_0 \tag{8.1}$$

On removal of the radiation, the moments will relax back toward alignment with the static magnetic field, and as they do so will emit radiation at a frequency given by

$$\omega = \gamma B \tag{8.2}$$

where B is the magnetic field at the relaxing spin. It is important to notice that in the case of a volume of spins in an inhomogeneous magnetic field, a spectrum of frequencies will be emitted as each spin emits radiation proportional to its local magnetic field. Many of the atoms in biologic tissue have large magnetic moments, especially hydrogen, sodium, and phosphorus, and can serve in magnetic resonance experiments. Hydrogen, because of its great natural abundance (found in water and fat—the major constituents of most living tissue)

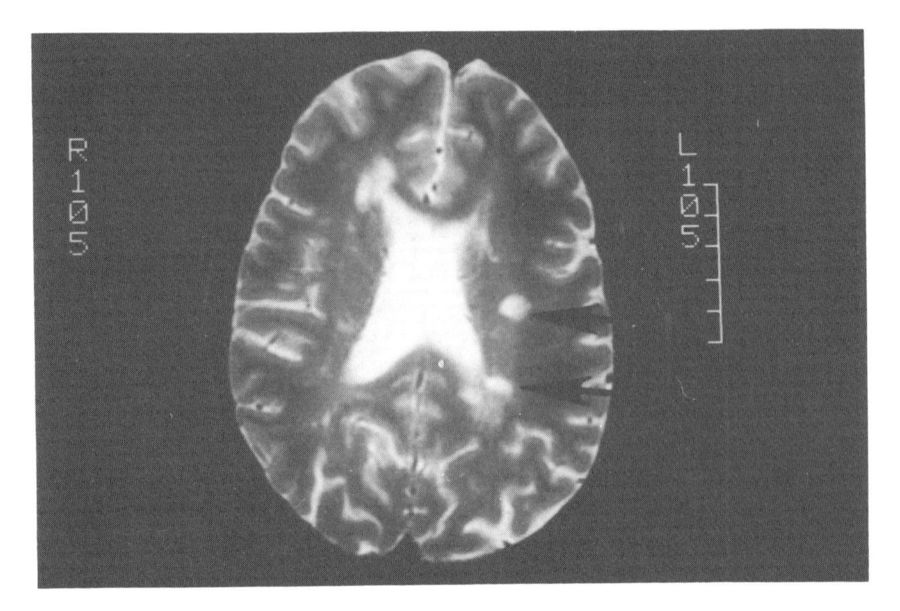

FIGURE 8.3 Axial view using the T_2 technique, taken through the lateral ventricles of a patient with multiple sclerosis. Arrows point to two of the white-matter lesions, which are characteristic of but not pathognomonic of this disease. Other lesions are seen adjacent to the frontal horns of this patient, and many others were seen at adjacent levels through the brain. A differential diagnosis for these lesions would also include ischemic changes, trauma, vascular diseases such as lupus, or infections such as Lyme disease. In this patient other tests confirmed the diagnosis of multiple sclerosis.

and high magnetic resonance sensitivity, is the atom of interest in nearly all clinical applications of nuclear magnetic resonance.

There are two relaxation processes for hydrogen in biologic tissue. The first (T_1, or spin–lattice relaxation) can be viewed as a measure of how well the atoms in a tissue "communicate" or share energy with their surroundings. The hydrogen spins excited by application of the appropriate electromagnetic energy relax (or return to their ground state) by transferring the energy to the lattice around them. The better coupled the atoms are to the lattice, the faster they will relax (or the shorter the T_1). What is so interesting about this from a medical viewpoint is that the degree of coupling varies from tissue to tissue, and furthermore varies with disease!

The second relaxation process (T_2, or spin–spin relaxation) is a measure of intrinsic magnetic field uniformity. Atoms in a perfectly uniform magnetic field will emit radiation of a single frequency, with the same phase as they relax after excitation. In an inhomogeneous environment there will be local, possibly time-varying, differences in phase and frequency of the signals from excited atoms. The net signal, summed over all the excited atoms in the volume, will decay as the various individual signals "dephase," with rate of net signal decay proportional

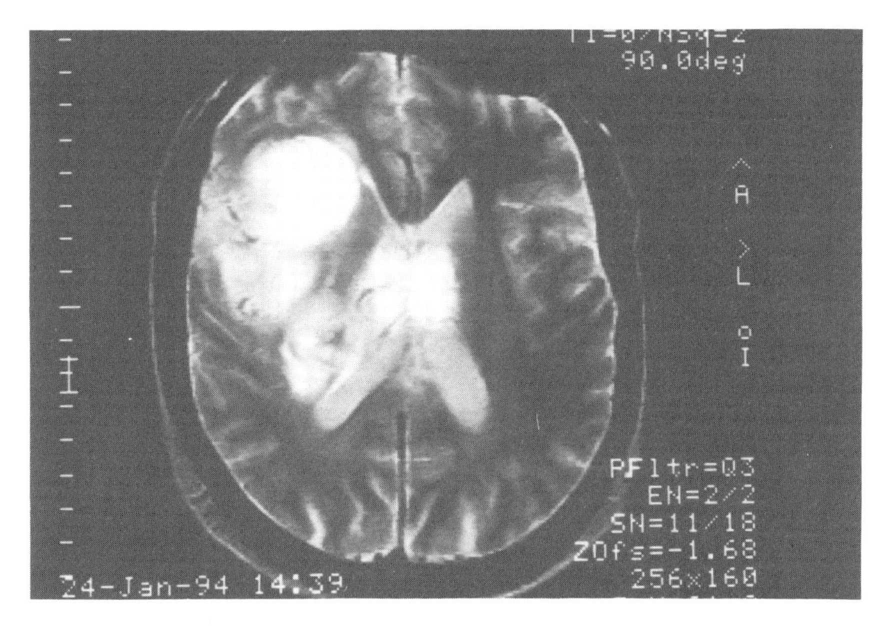

FIGURE 8.4 Axial view through the lateral ventricles of a patient with a large glioblastoma multiforme seen in the right side of the brain (left on the image). Notice the hyperintense signal of this malignant tumor. The multiloculated nature of the tumor attests to its high malignant potential.

to the degree of local field inhomogeneity. In real biologic material that is in an external magnetic field, the local field inhomogeneity will be proportional to local tissue variations (i.e., susceptibility) that are proportional to local tissue composition. Again, the interesting thing from a medical point of view is that these variations (and so the T_2 relaxation times) vary from tissue to tissue and with disease.

It is the variation in the relaxation parameters – from tissue to tissue and with disease that is the underlying reason for the medical applications of magnetic resonance. X-ray techniques, whether plain film or CT (computerized tomography), measure only one parameter of tissue, the electron density, which does not vary significantly between soft tissues (a maximum variation, except bone and fat, of approximately 5% is typical). The differences in relaxation times, on the other hand, are typically hundreds of percent between different tissues.

8.1.2.2 MRI

In the early 1970s, several workers, most notably Dr. Paul Lauterbur and Dr. Ray Damadian recognized that it is possible to take advantage of the relationship of Equation (8.2) and derive information about where in a volume of tissue the signals emitted by excited spins might be coming from; that is, it could be possible to make an image proportional to spatial variations in degree of spin

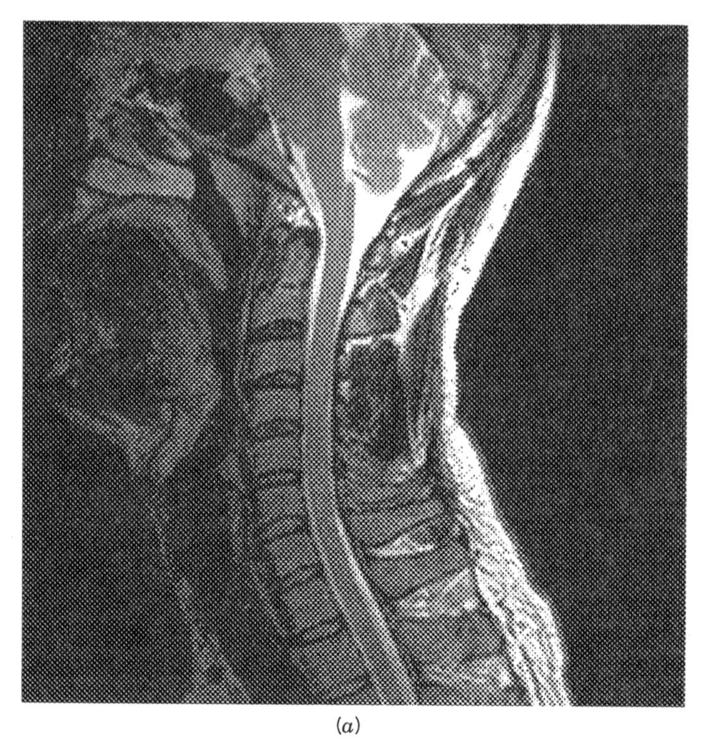

(*a*)

FIGURE 8.5 Sagittal views using T_2 (*a*) and T_1 (*b*) techniques. Note the good visualization of the spinal cord arising from the medulla and extending caudally through the spinal canal. Note that the intravertebral disks are well visualized and are relatively bright on the T_2-weighed images and relatively dark on the T_1-weighed images. No abnormalities were seen on this patient.

excitation. The basic idea is the purposeful application of an extrinsic magnetic field gradient. If a linear magnetic field gradient is applied so that the net magnetic field is greater at one side of a volume than at the opposite side (say, along the x axis), then by Equation (8.2) there will be a proportional linear variation in the signals emitted by excited spins. Similarly, it is possible to limit the number of spins excited in the first place by applying a magnetic field gradient simultaneously with the exciting electromagnetic energy. If the energy is at a single (or at most a narrow band of frequencies about a frequency), then only a plane of spins, corresponding to those that satisfy the relationship of Equation (8.1) will be excited. The remaining moments in the volume will be unaffected (since they are not "resonant" with the applied frequency). Using combinations of magnetic field gradients applied at appropriate times (i.e., simultaneous with excitation or during relaxation) and in appropriate directions (usually along orthogonal directions), an image of a volume can be generated.

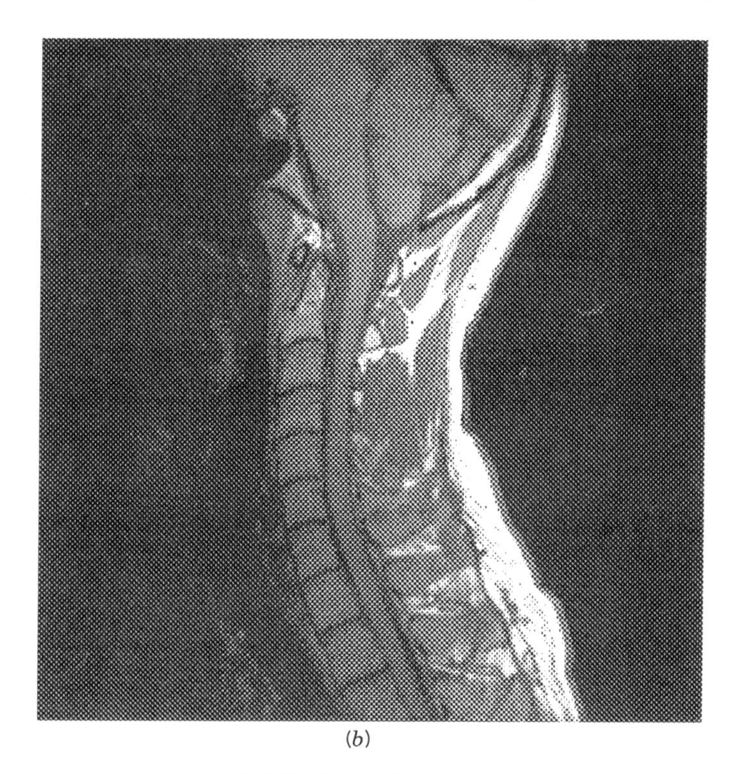

(*b*)

FIGURE 8.5 (*Continued*).

The resolution is limited primarily by the strengths of the applied gradients (which determine the finesse with which planes of atoms can be excited or different frequencies discriminated) and by the number of atoms in a "voxel" (which determines the signal to noise ratio of the received signal). At present, voxels measuring considerably less than 0.1 mm^3 can be resolved.

The speed of image acquisition depends on the number of lines in an image, the number of planes imaged at one time, the number of signal averages taken (signal–noise ratios can be enhanced by summing over multiple signal acquisitions), and by the rate of tissue excitation (data for an image plane are obtained by successive excitations, with the number of excitations proportional to the number of lines in the image. The time between excitations, relative to the tissue relaxation times, is an important determinant of image characteristics—because of the relative weight given to the signal from tissues of different relaxation times—and its choice is often dependent on the type of tissue being examined and the pathology being sought). The total image acquisition time is given by

$$T = T_r N_x N_{av} \tag{8.3}$$

where T_r is the time between excitations (repetition time), N_x is the number of lines in an image, and N_{av} is the number of signal averages. It should be noted that if T_r is long enough, more than one image plane can be acquired at one time by using methods that are akin to time-division multiplex. Thus, if T_r is 1000 ms, and the time to acquire the data due to an excitation is 100 ms, up to 10 adjacent planes could be excited, each with the same T_r of 1000 ms. Each plane, or course, would be excited at a slightly different frequency and in the presence of an appropriate gradient.

8.1.2.3 Flow

The appearance of flowing fluid (either blood or CSF) depends directly on the details of the image acquisition technique and on whether the phase or amplitude of the received MR signal is measured and displayed.

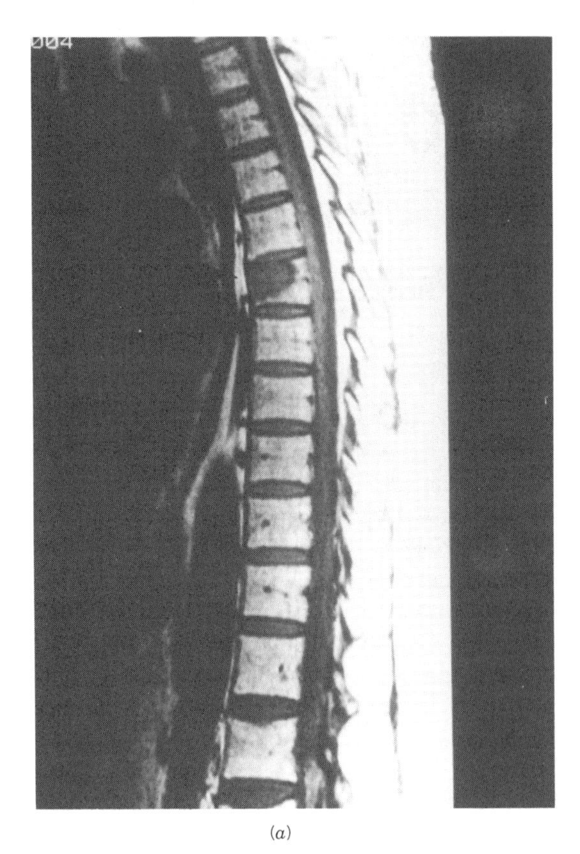

(*a*)

FIGURE 8.6 Sagittal views of thoracic spine (*a*) and lumbar spine (*b*) showing metastases as indicated by hypointense signal. The tumor in the lumbar vertebra extends into the spinal canal, causing a partial block. No contrast was used or needed to diagnose the location of this significant lesion in the lumbar spine and is sufficient to guide radiation therapy.

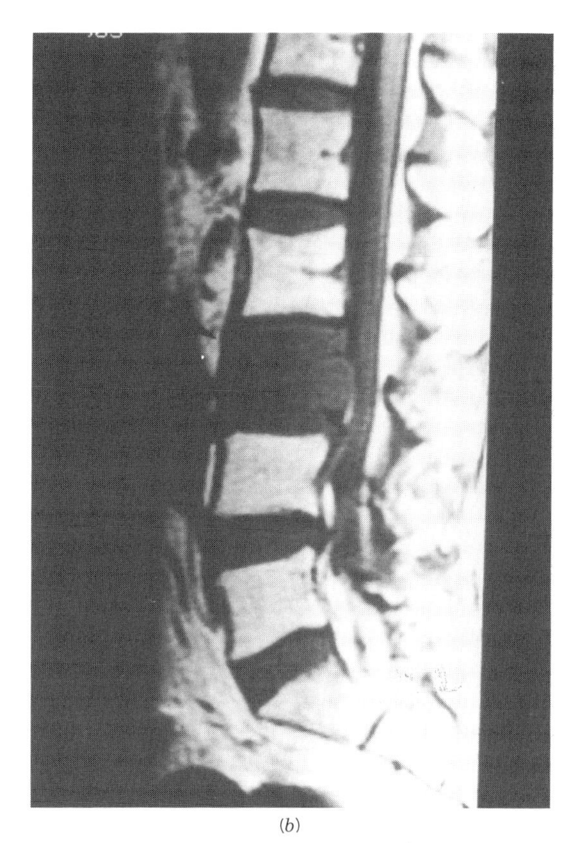

(*b*)

FIGURE 8.6 (*Continued*).

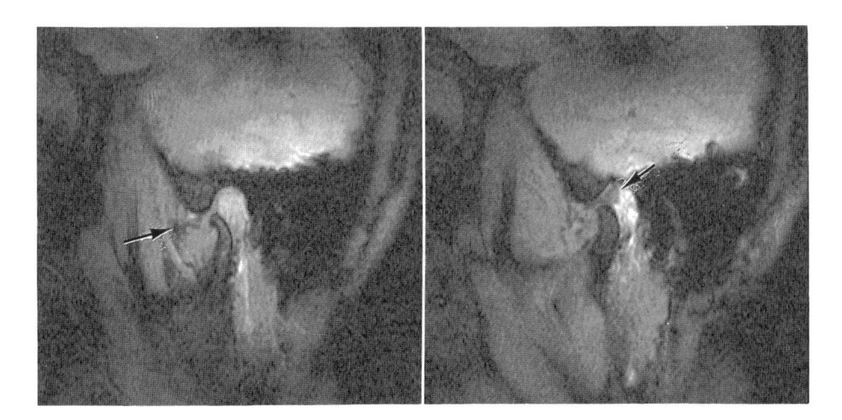

FIGURE 8.7 Sagittal views through the temporal mandibular joint showing the menisci (*arrows*) on the left and right sides of the patient. This patient has normal meniscal position on the right and anterior displacement on the left. This type of study is the most effective means of diagnosing the true extent of the disease in patients presenting with clicking and pain while chewing or talking. Depending on the diagnosis, conservative treatment with splints or else surgery is used for treatment.

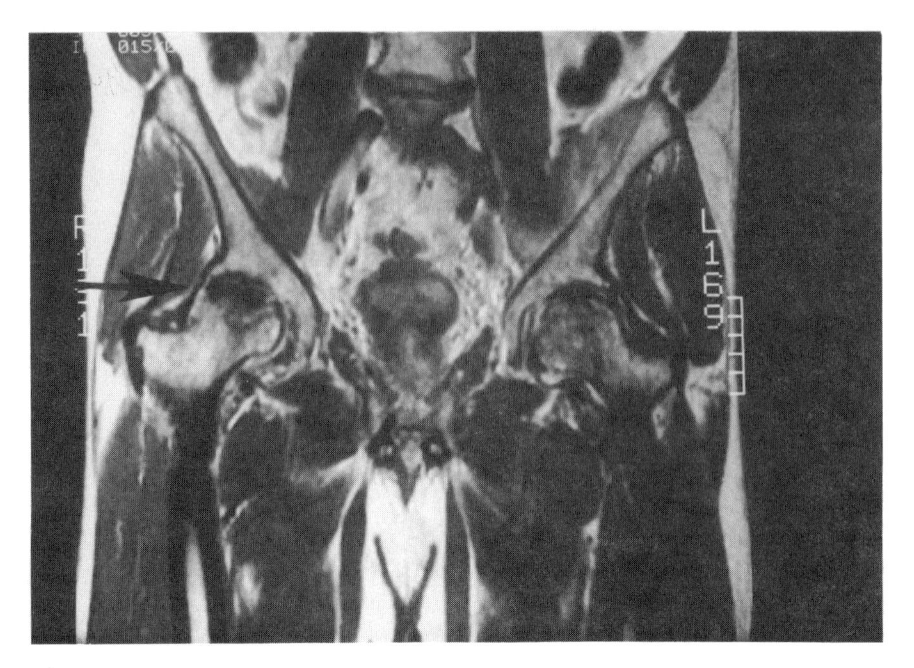

FIGURE 8.8 Coronal view of a middle-aged asthmatic who has had to take steroids to treat his disease. He now presents with hip pain. The abnormal signal in the right hip (*arrow*) is pathognomic for avascular necrosis (AVN) and the changes in the left hip are also consistent with that disease. Avascular necrosis is a result of impaired blood supply to the hips that can occasionally be seen in patients taking high doses of steroids, in caisson workers or deep-sea divers, in alcoholics, or in some patients with hip fractures. An MRI is the most sensitive test for this disease and can detect disease weeks to months before it becomes apparent on other tests.

In the case where multiple planes are being imaged within a volume, flowing blood may not remain within a plane for the interval between spin excitation and signal acquisition and will appear dark.

In the case where only a single plane is being imaged, blood flowing into that plane may appear bright relative to the stationary tissue of the plane. In the imaging process the stationary tissue is repeatedly excited and the spins become partially saturated, while the blood flowing into the plane contains "fresh" spins and has a relatively stronger net magnetic moment. This method has been dubbed "time of flight."

Finally, for blood flow along an applied magnetic gradient, the phase of the signal from excited spins will vary in proportion to the velocity of the blood and the strength of the gradient. Signal acquisition with quadrature techniques can be used to measure this phase change and directly display blood (or CSF) velocity. This method is usually referred to as the *phase-contrast* method.

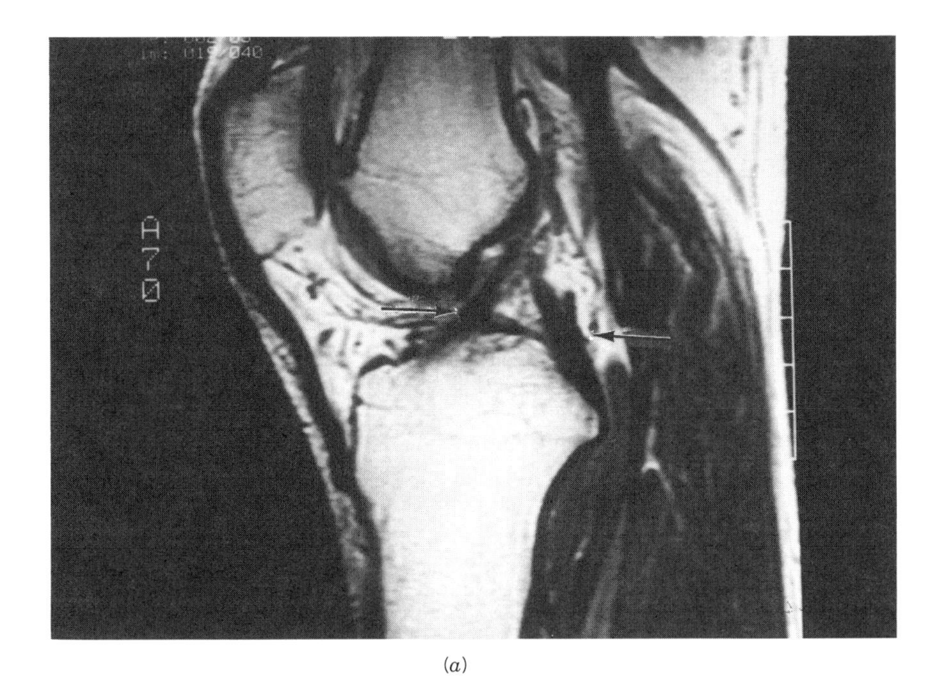

(*a*)

(*b*)

FIGURE 8.9 Sagittal (a) and coronal (b) views of a normal knee. The menisci can be well seen on the coronal views (b) as a triangle-shaped dark structure (*arrows*), and the cruciate ligaments are well seen on the sagittal views (*arrows*) (a).

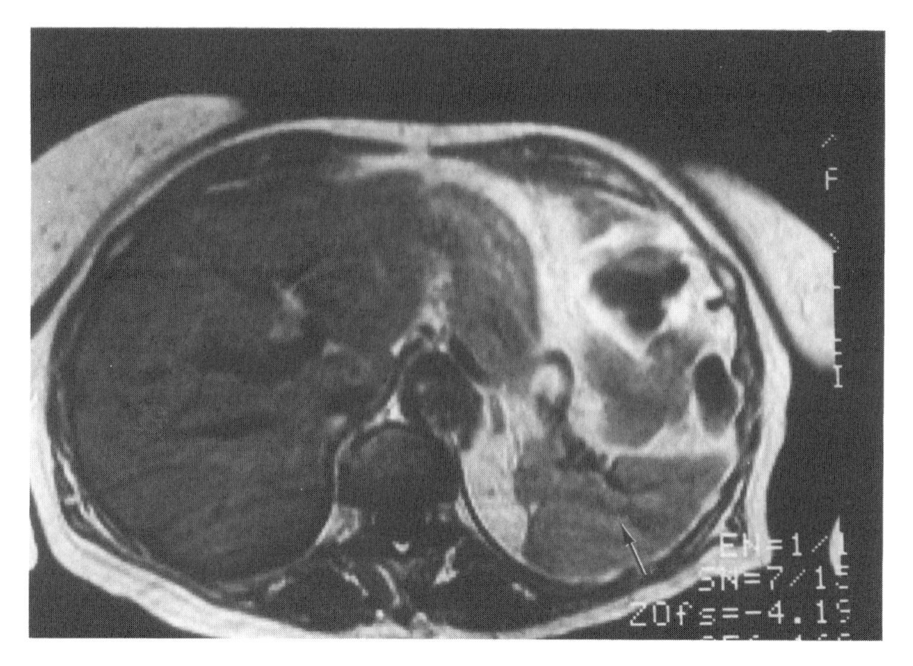

FIGURE 8.10 Axial view through the liver (*asterisk*) and spleen (*arrow*).

8.1.3 Clinical Applications

Because of its high resolution and high-soft-tissue contrast characteristics, MR has become the primary modality for the study of central nervous system (CNS) disease (which includes both the brain and the spine) and is becoming accepted as the preferred method for the study of joint abnormalities [especially the hip, knee, shoulder, and temporomandibular joint (TMJ)]. Recent studies have indicated that MR may have important uses in evaluating breast masses—distinguishing benign from malignant lesions (Fig. 8.11). The clinical literature is replete with other examples of the utility of MR, and several references are provided for the interested reader.

8.1.4 New Capabilities

In the past several years a number of new techniques have been developed that are now being introduced into clinical practice.

8.1.4.1 MR Angiography

Both time-of-flight and phase-contrast flow visualization methods have been incorporated into angiographic display techniques. In both methods, flow data from a collection of planes within the volume of interest (say, the neck or brain) are collected and processed for display. With the proper choice of acquisition

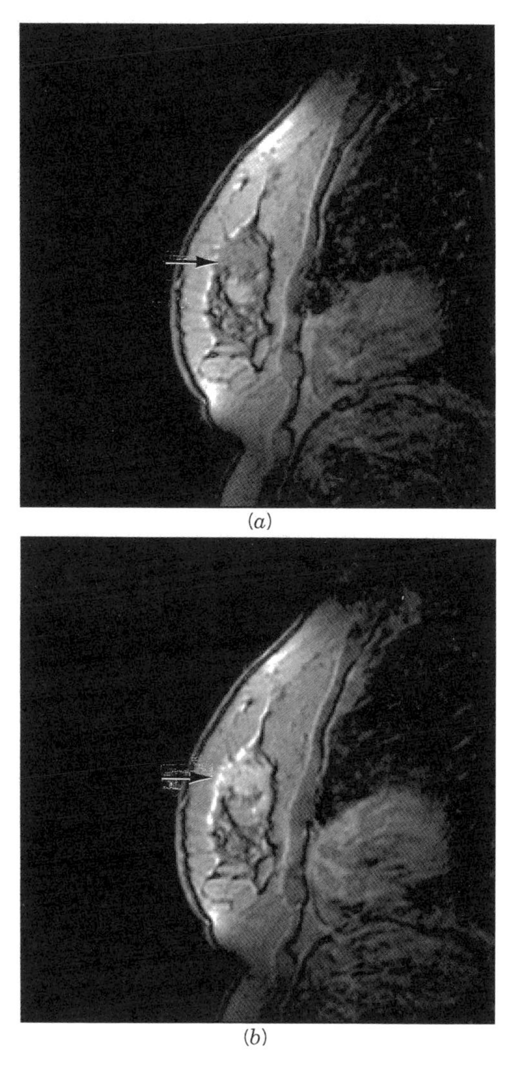

FIGURE 8.11 Sagittal views through the breast of a patient with a malignant cancer (*arrows*). The tumor is relatively dark on the precontrast image (a) and becomes hyperintense after contrast injection (b). While too expensive for regular screening, MR mammography shows promise as a means of identifying which lesions seen on an x-ray examination are truly malignant, perhaps even determining preoperatively the degree of malignancy.

parameters the signal intensity from the blood flowing within vascular structures is considerably greater than that of surrounding tissue, even for small vessels, and relatively straightforward thresholding techniques can be used to separate (or segment) the structures. Rendering techniques are then used to display the vascular anatomy in three dimensions.

Clinical trials, which have been continuing over the past few years, have demonstrated that the techniques have high anatomic fidelity, with high pathologic correlation with conventional angiographic techniques. The great advantage of the MRI technique, of course, is that no catheters need be placed into the patient's body, no contrast need be injected, and no ionizing radiation need be used. The methods are now being incorporated into clinical practice, both for screening and diagnostic use—especially when the patient cannot tolerate the iodinated contrast material used in conventional angiography (Fig. 8.12).

8.1.4.2 Cardiac Cine

By synchronizing the MR acquisition sequence with the cardiac cycle, very-high-resolution images of the heart can be obtained, at any phase of the cardiac cycle and at any plane through the heart. Using repetition times (T_r) that are multiples of the R–R times of the cardiac cycle, the relaxation characteristics of the myocardium, and their changes with ischemic or hypertrophic disease can be examined in detail. In very recent experiments acquisition with very short relaxation times have been used which add the capability of blood flow visualization to the high anatomic imaging capabilities of cardiac MRI. Valvular disease and septal abnormalities are directly seen and quantified, regardless of cardiac orientation or body habits. The potential for interactive three-dimensional display of dynamic cardiac activity is now being examined and may become a clinical tool for better appreciation and diagnosis of heart disease (Figs. 8.13, 8.14).

8.1.4.3 Turbo (Very-High-Speed) Techniques

Gradual improvements in MRI technology, particularly in the generation and control of the magnetic gradients, coupled with innovative approaches to signal acquisition and data manipulation (e.g., realizing that for a "real" structure only half the spatial frequencies need be sampled in order to generate the final image —halving the image acquisition time) have led to dramatic reductions in the times needed to acquire an MR image. Reasonably good, high-resolution images can now be acquired in less than 0.5 s. These techniques are being clinically applied in order to image abdominal and thoracic structures, eliminating the motion artifacts that plagued earlier MRI techniques.

These high-speed techniques are also being used in conjunction with contrast agents in order to study organ perfusion (especially in the heart and brain), with the hope of adding functional information to the high-resolution anatomic capabilities of MRI.

8.1.4.4 Contrast Agents

Although MRI has intrinsically high soft-tissue contrast capabilities, the use of exogenous contrast agents enhances the visibility and conspicuity of pathology and can sometimes allow demonstration of function. Paramagnetic contrast agents, (chelated compounds of Gadolinium) reduce the T_1 times of nearby protons, increasing signal intensity at sites of contrast agent accumulation. Contrast agents

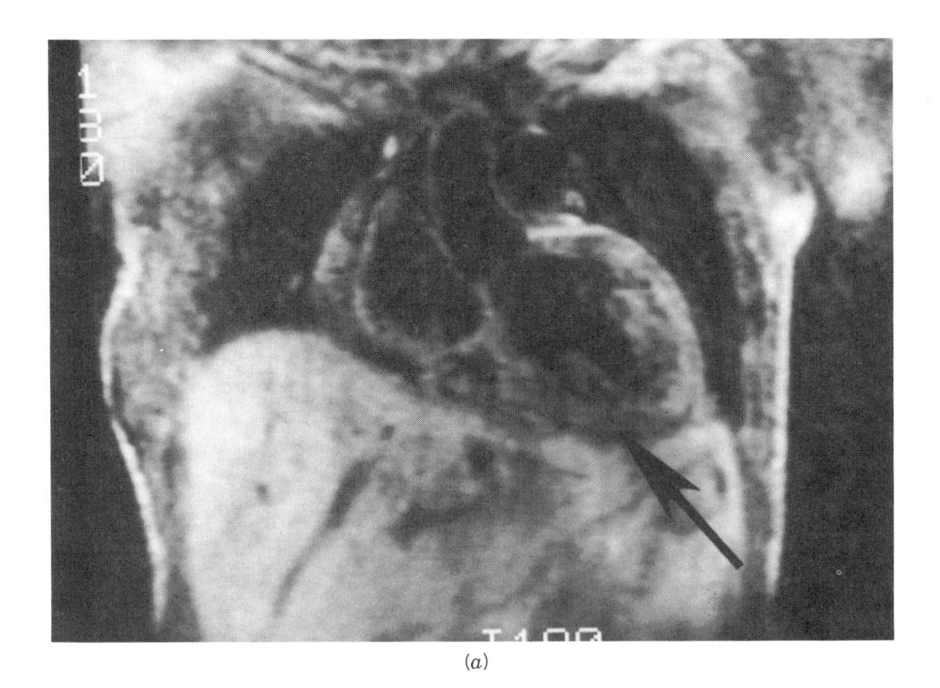

(*a*)

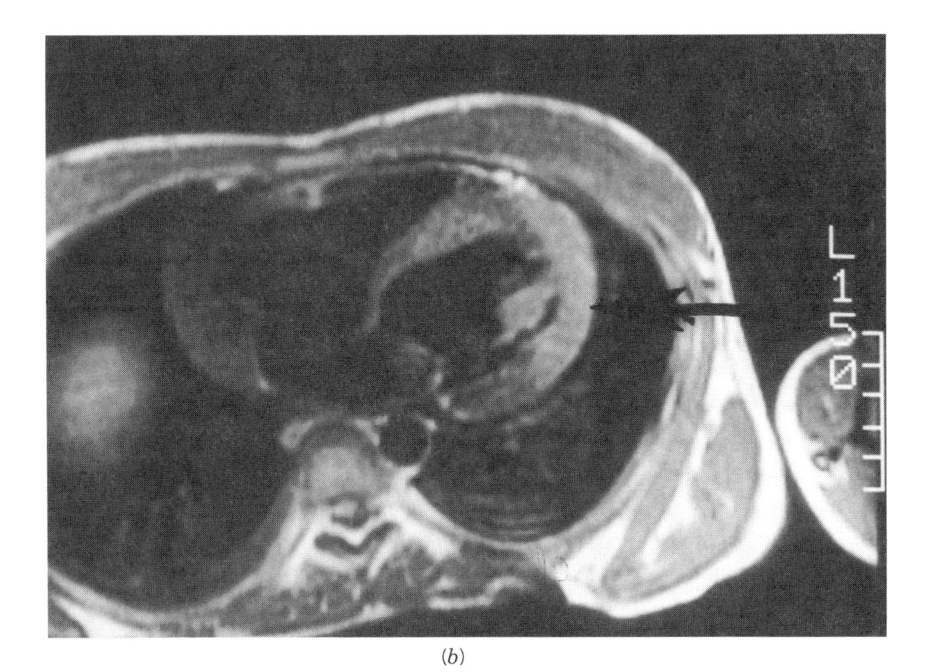

(*b*)

FIGURE 8.12

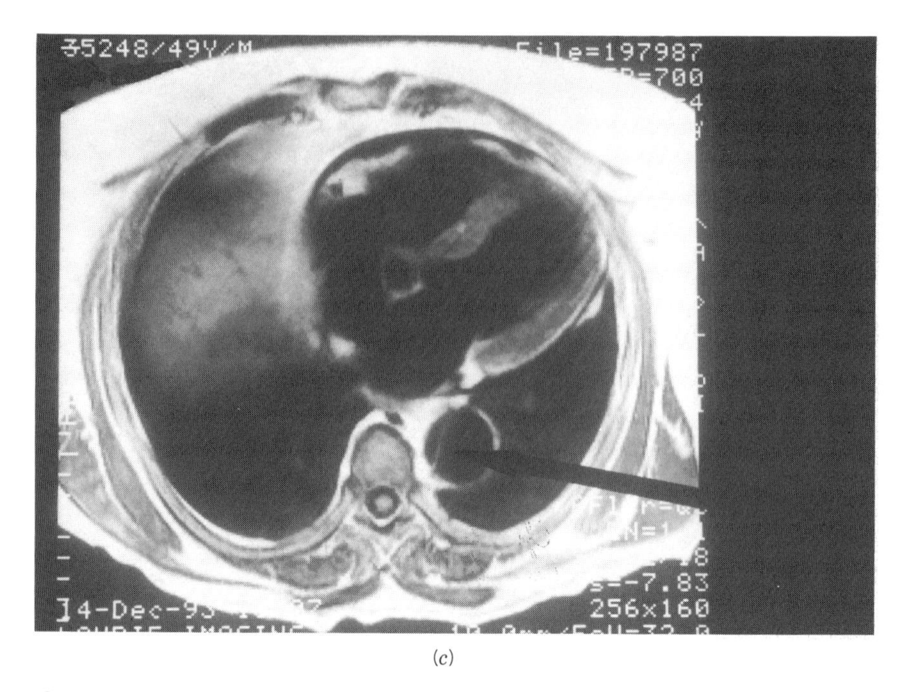

(*c*)

FIGURE 8.12 Coronal (*a*) and axial (*b*) views of the heart. The images are obtained synchronized to the cardiac rhythm to provide stop-action views. The large papillary muscle in the left ventricle in this young patient is clearly seen on both axial and coronal views (*arrows*). (*c*) Aorta dissection. Axial view synchronized to the cardiac rhythm shows the heart and immediately behind it the descending aorta. The thin line extending through the aorta (*arrow*) is one section of the flap where the inner wall has separated from the outer wall of the aorta. If untreated, this could lead to rupture of the aorta and exsanguination of the patient. The MRI is an effective way to illustrate these intimal tears.

have so far found greatest application in the study of neurologic disease, where pathologic destruction of the usual blood–brain barrier (by either stroke or tumor) leads to contrast accumulation in areas where contrast would normally not be found.

Newer agents, based on ferromagnetic agents, show promise in other soft-tissue application (especially the liver), and recent experiments suggest that small iron-based compounds can be bound to monoclonal antibodies for enhanced visualization of tumors and metastases.

8.1.4.5 Functional Imaging

All the MRI techniques described above display internal anatomy with exquisite detail, and all the research and development efforts focused on these techniques are devoted to improving spatial and temporal resolution. Recently techniques have been proposed and demonstrated that allow visualization of function . . . in particular brain function. The approaches take advantage of the fact that as brain

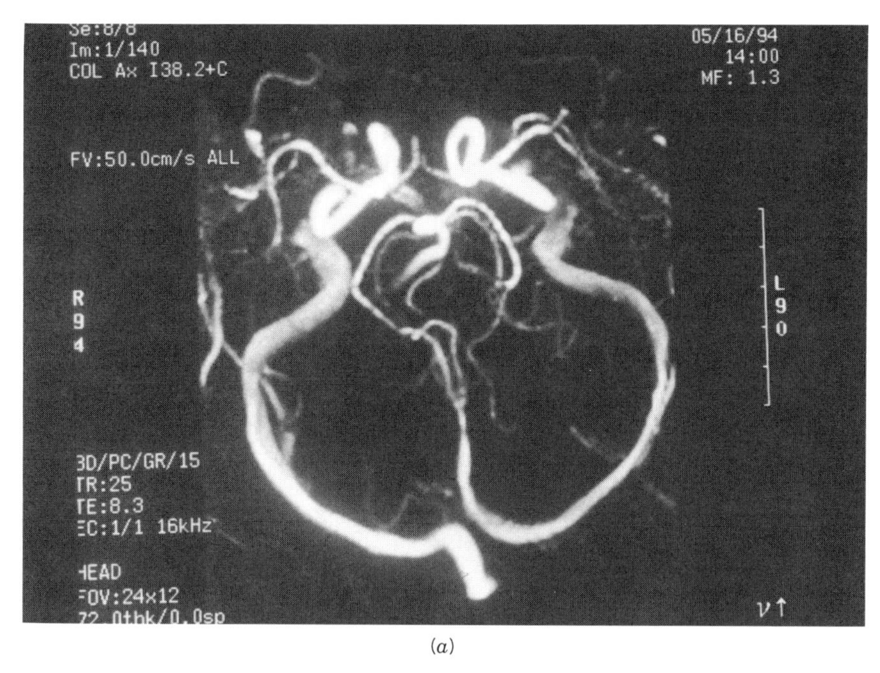

(*a*)

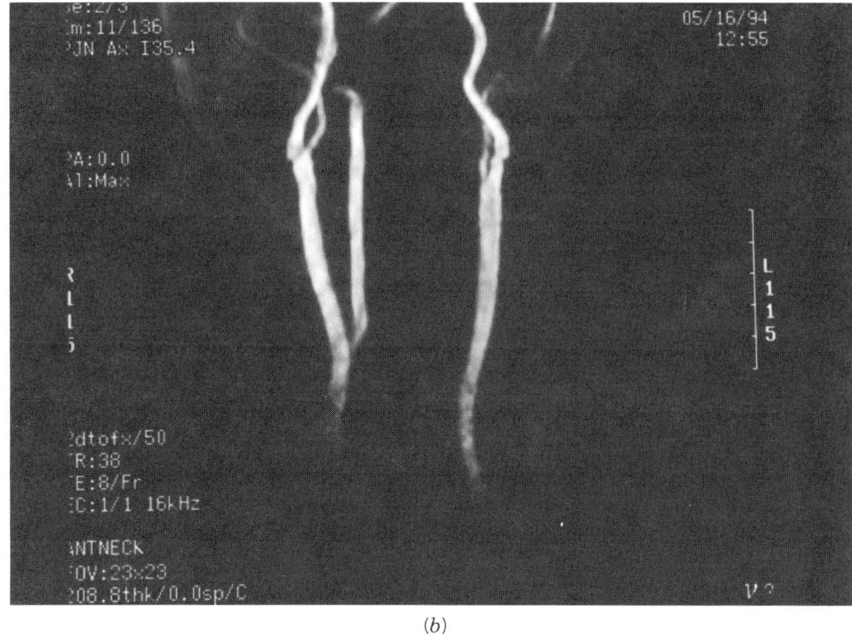

(*b*)

FIGURE 8.13 MR angiograms of the brain (*a*) and carotid arteries (*b*) of the neck. These images are acquired without injection of contrast and have been demonstrated to be an accurate means for screening the circulation in the neck and brain for the presence of stenosis.

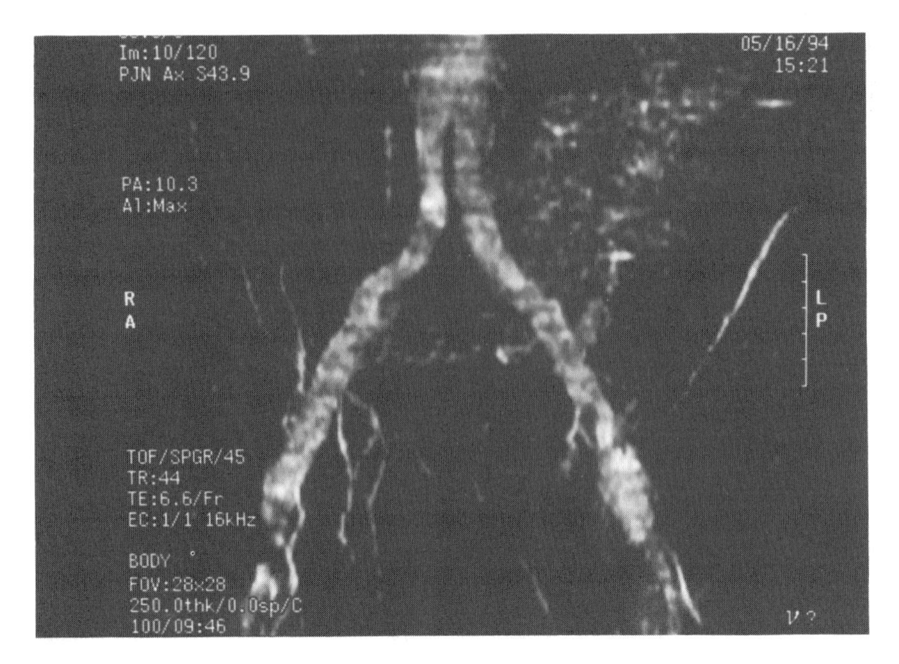

FIGURE 8.14 Frontal view of an MR angiogram taken of a patient's pelvis showing the aorta bifurcating into the common iliac vessels. There are high-grade stenoses of both iliac vessels. This examination is done without the use of intravenous contrast of any sort. It is an ideal method to screen patients and a first step in planning surgery.

cells become active their metabolism increases, raising local blood perfusion and increasing the use of oxygen. One method for brain function visualization acquires images before and after intravenous contrast injection (Gd–DTPA is commonly used). By subtracting the two images, made while the patient is performing some task such as tapping two fingers together, allows direct display of differences in perfusion, which are proportional to brain activity associated with the task. A more subtle but potentially more powerful technique takes advantage of the fact that the magnetization of oxygenated hemoglobin is different from the magnetization of deoxygenated hemoglobin. Since active brain cells avidly take oxygen from the red blood cells perfusing the brain, there will be relatively increased concentrations of deoxygenated hemoglobin, and therefore differences in local magnetization, at sites of increased brain activity. Use of acquisition protocols sensitive to small magnetization differences (such as gradient echo techniques) allows direct and continuous visualization of brain activity, without the need for contrast injection. These methods are attracting great interest as clinical and research tools with potential applications for study of patients with seizures, psychologic disturbances, and changes in mental status. It is too early to tell what the limits of applicability of this technique will be.

BIBLIOGRAPHY

Atlas S: *Magnetic Resonance Imaging of the Brain and Spine.* Raven Press, New York, 1991.

Belliveau JW, Dennedy DN, McKinstry RC et al: Functional mapping of the human visual cortex by magnetic resonance imaging. *Science* 254:716, 1991.

Edelman RR, Hesselink JR: *Clinical Magnetic Resonance.* Saunders, Philadelphia, 1990.

Higgins CB, Hricak H, Helms CA: *Magnetic Resonance Imaging of the Body.* Raven Press, New York, 1992.

CHAPTER NINE

Medical Ultrasound Imaging: State-of-the-art and Future

DAVID VILKOMERSON, Ph.D., *EchoCath, Inc., Princeton, NJ*

9.1 INTRODUCTION

Ultrasound imaging commands the largest share of the diagnostic medical imaging market; as shown in Figure 9.1, since 1990 it has led all other advanced forms of diagnostic medical imaging, responsible for more than one-third of all expenditures for medical imaging instrumentation.

The rapid rise in importance of ultrasound imaging reflected in Figure 9.1 is the result of the technical improvements in ultrasound images that have made ultrasound-based diagnoses definitive; as ultrasound examinations cost less and do not require either radiation or contrast agents, ultrasound imaging has become the modality of first choice.

Ultrasound imaging is a rich and diverse medical diagnostic field (see, e.g., Ref. 1), the medical aspects of which we do not even attempt to review here; rather, we survey the present technology used in medical ultrasound imaging, and discuss the areas where we believe future developments will take place. For these developments, we sketch the general approach and provide recent references (not the usual original reference but one that shows the current state of the art) for details of these developments.

New Frontiers in Medical Device Technology, Edited by Arye Rosen and Harel Rosen
ISBN 0-471-59189-0

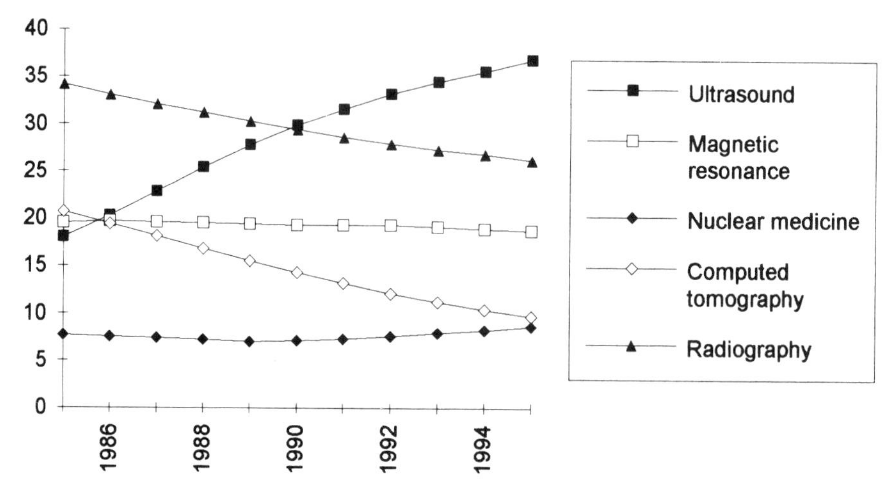

FIGURE 9.1 The percentage share of the different imaging modalities of total domestic purchases of diagnostic imaging systems, 1985–1995. After *Hospitals*, Nov. 5, 1990.

9.2 IMAGING

9.2.1 Propagation Considerations

Medical ultrasonic imaging uses ultrasound frequencies of 1–30 MHz, corresponding to wavelengths of 1.5–0.05 mm. (In the body, the ultrasound wavelength in millimeters is equal to 1.5 divided by the frequency in megahertz.) The reflectivity of human tissue in this frequency range is low, on the order of −30 dB per centimeter; while the low reflectance means reflected signal levels are low, the low reflectivity also allows an ultrasound pulse to propagate through the body. How deeply the pulse can penetrate depends on the frequency used: tissue absorption plus scattering produces attenuation of about 0.5 dB per centimeter per MHz, for example, 5 dB/cm at 10 MHz.[2] We cannot use an arbitrarily low frequency to ensure penetration of the body, however, because the higher the frequency used, the better the resolution. In fact, like other diffraction-limited imaging systems, medical ultrasound imaging systems have (optimally) resolutions equal to a few wavelengths, and the wavelength is inversely proportional to the frequency. Therefore, for the diagnostic range of 1–30 MHz noted above, the resolution ranges from a few millimeters at 1 MHz to a few hundred micrometers at 30 MHz. *Medical diagnostic ultrasound imaging systems use the highest possible frequency that still allows the reflected signals from the deepest tissue needed to be imaged to reach the system with adequate signal–noise ratio.* The limit on peak transmitted ultrasound power, which limits the intensity of the ultrasound pulse that is sent out, is set by the U.S. Food and

Drug Administration (FDA), and depending on the exact imaging situation, is on the order of 100 W peak. (See Ref. 3 for a fuller discussion of output power constraints.)

9.2.2 Image Formation

As in radar and sonar, the information about the image is obtained by transmitting a pulse and receiving the reflected pulses ("echoes"). The energy in the pulse is spatially constrained in both width and length: its width is described by the *beam pattern* of the transmitter; its length, by the duration of the pulse. Imagine a pulse transmitted into the body at time $= 0$ and at an angle θ from the perpendicular. The pulse travels at a known velocity in the body, 1.5 mm/μs, so the time that an echo appears gives the distance r to that reflector; the angle of the beam is known, so the reflector can be mapped into the image at the cylindrical coordinates r, θ. Then another pulse is sent at $\theta + \Delta\theta$, where $\Delta\theta$ is the width of the beam, and another image "skinny pie slice" is obtained. The beam scans progressively across the region of interest, forming an image of the reflecting structures the beams have encountered.

The number of such beams required depends on the angle (or size) of the region of interest; the diffraction-limited angular extent (in radians) of an ultrasound pulse is approximately λ/a, where λ is the acoustic wavelength and a is the size of the transducer being used (discussed below), so the number of beams must be the angle divided by λ/a. For example, if a 60° region is to be imaged using 0.3 mm ultrasound (5 MHz) and a 5-cm scanhead, a total of 166 beams would be required. If a depth of 12 cm were to be reached (typical for a 5-MHz scanhead), each transmission would require 160 μs (total path length of 240 mm out and back divided by 1.5 mm/μs), and the total image formation would require (at the very least) 27 ms (i.e. 166 transmissions of 160 μs each). Formation of images at 27 ms each would allow 37 images per second, "real-time" operation.

9.2.3 Beam Forming and Scanning

Until the mid-1980s, most beam forming and scanning was accomplished by mechanically sweeping a focused ultrasound transducer over a 60°–110° sector. Although this produced an excellent image in the region of focus of the transducer, the limited region-of-focus of any particular scanhead meant that a complete examination required changing scanheads a number of times. (The lower the transducer's *f*-number, the ratio of its focal length to its aperture, the better the resolution, proportional to *f*-number multiplied by λ; however, the smaller the *f*-number, the shorter the region-of-focus, requiring a tradeoff of resolution for examination time.) Phased-array transducers have displaced mechanically scanned focused transducers in general-purpose medical ultrasonic imaging systems (as phased-array antennas have replaced mechanically scanned antennas for high-performance radar systems). As shown in Figure 9.2, a phased array consists

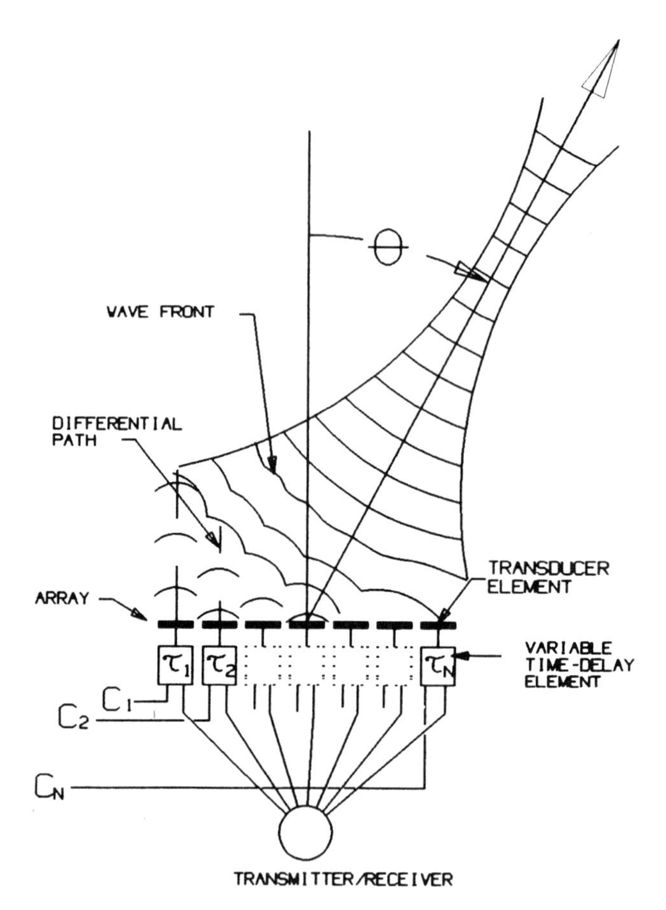

FIGURE 9.2 A phased-array system producing a focused beam at angle θ. The time-delay elements provide the differential delays that form the wavefront.

of individual transducers, each connected to an electronically controlled variable time-delay element. By setting the time delay for each transducer, the send/receive direction θ and wavefront curvature ("focus") of each pulse is controlled, as the wavefronts from each of the array elements constructively combine in a synthesized beam, which is an embodiment of Huygen's principle.[4] In addition, the elements can provide varying attenuation for the transducer elements in the array, improving the beam's profile, a process called *apodization*.[5]

The great advantage possessed by phased-array scanning is its capability to vastly increase the region in focus. Because the velocity of the transmitted pulse into the body is known, the *receive* focus can follow the transmitted pulse into the body, achieved by changing the variable-delay elements. As a result, the whole image depth is "in focus," rather than the limited region-of-focus of a single element transducer.

While the *transmitted* pulse can be focused at one depth only, by taking subimages, with the *transmit* focus at the center of each subimage, the full advantages of a double-focused imaging system (which can be considered to roughly double the resolution of a one-way focused system) can be obtained, at the price of a reduced frame rate. Figure 9.3 shows what a state-of-the-art system can achieve.

The phased-array beam direction and focus can be controlled by electronically setting the variable-delay elements. Under control of a microprocessor, the phased array has great agility in flexibly changing its sector size and scan pattern. Another advantage is the greater reliability of an all-electronic system compared to a mechanical one.

The result of these advantages is that high-end general-purpose radiology ultrasound imaging systems employ phased arrays. While mechanically scanned annular array transducers (which can be considered to be a phased array rotated

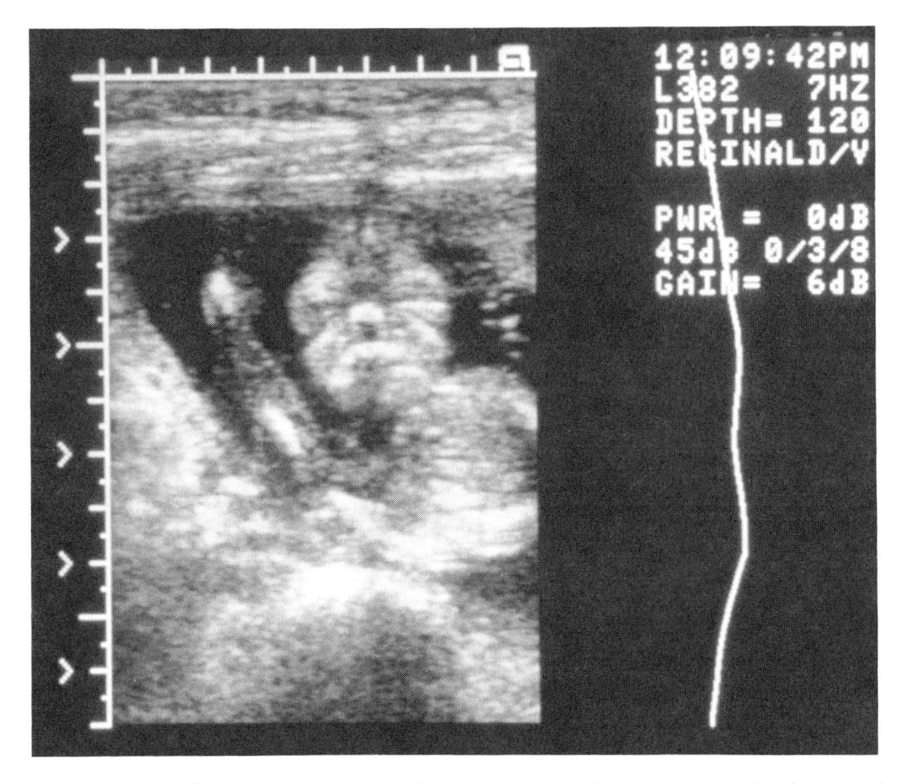

FIGURE 9.3 A 20-week-old fetal face (the nose is near the center, opposite the second arrow on the left; the eye sockets are above the nose, the developing dental arch below the nose) and arm (to the left). This image is a multizone focus composite, the arrows on the left indicating the points of transmit focus of the subimages that compose the image. Courtesy of Acuson.

around the axis of a single transducer) have an extended region-of-focus and – an advantage over a phased array – a symmetric beam profile, such systems still have limited flexibility in pulse and scan patterns and the lowered reliability of mechanical systems; they are generally used for special-purpose and lower-end medical systems.

9.2.4 Specialized Scanheads

Closer proximity to the region to be imaged, allowing shorter propagation lengths, enables the use of higher frequencies and lower transducer f-numbers; these factors produce higher-resolution images.

To gain this advantage, in recent years scanheads have been developed to get closer to certain important regions of the body: endovaginal (*endo* means within) scanheads to get closer to the uterus, endorectal scanheads to get closer to the prostate and rectum, and transesophageal echocardiographic (TEE) scanheads to get closer to the heart (by imaging through the wall of the esophagus that runs directly behind the heart). These scanheads may be phased arrays or mechanical scanners. They have improved the diagnostic capability of ultrasonic imaging for these important body structures. Figure 9.4 shows the various scanhead designs available for a modern ultrasound medical imaging system.

The present maximal extension of this principle of getting close is the endoluminal (*lumen* is the inside of a tubular structure, such as a duct or blood vessel) scanhead; such scanheads have diameters of as little as 0.9 mm and are mounted on the end of flexible tubes (catheters). Such catheters can be slid inside blood vessels, so that a catheter, inserted into an artery where it is close to the skin (e.g., near the groin), can be pushed through the circulatory system so that it can image regions of partial blockage in the arteries of the heart. The propagation paths for such transducers is typically one-tenth that of those used in abdominal imaging, and frequencies as high as 30 MHz are used with such transducers. Figure 9.5 is an image from such an endoluminal system, and shows a cross section of a heart artery with plaque (a deposit within the artery that interferes with circulation) at the 7 o'clock position. Such ultrasonic imaging of the inside of the heart or its arteries was previously impossible, and so opens up whole new areas for the use of diagnostic ultrasound.[6]

These small imaging systems have been used to image inside the abdomen as well, using not a natural opening but the small channels available through trocars. Such laparoendoscopic imaging[7] is a natural complement to optical laparoscopy, as ultrasound can penetrate into and allow visualization of the organs rather than only the organ surfaces as is possible with optical laparoscopy.

Figure 9.6 shows the scanhead that produced Figure 9.5. It consists of a transducer angled so that the resulting beam is almost perpendicular to the axis of the catheter; the counterwound cable rotates the transducer, producing the cross-sectional (or PPI in radar terms) image. The outer case protects the rotating core. A thin coaxial cable (coax) connects the transducer with the electronics outside the body.

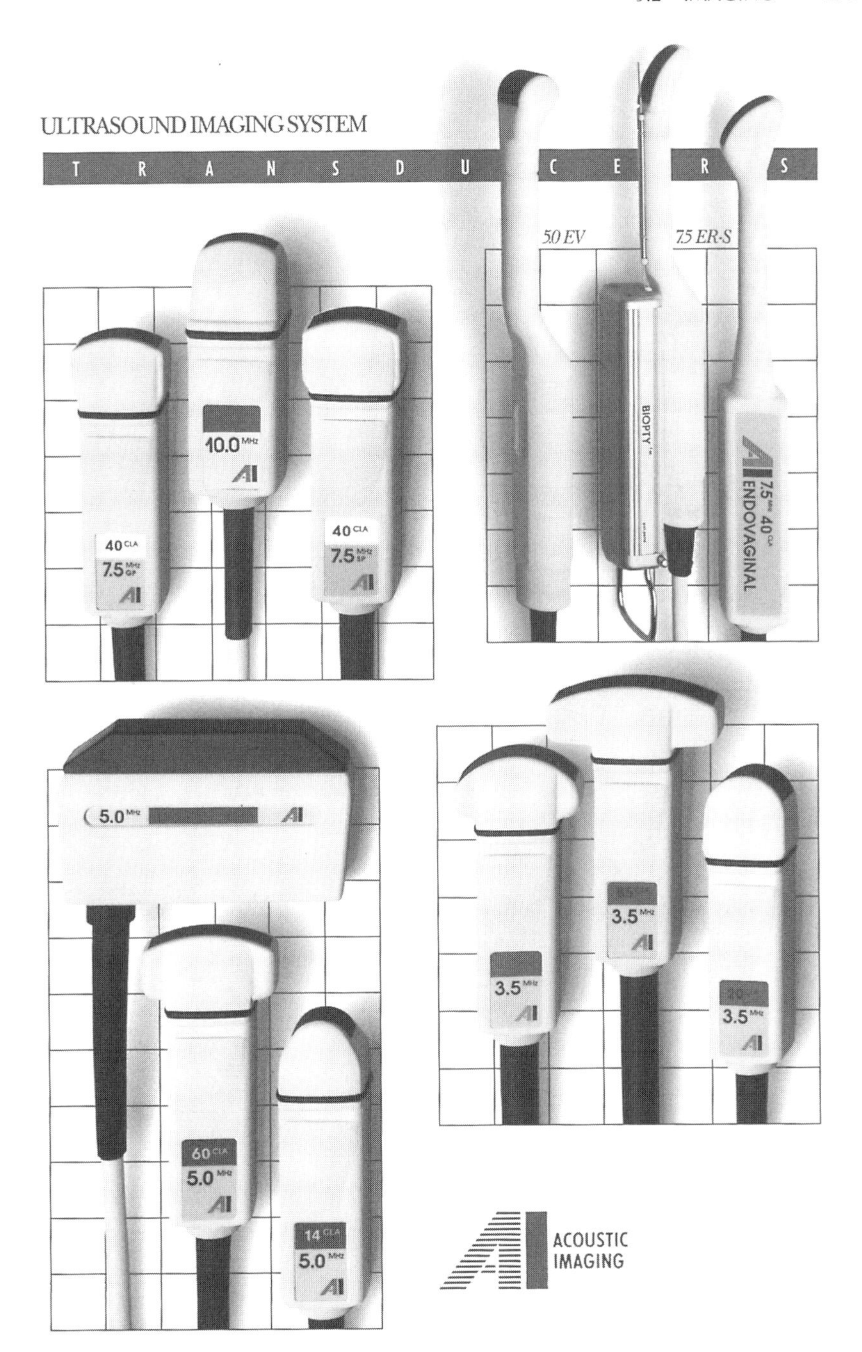

FIGURE 9.4 A variety of scanheads available for different imaging situations. Courtesy of Acoustic Imaging.

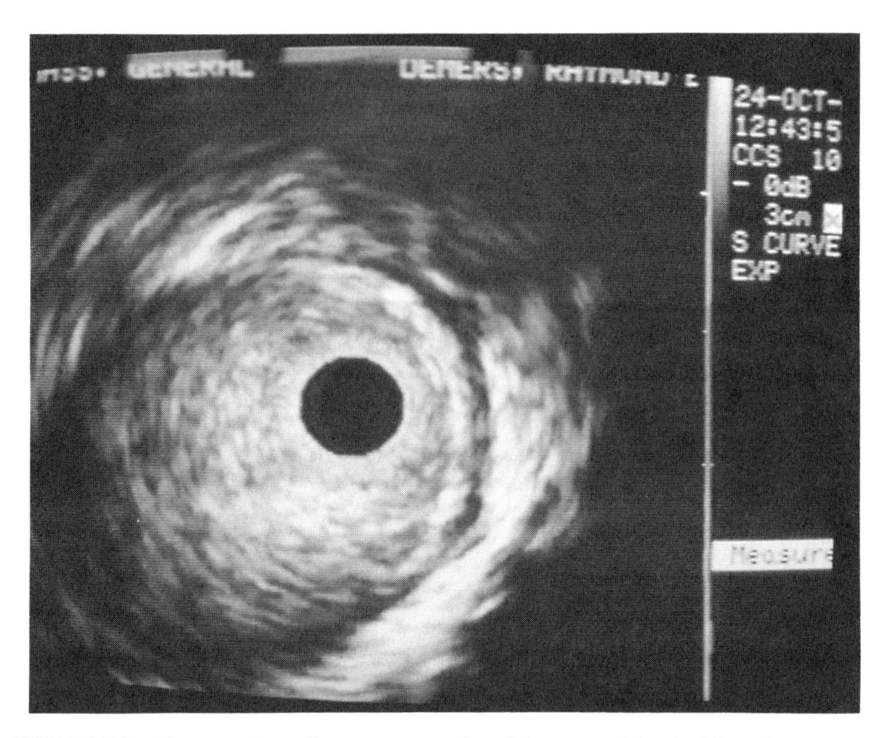

FIGURE 9.5 Cross section of an artery, produced by an endoluminal imaging system; note the reflective plaque in the upper right. Courtesy of Robert J. Crowley, Boston Scientific Corp.

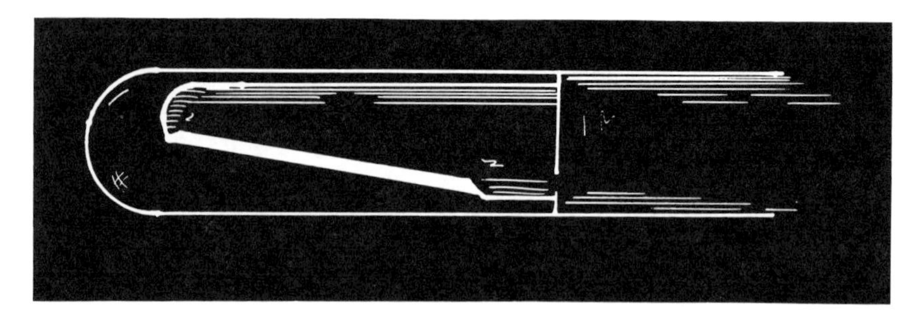

FIGURE 9.6 The transducer end of the Boston Scientific/HP Endoluminal Scanner; the transducer is the plate near the end that points almost perpendicularly to the axis of rotation. Courtesy of Robert J. Crowley, Boston Scientific Corp.

Other endoluminal scanheads use fixed transducers and rotating mirrors, or phased arrays. Some of the advantages of phased arrays are lost in endoluminal scanning because the cylindrical shape of the transducer surface limits the number of transducers that can be active to form a beam; the directivity of each transducer limits how many elements can contribute to the synthesized wavefront, and therefore the resolution obtainable is limited.

9.3 DOPPLER

Moving blood cells affect the frequency of the ultrasound reflected from them, just as the speed of a moving train affects the frequency of its whistle. The change in frequency caused by motion is known as the *Doppler shift* in frequency. Ultrasonic imaging systems have been used to locate the best point in a blood vessel to measure the velocity by measuring the Doppler shift, what is known as *duplex imaging*, but the Doppler ultrasound signal path was separated from the image ultrasound signal path. Now in state-of-the-art systems Doppler information about the entire field of interest can be obtained, using the same signals for Doppler as used for imaging.

The Doppler shift can be measured by determining how the phase of the ultrasound waves in the backscattered pulse changes as pulses are sent along the same beam direction and observed for the same travel time. As a reflecting red blood cell moves toward us, we can imagine that the phase of the reflected wave decreases with each successive pulse. By measuring how rapidly the phase changes, we can determine how fast the red blood cell is moving. (Technically, by quadrature-detecting time samples of the backscattered signal, equivalent to homodyne detection of the signal with sine and cosine multiplication at the transmitted frequency, we can detect the real and imaginary parts of the signal vector at the sampling time.) The actual signal we observe is the sum of the individual signals from each blood cell, and its amplitude changes slowly (assuming that the blood cells stay in the same relative positions) but advances or retreats in phase depending on whether the blood cells are moving toward or away from the transducer. Figure 9.7 shows how we can measure the speed of blood with two measurements in time: the rotation around the circle is proportional to the phase of the reflected ultrasound. If we obtain a signal vector at time t_1, that is, pulse echo at time t_1, and after a time t obtain a signal vector from the same blood sample at t_2, the signal vector will have advanced, as shown, from $\theta(t_1)$ to $\theta(t_2)$. As we continue to measure the signal vector, we find it rotates at a rate f_d, which can be shown to be the Doppler frequency as calculated by the standard physics formula. Note, however, that the two measurements shown in Figure 9.7 are sufficient to determine the velocity: as we know the interval between measurements, and that the phase shift is proportional to the distance (in wavelengths) that the backscattering cells have moved, we can calculate the velocity, specifically, the distance divided by time.

If the velocity is such that the signal vector moves more than π degrees, the direction of flow may be interpreted incorrectly, as shown in Figure 9.7 for $\theta'(t_2)$. This graphical ambiguity can be rephrased as saying that the pulse repetition frequency (PRF) must be higher than half the Doppler frequency to be measured (in either direction) if the ambiguity is to be avoided. If the Doppler

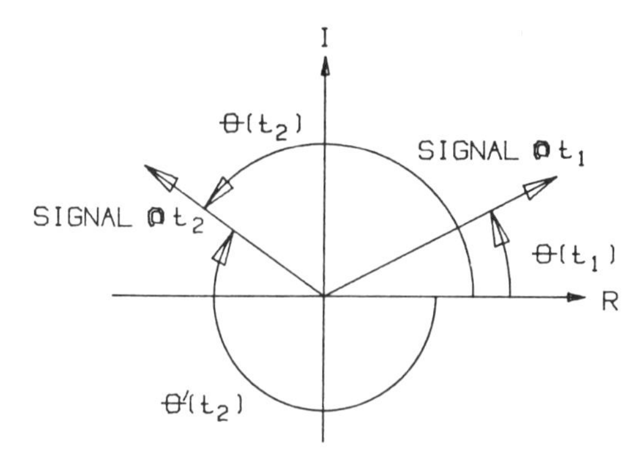

FIGURE 9.7 The quadrature-detected signal from moving blood at times t_1 and t_2; from the rate of rotation of the vector, the Doppler frequency can be calculated. The vector at $\phi'(t_2)$ represents an aliased signal.

signal is so high that the blood appears to be moving in the direction opposite to its actual direction of flow, the signal is considered *aliased.*

In the 1980s, workers at Aloka recognized that the two samples at a point (or a few more, to increase the signal–noise level) were sufficient to calculate the velocity.[8] Therefore, using the imaging pulses, the velocity of backscattering tissue at every point in the ultrasonic image could be determined. This velocity information was shown by coloring pixels in the image according to their velocity. This method of showing the velocity of tissue over the entire image by means of color is variously called *color Doppler*, *colorflow mapping*, or *colorflow imaging.*[9] (It should be pointed out that what is actually shown is the vector component of *velocity* along the ultrasound beam, not the *flow.*)

To keep publication costs down, it is not possible to show a color image as produced by this method. Figure 9.8, a black-and-white image in which you can see the shades of gray for color, shows how the color information adds information about blood flow.

Today, all advanced ultrasonic imaging systems include color Doppler. In addition, a correlation-shift method of measuring the velocity from the backscattered ultrasound has become a commercially available technique.[10] This method also measures the velocity of the blood by seeing how much the signal changes from one pulse to the next, but instead of measuring the phase shift between the two samples, as in Figure 9.7, it measures the time shift between the two signals as shown in Figure 9.9. By finding how much the second signal must be shifted in time to best match the first signal (technically, by finding the maximum in the cross-correlation between the two signals), and the time interval between samples, one can calculate the velocity.

The advantage of this method is that it overcomes the Doppler method's limited peak frequency range. As discussed, the signal at t_2 could also have resulted from counterclockwise change in phase, shown as $\phi'(t_2)$, corresponding

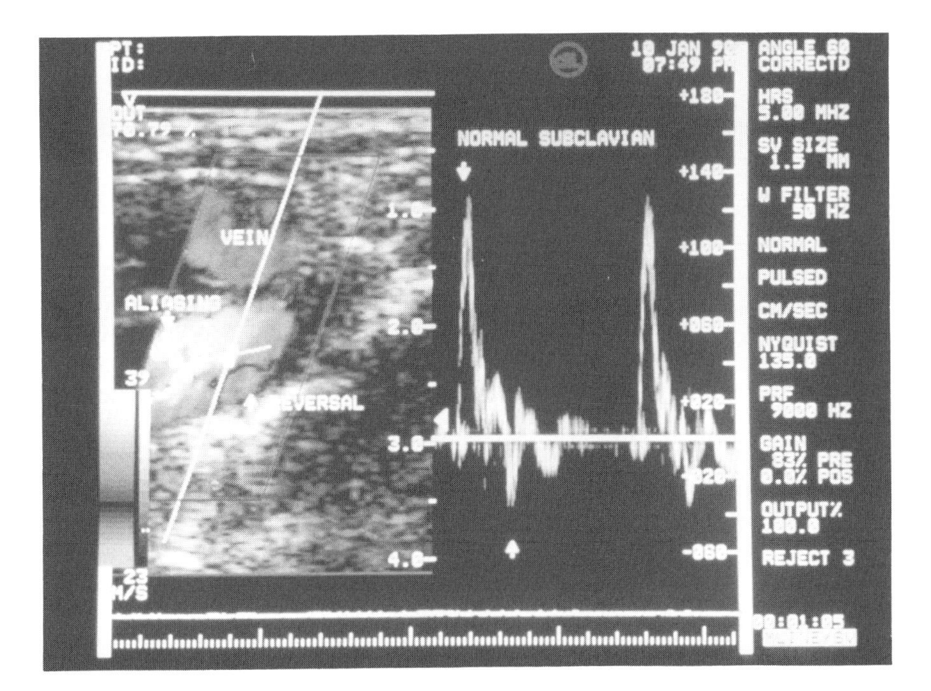

FIGURE 9.8 A colorflow image with spectral display (color here shown by shades of gray); the colorflow gives the overall pattern of flow, while the spectral display, to the right, shows the velocity as a function of time at the point indicated by the crossing of the diagonal line and short horizontal line in the colorflow image. Courtesy of Advanced Technology Labs, Bothell, WA.

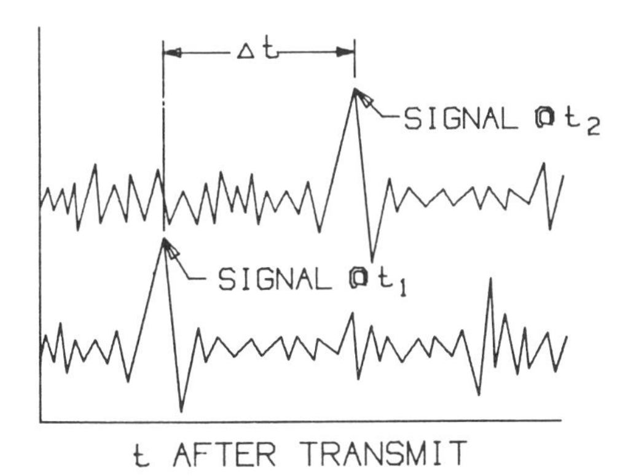

FIGURE 9.9 Correlation method of determining motion. The amount of shift needed to make the two signals the most similar depends on the velocity of the moving tissue and the time between t_1 and t_2; from this the velocity can be calculated without concern for aliasing.

to velocity away from the transducer, rather than toward it. With the time-shift algorithm, there is no ambiguity about direction, and no concern for aliasing.

The development of a number of improved algorithms to derive velocity from backscattered ultrasound has been an active field of endeavor; see the references cited in Ferrara and Algazi.[11]

9.4 FUTURE DEVELOPMENTS

We have tried to describe below the research and development activities that, while not yet in clinical use, will, we believe, become important parts of ultrasonic imaging by the end of the decade.

9.4.1 Tissue Characterization

The pathologist distinguishes between abnormal and normal tissue while examining it under a microscope; and, with the help of stains, can characterize tissues based on their structure and composition. Differences in structure and composition change the interaction between tissue and ultrasound and could serve to differentiate between normal and abnormal tissue, allowing ultrasound to characterize tissue while imaging it.

The key to such characterization is understanding the effect of tissue structure on the backscatter of ultrasound. In analogy to this problem, consider optical filters, which are combinations of thin films that transmit or reflect light depending on the color (wavelength) of impinging light; by shining a white light onto such a filter and seeing which colors are reflected, one can characterize the film structure of the filter. Similarly, by measuring how strongly the different frequencies that make up a short ultrasound pulse (typically an octave in span) are reflected from tissues, one can infer the structure of the tissue. The first use of ultrasonic tissue characterization for clinical decisions was achieved by Lizzi and coworkers:[12] on the basis of spectral analysis of the backscatter of a broadband pulse, eye malignancies are identified and treated. The ultrasound system extracts the needed information about the structure and composition of tissue by recognizing that the power spectrum of the backscattering from a collection of cells is the Fourier transform of the autocorrelation of the cellular arrangement; therefore the *slope* of the spectrum is inversely proportional to the average autocorrelation length, that is, the average size of the cell. The magnitude of the backscattering is proportional to the number of scatterers and their acoustic contrast. The slope of the spectrum then gives the size of the backscattering cells, and the intercept of that spectrum (extrapolated down to zero frequency) gives the number and acoustic impedance variation of cells. Figure 9.10 shows tissue characterization in operation.

Another area where tissue characterization is showing promise is in differentiating healthy from unhealthy heart muscle.[13] This is accomplished by monitoring the backscattered ultrasound and noting whether the backscattered

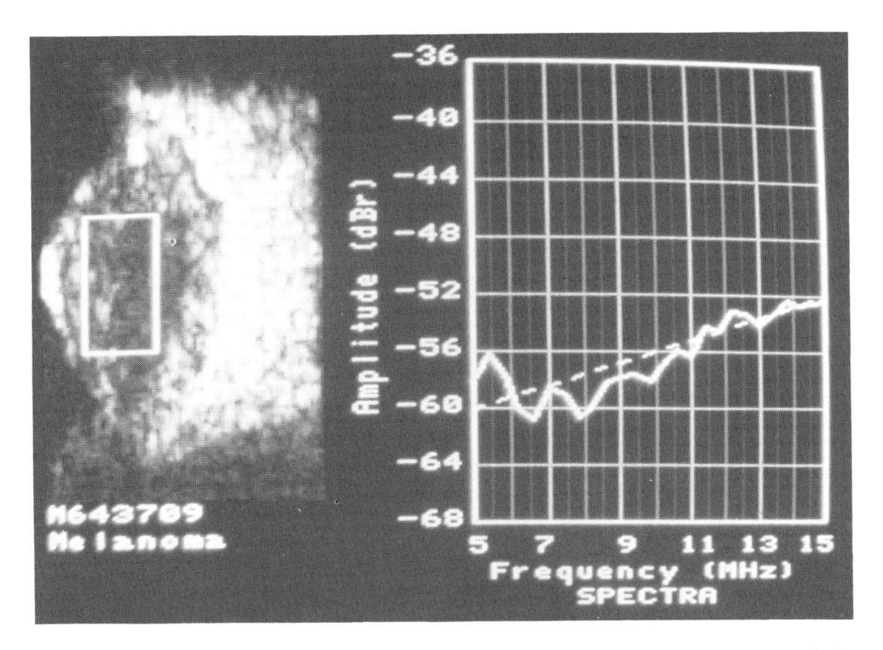

FIGURE 9.10 A melanoma in the retina of an eye, characterized by the spectral slope method. Courtesy of Fred Lizzi, Riverside Research Institute.

energy changes as the heart beats. If the heart muscle tissue is contracting, as healthy tissue should, the backscattering characteristics will change as the heart beats; if there is no change, the tissue is not contracting and the heart muscle is not healthy.

With continued research, tissue characterization will become an integral part of medical ultrasonic imaging.

9.4.2 Elastography

Elastography[14] is an ultrasonic imaging technique that displays the elastic properties of the tissue traversed by the ultrasound. Ultrasound pulses traverse tissue before and after a mechanical force (stress) is applied to the tissue; the change in phase of the sound reflected by the tissue shows the amount of motion (strain) the stress caused. The ratio of stress to strain is the elasticity.

Elastography may be considered another way of using ultrasound to characterize tissue. The presence of many diseases, for example, breast cancer, is indicated by a change in tissue elasticity, which is why palpation serves to find these diseases. Elastography would provide a quantitative visual indication of such diseased tissue.

While as yet no clinical use has been established for elastography, it is an area of keen interest by ultrasound researchers.[15]

9.4.3 Ultrasound Contrast Media

By injecting materials with special ultrasonic characteristics, the information content of the ultrasonic image can be increased; for example, a suspension of bubbles formed with an albumin shell (Albunex®) can backscatter more than 100 times as much ultrasound as can blood. Injected into the blood, the contrast agent shows where the blood is going (perfusion) by the "brightness" in the ultrasonic image.[16] The effect of ultrasound contrast media is analogous to the contrast media used in x-ray imaging, where radioopaque dye is injected to show otherwise invisible blood vessels in the x-ray image. Contrast media are extensively used in x-ray and MRI medical imaging, and by providing physiologic data, that is, which tissues are taking up the contrast media, contrast media can provide critical data to the physician.

Combined with colorflow imaging, contrast media make it possible to see the physiologic state of tissues, such as the increased blood flow to inflamed tissues or possibly the neovascularity characteristic of malignant growth.[17] Use of contrast media in this manner is another method of characterizing tissue.

Although ultrasound contrast media are still experimental, FDA allowance for general usage is expected within a short time.

9.4.4 Propagation Aberration Correction

Achievement of diffraction-limited resolution normally requires a uniform propagation medium. The human body has layers of fat and muscle whose velocity, analogous to the refractive index of optics, may vary from point to point by 5%. The quasirandom variation in propagation velocity means that a smooth wavefront is distorted in shape as it propagates through the body; this degrades the acoustic resolution, as can be recognized when the ultrasonic images from some people are much worse than others. An example of even more severe aberration is the effect of the skull: the aberration it produces prevents good ultrasound imaging of the adult brain.

To remove the "bumps" in the wavefront caused by the changes in acoustic velocity, compensating bumps in the opposite direction must be added in the detector. Such techniques have been developed so that optical telescopes are less affected by velocity of propagation in the atmosphere, and similar methods can be applied to ultrasonic imaging.[18] These techniques correct the distortion in phase by adding a corrective phase to the received wavefront. To do this optically one uses a deformable mirror, to compensate on the mirror surface for the different optical path lengths that cause the bumpy wavefront. In acoustic imaging, the detectors are phased arrays, so the wavefront correction takes place on a point-by-point basis. As described before, a phased array already has phase shifters built into its circuitry, so aberration correction is particularly well suited for phased-array imaging.

Present phased arrays are one-dimensional, which makes correcting the distortion caused by aberrating tissue—which distorts in two dimensions—much more difficult. The two-dimensional arrays discussed next will help solve the difficulties of aberration correction, and we expect aberration correction to be applied increasingly in clinical ultrasound.

9.4.5 Two-Dimensional Arrays and Their Advantages

Making a two-dimensional array of ultrasonic transducers has been a significant technical challenge; indeed, even making high-quality one-dimensional arrays is a combination of science, technology, and art that only a handful of manufacturers have mastered.

However, the advantages of two-dimensional arrays make meeting the technical challenge worthwhile. In addition to making it more feasible to correct for aberrations in the ultrasound wavefront alluded to above, the two-dimensional array makes it possible to significantly increase the amount of image information an ultrasound system can obtain. Multidirectional simultaneous scan beams can be generated, analogous to phased-array radars that have been developed to generate thousands of simultaneous beams. The intrinsic limitation on acquiring acoustic information by pulse-echo imaging is the need to wait for a pulse to propagate to the end of the volume of interest and back; the velocity of sound in the medium and size of the volume determine the time it takes to acquire the image. Using n beams simultaneously will reduce the acquisition time by n.

Although multiple beams theoretically may be generated by one-dimensional arrays, the technical difficulties of doing this, even for small n, have proved almost insurmountable. With two-dimensional arrays, even of limited size, up to 16 simultaneous beams have been used.[19] One important use of higher data rates has been the generation of three-dimensional images of moving structures, such as the heart.[20]

9.4.6 Three-Dimensional Ultrasonic Imaging

Three-dimensional visualization of body structures by computer reconstruction of multiple scan planes of computerized tomographic (CT) and MRI images is of rapidly increasing usefulness; similar activities in ultrasonic imaging are appearing, with the same aim as those for CT and MRI, making the images contain more useful information.

The first commercially available three-dimensional ultrasound system[21] (which is not yet available in the United States, pending FDA permission to market) straightforwardly uses a linear array to generate a conventional two-dimensional (2-D) image, and mechanically sweeps the array in the third dimension; the sum of the 2-D images produces a three-dimensional image.

In addition to this system, almost every ultrasound manufacturer has some sort of three-dimensional visualization system "in the works." One problem all these systems share is the length of time it takes to produce the three-dimensional image: waiting for the image is not the problem; rather, it is the motion of the tissue, caused by the heart or respiratory motion, blurring the three-dimensional image as the tissue moves during the scan.

Three-dimensional images are also routinely generated by the intraluminal scanners discussed above. In the "pullback" technique,[22] the catheter is withdrawn at a regular rate while imaging; the series of round images so generated is stacked, using the same kind of software as used for stacking CT images, and a three-dimensional image is produced. The motion of the vessel is not a problem, because the position of the catheter is determined by the walls, so if the vessel moves, the catheter moves, eliminating relative motion. Unfortunately, for the same reasons, curves in the vessel are not portrayed, because from the perspective of the catheter moving along the wall, there is no curve.

We have recently demonstrated a three-dimensional (3-D) forward-looking catheter-based imaging system that combines intraluminal imaging with real-time 3-D visualization.[23] It mechanically scans a focused transducer in a spiral fashion. At each point on the spiral the transducer is pulsed, and the resulting bundle of scan lines produces an image, as depicted in Figure 9.11. The high resolution and small image volume required for endoluminal imaging dictate high frequencies, which, in turn, means short propagating paths. Because the path is so short, propagation times for each image line is short, so three-dimensional images, constructed from 2000–3000 such lines, can be generated in real time (15 frames per second). Figure 9.12 shows image acquisition and the resulting 3-D image. Note that the total occlusion shown could not be made visible by

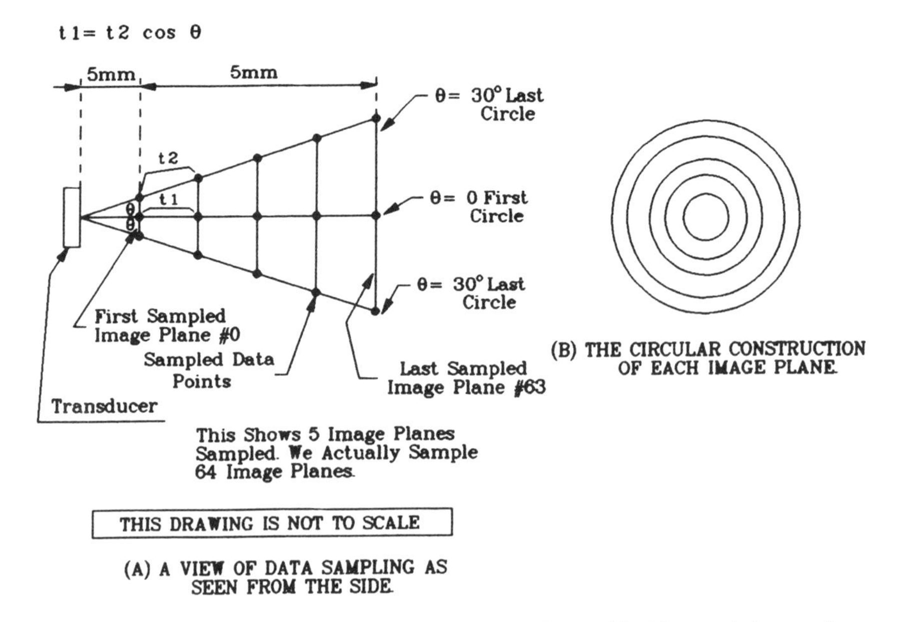

FIGURE 9.11 The spiral scanning pattern of the forward-looking real-time catheter-mounted imaging system.

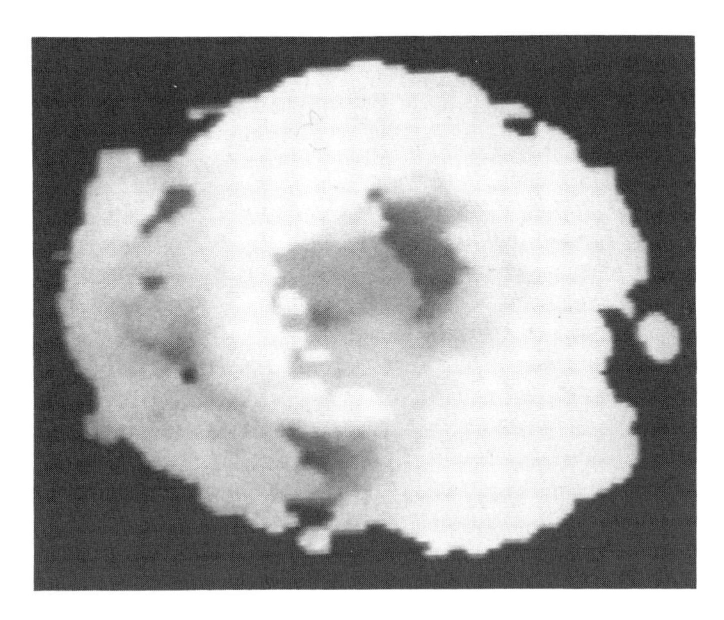

FIGURE 9.12 The image of an occluded femoral artery produced by the system depicted in Figure 9.11.

either x-ray techniques (as no contrast media could outline the lumen) or side-looking ultrasound (as the catheter could not penetrate the occlusion). Such forward-looking systems could be used to guide therapy for such occlusions.

9.4.7 Ultrasound-Guided Interventional Procedures

Interventional radiologists were those who, rather than only diagnose, would "intervene" in the situation. A perhaps better term than "interventional medicine" for their practice is "image-guided-therapy" (IGT), which utilizes imaging capabilities to guide various devices to perform therapy. The most frequent image-guided therapy is balloon dilatation of narrowed arteries (balloon angioplasty) in the heart or limbs. Moving therapeutic devices under imaging guidance to the tissue to be treated, rather than cutting through the overlying tissue to expose the area, minimizes the trauma to the body. Such treatment is much less expensive, has a much faster recovery time, and is safer than traditional techniques. For these reasons, balloon angioplasty has grown from its inception a decade ago to over a million procedures a year.

We have devised a method to guide catheters with ultrasound, rather than by x-ray imaging.[24] By eliminating the need for an expensive radiographic facility and the need for expensive contrast media and cut film, ultrasound guidance can still further reduce the cost and increase the safety of catheter procedures.

Figure 9.13 shows the principle of operation. A small, omnidirectional transducer is placed at the point on the catheter that is important, such as the center of the balloon of an angioplasty catheter. An antenna on the scanhead cable detects the transmission of each ultrasound beam; the reception of the ultrasound energy on the catheter-mounted transducer determines which acoustic beam (ray 54 in Fig. 9.13) hits it, and the distance, by its time of arrival, of the catheter transducer from the transmitting scanhead. An arrow icon is injected into the scanhead by an antenna wrapped around the scanhead. The icon points at the spot in the image corresponding to the position of the catheter-transducer, as shown in Figure 9.14. This procedure has proven itself effective.[25]

Another procedure that will benefit from ultrasound guidance is insertion of needles. There is a frequent need for placement of a needle in a lesion to sample (biopsy) the tissue to determine the cause of the lesion. There is also a growing use of needles to place therapeutic agents – whether a laser fiber to ablate the lesion, or to destroy the lesion by injecting an ablating chemical, freezing it, or using RF energy to burn (cauterize) it out.[26]

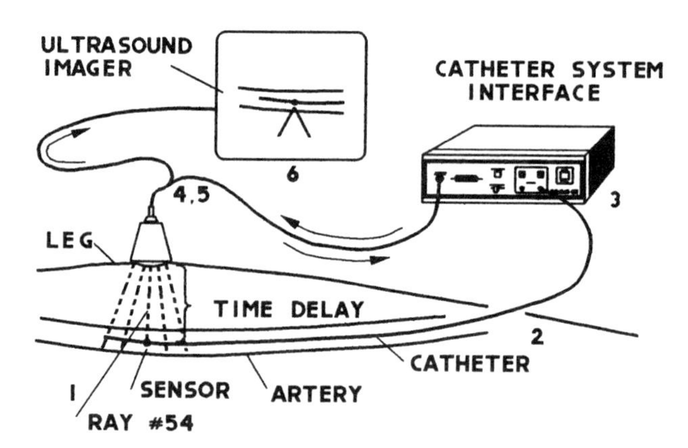

FIGURE 9.13 The EchoMark® system. An imaging pulse strikes (1) the omnidirectional receiver attached to the midballoon region of the catheter; the signal is carried via a wire (2) to the (3) catheter–system interface (CSI); the CSI determines the ray number in the frame (54 in the figure shown), and the time delay between the emission of the pulse and its reception at the sensor, by detecting the transmitted pulses from an antenna that picks up the electronic leakage from the ultrasound system (4); on the succeeding frame, a series of pulses are injected (5) into the scanhead by electromagnetic coupling to form a flashing arrow in the image (6) at the position of the sensor, as calculated from the information gathered in the previous frame.

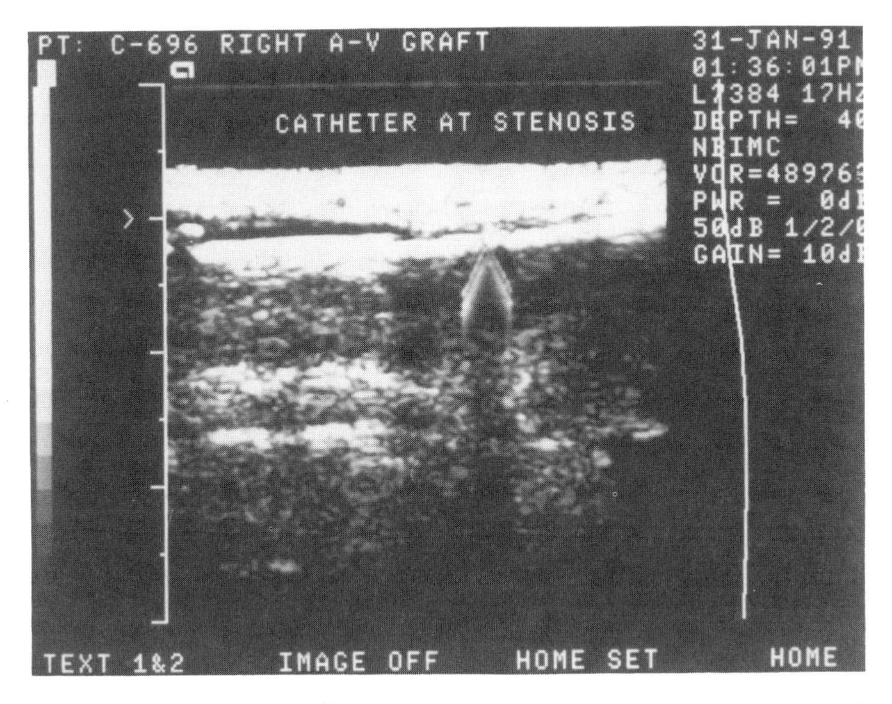

FIGURE 9.14 An EchoMark® image showing the arrow icon pointing to the position of the sensor.

To guide needles for biopsy or therapy, we have developed a means of enhancing the visibility of needles under ultrasound, called the *ColorMark*® system. As shown in Fig. 9.15, by exciting flexural waves of a few tens of micrometers of motion on the needle by means of an exciter near the needle hub, a colorflow imager will show the needle in color against a background of black-and-white tissue, making it easy to guide to its target.[27]

By combining several ultrasound guiding techniques with a trocar-mounted forward-looking three-dimensional real-time ultrasound imaging system described above, it is possible to perform operations deep within the solid organs, such as the liver or kidney. This would allow the benefits of IGT to be extended to many of the operations that today require open abdominal surgery with all the attendant pain and recuperative time.

Time will tell whether this as well as the other developments presented will fulfill their promise; what seems without doubt is that the role of ultrasound medical imaging in medical diagnostic and therapeutic practice will increase.

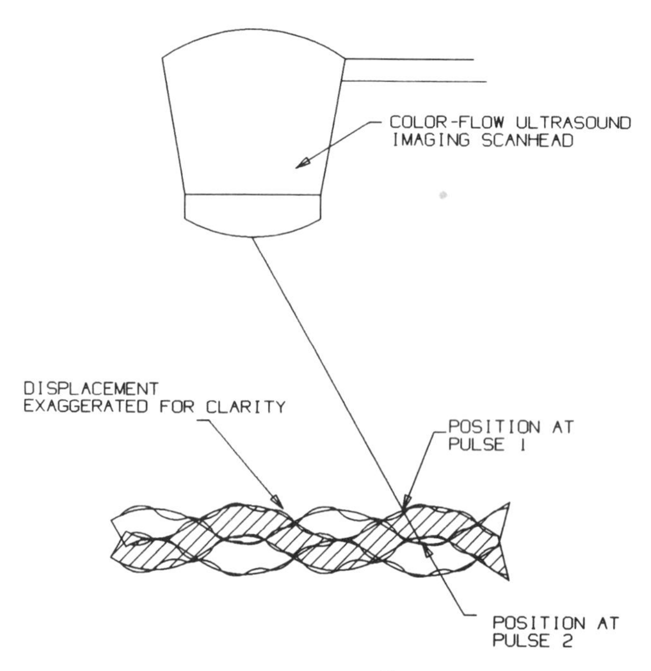

FIGURE 9.15 The principle of the ColorMark® system: a flexural wave, snake-like undulations on the needle, will cause successive pulses to appear to have different path lengths if the flexural frequency is half (or odd multiple of half) of the pulse-repetition frequency, making the needle appear colored in a colorflow imaging system.

REFERENCES

1. Goldberg, BB (ed.): *Textbook of Abdominal Ultrasound.* Williams & Wilkins, Baltimore, 1993.
2. Kremkau FW: *Diagnostic Ultrasound: Principles, Instruments & Exercises,* 3rd ed. Saunders, Philadelphia, 1989.
3. American Institute of Ultrasound in Medicine: *Acoustic Output Measurement and Labeling Standard for Diagnostic Ultrasound Equipment.* AIUM, Rockville, MD, 1992.
4. Kino GS: *Acoustic Waves.* Prentice-Hall, Englewood Cliffs, NJ, 1987, pp. 227–229.
5. Wright JN: Resolution issues in medical ultrasound. *Proc 1985 IEEE Ultrasonics Symposium.* pp. 137–140.
6. Cavaye DM, White RA: *Intravascular Ultrasound Imaging.* Raven Press, New York, 1993.
7. Goldberg BB et al: Sonographically guided laparoscopy and mediastinoscopy using miniature catheter-based transducers. *J Ultrasound Med* 12:49, 1993.
8. Kasai C, Nemekawa K, Koyana A, et al: Real-time two-dimensional blood flow imaging using an autocorrelation technique. *IEEE Trans Sonics Ultrasonics* 32:458, 1985.

9. Mitchell DG: Color Doppler imaging: Principles, limitations, and artifacts. *Radiology* 177:1, 1990; or Ralls PW, Mack LA: Spectral and color Doppler sonography. *Semin Ultrasound CT MR Computerized Tomographic Magnetic Resonance* 13:355, 1992.

10. Bonnefous O, Pesque P: Time-domain formulation of pulse-Doppler ultrasound and blood velocity estimators by cross correlation. *Ultrasonic Imag* 7:73, 1986.

11. Ferrara K, Algazi VR: Comparison of estimation strategies for color flow mapping. In *Acoustic Imaging*, Vol. 19. Plenum Press, New York, 1993.

12. Coleman DJ, Lizzi FL, et al: Ultrasonic tissue characterization of uveal melanoma and prediction of patient survival after enucleation and brachytherapy. *Am J Opthalmol* 112:682, 1991.

13. Waggoner AD, et al: Differentiation of normal and ischemic right ventricular myocardium with quantitative two-dimensional integrated backscatter imaging. *Ultrasound Med Biol* 18:249, 1992.

14. Ponnekanti H, Ophir J, Cespedes I: Axial stress distributions between coaxial compressors in elastography: An analytical model. *Ultrasound Med Biol* 18:8667, 1992.

15. Cespedes I, Ophir J, Ponnekanti H, et al: Elastography: Elasticity imaging using ultrasound with applications to muscle and breast *in vivo*. *Ultrasonic Imag* 15:73, 1993.

16. Goldberg BB: Ultrasound contrast agents. *Clin Diagn Ultrasound* 28:35, 1993.

17. Dock W, et al: Tumor vascularization: Assessment with duplex sonography. *Radiology* 181:241, 1991.

18. Karaman M, Atalar A, Koymen H, et al: A phase aberration correction method for ultrasound imaging. *IEEE Trans UFFFC* 40:275, 1993.

19. Mallart M, Fink M: Improved imaging rate through simultaneous transmission of several ultrasound beams. *Ultrasonic Imag* 15:175, 1993.

20. Pavy HG, Smith SW, von Ramm OT: Real-time volumetric ultrasonic imaging with stereoscopic display. *Ultrasonic Imag* 15:178, 1993.

21. Kretz Technik, Combison 550.

22. See Ref. 6 (above), pp. 64–66.

23. Gardineer B, Lyons D, Vilkomerson D: A forward-looking 3-D imaging endoluminal scanner. *J Ultrasound Med* 12(Suppl):9, 1993.

24. Vilkomerson D, Gardineer B, Lyons D: Theory and practice of beacon-guided interventional ultrasound. *J Ultrasound Med* 11(Suppl):44, 1992.

25. Rice K, Hollier L, Ferrara-Ryan M, et al: Ultrasound-guided balloon angioplasty: A new therapeutic modality. *J Vasc Technol* 17:33, 1993.

26. See Proceedings VI Internatl Congress Interventional Ultrasound, Copenhagen, Denmark (in press).

27. Sinow R, et al: Color Doppler guided breast biopsies with a vibrating needle system. Proceedings 8th Internatl Congress Ultrasonic Examination of the Breast, Heidelberg (in press).

CHAPTER TEN

Thermography: Radiometric Sensing in Medicine

KENNETH L. CARR, D. Eng., *Microwave Medical Systems, Inc., Acton, MA*

According to the American Cancer Society,[1] approximately 180,000 American women were diagnosed with breast cancer in 1993. Approximately 45,500 died from the disease. In the United States, breast cancer continues to be the most common of nonpreventable cancer diagnosed among women. Much of the urgency in improving early diagnosis of breast cancer stems from the tragic and steady rise in the incidence... 1 in 16 women in 1962, to 1 in 9 women in 1993. The incidences of breast cancer have been steadily increasing in the United States since formal tracking of cases through registries began in 1930.[2] There has been no appreciable change in death rate during the same period (see Fig. 10.1).

It has long been known that early detection increases the chance of survival. Results of the 1976 National Institute of Health Survey[3] indicated a dramatic increase in survival (from 56 to 85%) as a result of early detection. Table 10.1 illustrates the importance of early detection with respect to 20-year survival rate.[4] It is an established observation that survival depends on the pathologic stage of disease at the time of treatment; that is to say, size and stage at the time of detection are the key to survival. For example, in 1930 cancer of the uterus was the leading cause of death due to cancer in women in the United States. Cancer of the uterus has declined steadily since that time, in part because of improved hygiene, but primarily because of the development of an early detection

New Frontiers in Medical Device Technology, Edited by Arye Rosen and Harel Rosen
ISBN 0-471-59189-0

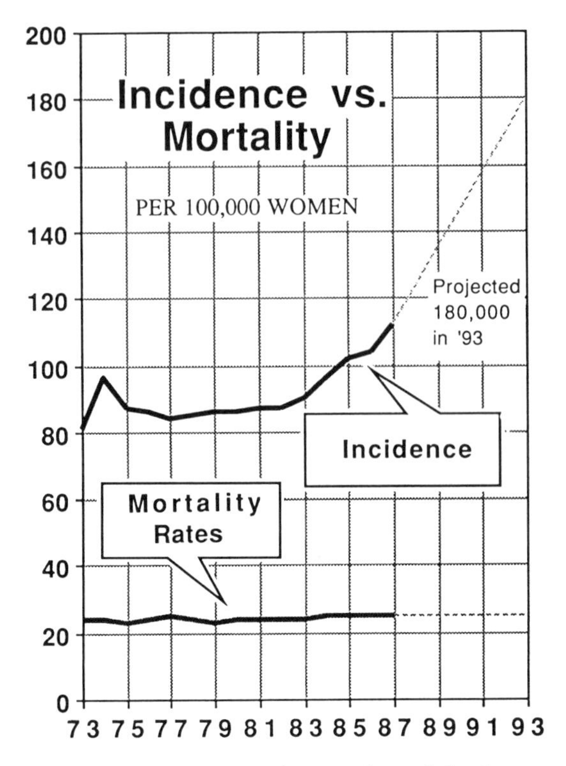

FIGURE 10.1 The outlook. The death rate has not changed, but breast cancer's incidence is rising. (*Source:* National Cancer Institute.)

TABLE 10.1 Twenty-Year Survival Rate—Breast Cancer

Size at Time of Detection (diameter, cm)	Approximate Survival Rate
3	50
2	65
1	80
<1	95

Source: From *Cancer* (February Supplement, 1984).

technique (i.e., the Papanicolaou test). Today, the vast majority of all breast tumors are found by physical examination by either the patient or the examining physician. The result is that long before a breast tumor can be detected by present technology, nodal involvement may occur.[5,6]

Figure 10.2 illustrates the long preclinical existence of breast carcinoma.[7,8] The curve was generated by measuring the growth over a given period of time and,

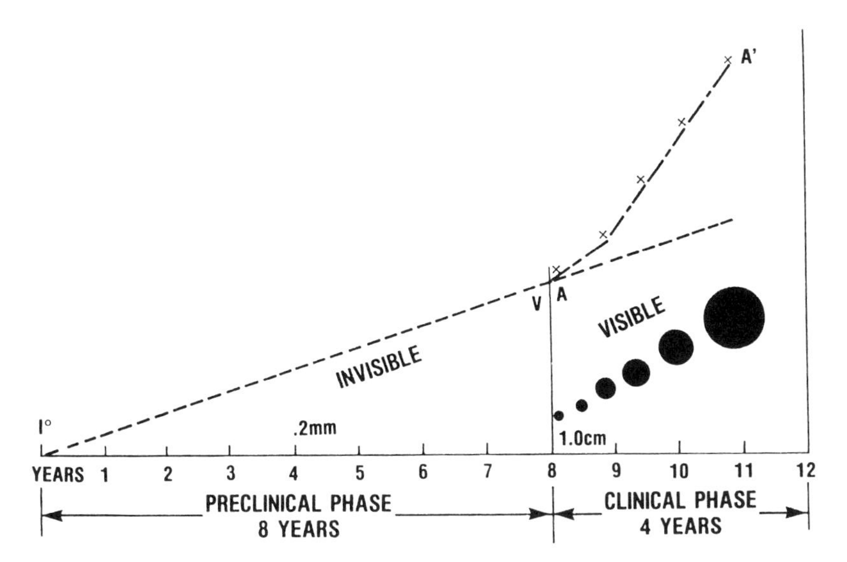

FIGURE 10.2 Doubling time of tumor in relation to clinical phase. Schematic representation of the life cycle of a breast cancer with a doubling time of 100 days.This demonstrates that when it reached the size of 0.2 mm (which may have taken 4 years) there are still 4 years to go before it becomes a 1.0 cm mass. The visible or clinical phase of a breast cancer may be only a short period in its life history.

assuming the growth rate to be constant, in this case using a tumor *doubling time* of 100 days and extrapolating to establish the time of inception. Accordingly, the visible or clinical phase (i.e., when a tumor diameter of 1 cm is achieved) occurs on the average 8 years after inception. Unfortunately, the average tumor diameter when first detected and diagnosed as malignant is approximately 2–2.5 cm and typically is not a localized disease.

Mammography will remain the standard against which new screening techniques will be compared. According to Lundgren,[8] however, the average diameter detected by mammography was 75% of the average diameter detected by palpation. This is not adequate lead time. (The time assumed to be gained in the diagnosis of breast cancer by screening a population of apparently well women is known as the *lead time.*)

Evidence suggests that metastases occur very early in the course of the disease. While cancer cells can be released at any time during tumor growth, the larger the tumor, the larger the number of cells released. According to Gullino,[9] "We know that the great majority of circulating neoplastic cells are destroyed, but the higher their number the higher is the frequency of metastasis. On this ground, early diagnosis and removal of the primary tumor is essential."

Present detection techniques, other than thermography, require that the tumor have mass and contrast with respect to the surrounding tissue (i.e., palpation or physical examination, mammography, ultrasonography, and diaphonography).

Thermography, on the other hand, is a passive, noninvasive, nonionizing procedure determining thermal activity rather than mass that, when used in conjunction with one or more of the other techniques, could provide early detection. Microwave and infrared radiometry should not be considered competing technologies.

The use of an adjunctive modality, together with radiographic mammography, is one way of improving the effectiveness of a breast cancer screening program. This observation was initially noted in two of the most widely publicized, large-scale screening projects, the Health Insurance Plan (HIP)[10] project from 1963 to 1966, and the Breast Cancer Detection Project (BCDDP)[11] conceived by the American Cancer Society from 1974 to 1981. These projects showed that the rate of success in detecting breast cancers was considerably greater when the modality of physical examination was combined with mammography.

Just as mammography should not be viewed as an examination competing with breast palpation in the diagnosis of breast cancer, other modalities,[12,13] as they become available as potential adjunctive screening techniques, may provide earlier detection. The determination of thermal activity is a measurement of tumor activity or growth rate,[14] providing data beyond the physical parameters (i.e., size and depth determined by mammography). As an early detection technique currently does not exist, approximately 85% of all determinations of breast disease result in extensive surgical procedures (i.e., discovery of a tumor usually means loss of breast and, with it, a negative attitude toward detection).

Early detection could lead to a conservative treatment and a positive attitude toward detection.

10.1 TECHNICAL DISCUSSION

Thermography, or more correctly radiometry, is the measurement of received radiation. *Radiometry* is defined as the technique for measuring electromagnetic energy considered as thermal radiation. Clinical thermography, in turn, is the measurement of natural emission from the human body. Any object above absolute zero will radiate electromagnetic energy to an extent governed by its radiant emittance. A body on which electromagnetic radiation falls may transmit, reflect, or absorb all the incident radiation or energy is known as a *black body*. To remain in equilibrium, a perfect absorber is also a perfect emitter, or radiator, and from black body theory, any perfectly absorbing body emits radiation at all frequencies in accordance with Planck's radiation law.[15,16]

The distribution of radiation is a function of both the temperature and the wavelength, or frequency. Figure 10.3 illustrates the intensity of the radiated signal with respect to frequency. It should be noted that, regardless of the method of heating, the intensity or radiation increases proportionately at all frequencies with increasing temperature.

Figure 10.4 demonstrates at 4.7 GHz (1 GHz $=$ 10^9 Hz) the emissivity of various materials at a common temperature. Notice that the amplitude of emittance

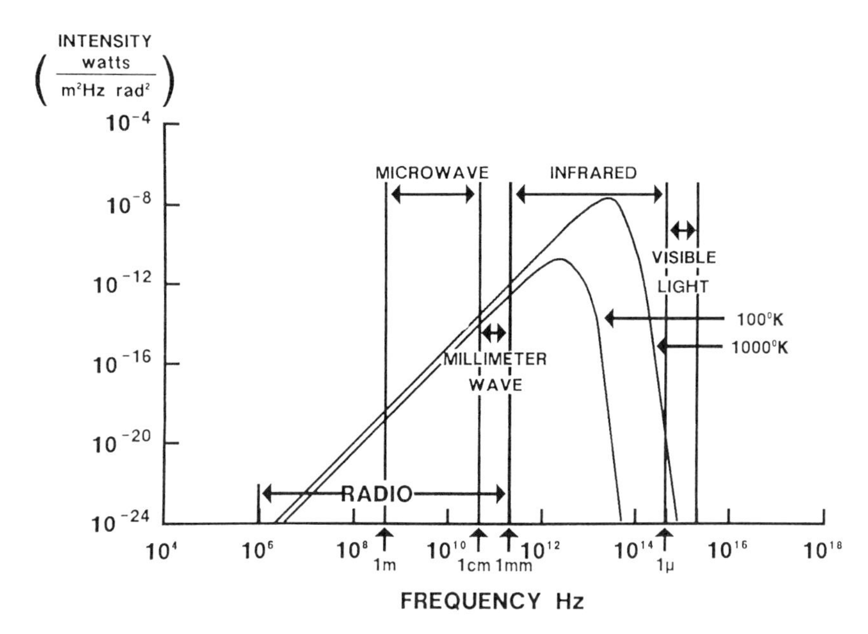

FIGURE 10.3 Black body radiation.

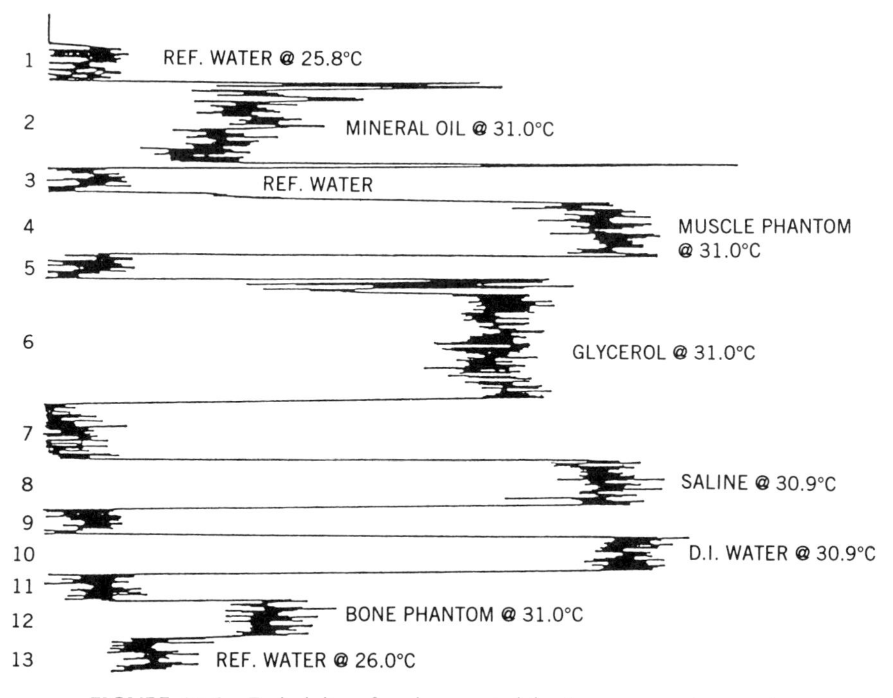

FIGURE 10.4 Emissivity of various materials at a common temperature

of muscle phantom is much greater than bone phantom at the same temperature. Correspondingly, transmission loss or attenuation of muscle (high-water-content tissue) is greater than that of fat or bone (low-water-content tissue) (see Fig. 10.5).

The use of thermography in cancer detection is based on the assumption that a temperature differential exists between a malignant tumor and the surrounding tissue.[17–19] Evidence of temperature elevations associated with the presence of carcinoma has been reported by several investigators examining patients with breast cancer.[20–26] One explanation for this is that the malignant cells may be more metabolically active and produce more heat. Blood flow[21,22] and angiogenesis[28] may also be factors. It is further recognized that tumors do not have the thermoregulatory capacity of normal tissues.

Gautherie and Albert[14] demonstrated (Fig. 10.6) that metabolic heat production was directly related to the doubling time of tumor volume. This would suggest that a tumor having a slow-growth factor would not be accompanied by significant thermal activity and, therefore, be difficult to detect using thermographic techniques. U et al.[29] measured tumor temperature with respect to surrounding

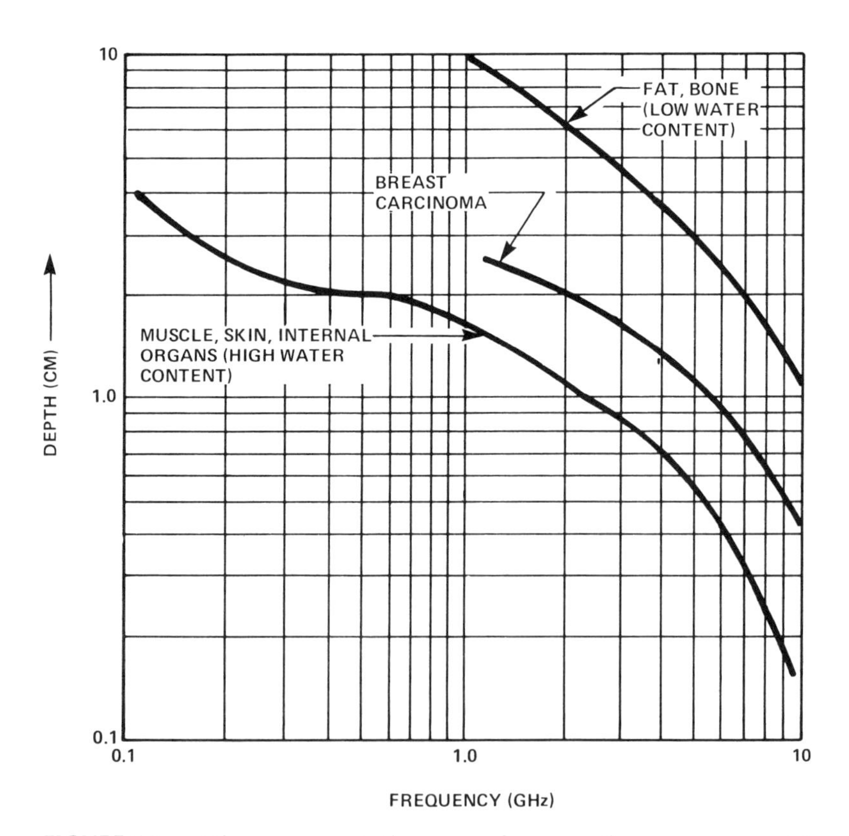

FIGURE 10.5 Microwave penetration versus frequency (homogeneous media).

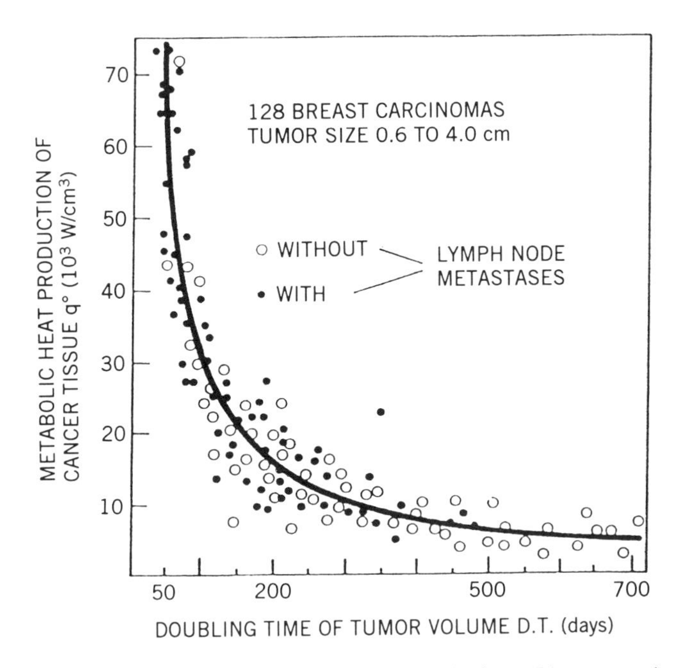

FIGURE 10.6 Growth rate and specific-heat production of breast carcinomas.

tissue prior to the application of microwave hyperthermia. In 16 of 17 patients at various sites, temperature differentials of 1–3 °C were observed.

There is considerable experimental evidence to indicate that unrestricted tumor growth is dependent on angiogenesis (i.e., the formation of new blood vessels).[27] Thus, after a new tumor has attained the small size of a few millimeters in diameter, further expansion of the tumor cell population requires the induction of new capillary blood vessels. The increase in vascularity, which can increase the opportunity for tumor cells to enter the circulation,[30,31] will also be associated with elevated temperature (i.e., inflammation), supporting the belief that thermal activity could precede the formation of significant mass providing earlier detection.

Development of early diagnostic thermography equipment occurred at the infrared frequency range taking advantage of the corresponding higher levels of emission, as evident in Figure 10.3. Recently the emissions at much lower frequencies have been investigated,[32–36] since improved transmission characteristics at the lower frequencies tend to offset the corresponding lower levels of emission, in essence shifting the apparent peak as viewed from the surface downward into the microwave region. The dielectric properties of various biologic tissues at the microwave frequencies have been presented in the literature.[37–41] Figure 10.5 illustrates the attenuation or depth of penetration with respect to frequency and water content. The "hot spot" or tumor can be viewed as a broad-spectrum signal generator coupled to tissue having absorptive characteristics equivalent to a low-pass filter (i.e., increasing attenuation with increasing frequency). Therefore, the shift to lower frequencies will allow detection at greater depths. The continued discussion will focus primarily on radiometry at the microwave frequencies. It should be noted that microwave radiometry is currently an experimental technique with studies being conducted by several teams around the world, with encouraging yet limited results.

Every component in the radiometer generates noise power that contributes to the overall noise of the system; therefore, the total system output contains not only noise received by the antenna but noise generated within the system. Also, gain variations within the system can produce output fluctuations far greater than the signal level to be measured. To overcome these system gain variations, Dicke[42] developed the common load comparison, or Dicke, radiometer. This configuration greatly reduces the effects of short-term gain fluctuations on the radiometer since the switch provides a mechanism to allow both the reference and the unknown signals to pass through amplification, essentially at the same time relative to expected gain drift in the amplifiers. Thus, any drift in gain will be applied equally to both signals. The receiver input is switched at a constant rate between the antenna and a constant temperature reference load, or reference antenna. The switched or modulated RF signal is, therefore, inserted at a point prior to RF amplification and as close to the antenna as possible. In turn, it is then amplified and coherently detected. The final output is proportional to the temperature difference between the antenna and the reference load. In the case where long integration times are involved, the long-term gain variations in the receiver must be considered. The long-term gain variations can degrade the minimum detectable temperature sensitivity (ΔT) in accordance with the following expression:

$$\text{Variation due to } \Delta T, \text{ long-term gain change} = \frac{\Delta G}{G} |T_1 - T_2| \quad K_{\mathrm{RMS}} \qquad (10.1)$$

where ΔG is the receiver gain change, G is the nominal receiver gain, T_1 is the temperature of the reference load in kelvins, and T_2 is the temperature of the antenna in kelvins. Obviously, if T_1 approaches T_2, the effect of long-term receiver gain variations becomes negligible. It becomes advantageous, therefore, to maintain the temperature of the reference load approximately equal to the temperature of the unknown.

Normally, for optimum system performance, one should consider reducing receiver noise temperature. This would involve the use of liquid nitrogen (77 K) or possibly liquid helium (4 K). This, however, would have a dramatic increase in cost, adding significant complexity. With the advent of low noise microwave transistors, it is possible to achieve 50 dB of RF amplification with less than 1.2 dB noise figure at microwave frequencies of less than 10 GHz, thereby eliminating the need for cooling for this particular application.

Consistent with current technology trends, the use of monolithic microwave integrated circuits will significantly reduce size, weight, and cost. The significant reduction in size will result in the direct attachment or integration, with the antenna forming a monolithic integrated subsystem having improved performance.

The radiometer can be further modified to take into account antenna mismatch with respect to the tissue or volume under test. If the receive antenna is noncontacting or remote, for example, the mismatch at the surface relative to the air can be significant, resulting in a dramatic reduction in surface emissivity. Ludeke and Kohler[43] have suggested the use of a radiation-balancing radiometer employing noise injection, thus making the receiver temperature equal to the object temperature to eliminate the error due to reflectivity. However, if the radiometer is designed for a specific application, the use of site-optimized contact antennas could eliminate the need for this added complexity.[44] In this situation, however, thermal drift results from prolonged contact between a microwave antenna at room temperature and a subject at a different temperature. Appropriate antenna heating (i.e., thermal matching of the antennas) can minimize thermal drift and realize a more accurate temperature measurement.[45]

The choice of radiometer frequency is based on several factors: intensity of emission, which increases with increasing frequency; resolution, which improves with increasing frequency; and transmission characteristics, which deteriorate with increasing frequency. Resolution is related to the wavelength with respect to the size of the object. Obviously, aperture size will be less at higher frequencies; however, this is offset by increased insertion loss or depth of penetration.

The Dicke radiometer design pertaining to the following discussion employs a low-noise RF amplifier in conjunction with a simple single-ended square-law detector rather than the common superheterodyne configuration involving a local oscillator and RF amplifier, thereby minimizing the potential drift and noise associated with the local oscillator in this approach.

Figure 10.7 represents a simplified block diagram of the radiometer. A switchable ferrite circulator, SW_2, has been developed to perform the load comparison, or Dicke[42] switch, function. The ferrite switch is preferred to the

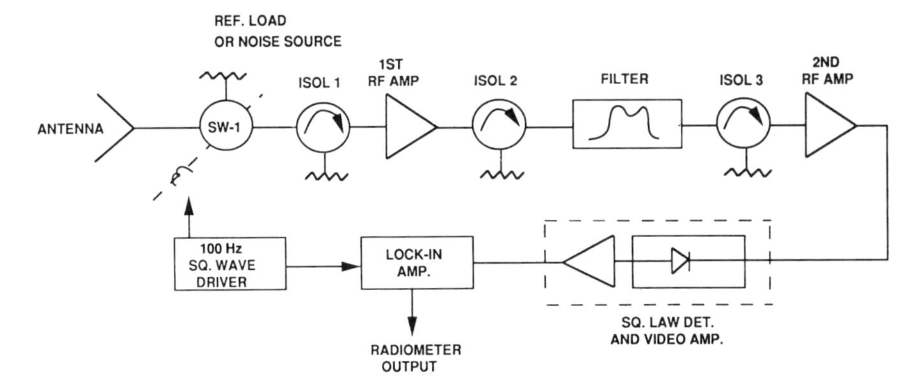

FIGURE 10.7 Radiometer block diagram. (The necessary transitions, cables, etc. are not shown.)

semiconductor approach primarily in view of the lower insertion loss—typically less than 0.2 dB. Briefly, the device is a switchable ferrite junction circulator[46] utilizing the remnant or latching characteristics of the ferrite material. The latching ferrite switch has been constructed in a waveguide having a single ferrite element contained within the microwave circuit. The insertion loss was measured and found to be less than 0.2 dB, having an isolation in excess of 20 dB. The Dicke switch (SW_1) is used to alternately connect the antenna and the reference stable noise source. The RF amplifier is a multistage field-effect transister (FET) device constructed in microstrip. The noise figure of the first RF amplifier is 1.2 dB with a gain of 35 dB. The noise figure of the second RF amplifier (which is less critical) is 2.2 dB. With the input and output VSWR at less than 1.5:1, the gain compression for signal levels of between −55 and −10 dBm (dBm = decibels above 1 mW) is less than 0.1 dB. The bandwidth of the microwave radiometer is basically determined by the bandpass characteristics of the filter. The filter characteristics were chosen to minimize possible interference due to nearby microwave communications or radar bands. The bandwidth is approximately 500 MHz, centered at 4.7 GHz. This is considered a quiet frequency range, used primarily for radio astronomy and troposcatter communication. The lock-in amplifier[47] enables the accurate measurement of signals contaminated by broadband noise, power-line pickup, frequency shift, or other sources of interference. It does this by means of an extremely narrow band detector that has the center of its passband locked to the frequency of the signal to be measured. Because of the frequency lock and the narrow bandwidth, large improvements in signal : noise ratio can be achieved. This allows the signal of interest to be accurately measured, even in situations where it is completely masked by noise. In addition, the lock-in amplifier provides the synchronous function associated with the Dicke switch (i.e., the unit supplies the 100-Hz reference clock frequency to drive the ferrite switch drive circuitry).

The minimum detectable temperature sensitivity ΔT is expressed as follows:

$$\Delta T = \frac{K[(FL-1)T_1 + T_2]}{(\beta\tau)^{1/2}} \quad K_{\text{RMS}} \tag{10.2}$$

In the case of the Dicke switch employing square-wave modulation, the value of K is 2.0, F is the noise figure (first amplifier stage, which in our case was 1.2 dB), and L is the sum of input losses expressed as a power ratio. The total loss is less than 2.0 dB. The effective noise figure, FL, is therefore $1.2 + 2$, or 3.2 dB, which represents the power ratio of 2.08. T_1 is the ambient radiometer temperature (microwave portion), namely, 290 K; T_2 is the source temperature (i.e., the temperature seen by the antenna), namely, 310 K. β is the receiver bandwidth (i.e., 500 MHz), and τ is the radiometer output time constant in seconds.

Utilizing a 3-s time constant, we obtain a minimum detectable temperature sensitivity of 0.03 K_{RMS}. This calculated temperature sensitivity is well within the design goal of 0.1 °C as were the actual results.

10.2 ANTENNA DESIGN

Early thermography was carried out at the infrared frequency range corresponding to frequencies at which the peak of intensity of black body radiation would occur (see Fig. 10.3). Thermographic diagnostic techniques utilizing electromagnetic emission at the infrared frequencies involving both thermographic cameras (remote) and liquid-crystal films (contact) have been in use for over 30 years[48–52] and have proved useful in measuring surface temperature distributions. Surface temperatures in the body vary continually in response to physical activity, menstrual cycle,[53] environment, substance intake, and so on; however, temperature differences from one side of the body to the other are quite small and reasonably stable in a normal healthy person. Since heat associated with a subcutaneous hot spot is transferred by radiation as well as convection and conduction, the thermal patterns seen at the surface can be significantly altered. This limitation associated with infrared thermography is due to the rapid absorption of electromagnetic energy at the infrared frequencies. Basically, the thermal pattern generated at the surface results from heat transmitted to the surface through a lossy nonhomogeneous layered media by conduction, convection, and radiation. Again, assuming the "hot spot" to be a broad-spectrum signal generator, electromagnetic energy transfer depends, to a great extent, on the absorption properties of the tissue. This would indicate that operation at millimeter frequencies would yield results similar to those experienced at infrared frequencies.[54] A shift, therefore, to the lower microwave frequencies would provide improved transmission characteristics[55–57] (i.e., greater depth of penetration). Furthermore, if the antenna is matched to the tissue, the tissue–air interface reflection is minimized, providing maximum coupling of the emitted signal to the generator.

It had been determined by Guy[58] that the optimum aperture size to achieve effective coupling of microwave energy associated with biologic tissue is the simple TE_{10}-mode aperture in direct contact with the emitting surface. To reduce the physical size of the aperture, dielectric loading was used.[59,60]

The antenna (Fig. 10.8) used in conjunction with the microwave radiometer described was reasonably well matched over the band of the radiometer when site optimized. A return loss of 10 dB corresponds to a transmission loss of 0.45 dB, or 90% power transmitted. Correspondingly, a return loss of 15 dB would yield a transmission loss of less than 0.2 dB and 96% power transmitted. A low-loss dielectric having a relative dielectric constant of 12 was employed, providing an aperture size of 1.58×0.79 cm. A heater and proportional thermostat are provided to maintain a constant temperature approximately equal to the surface temperature of the subject, thus providing thermal matching.

Antenna design operating in the near-field region in a layered, inhomogeneous medium is complex and difficult, with results determined generally by test rather than design. Further, antennas are normally evaluated in their radiate (transmit) mode rather than the receive; however, reciprocity dictates that the transmit and receive antenna patterns be identical.[61]

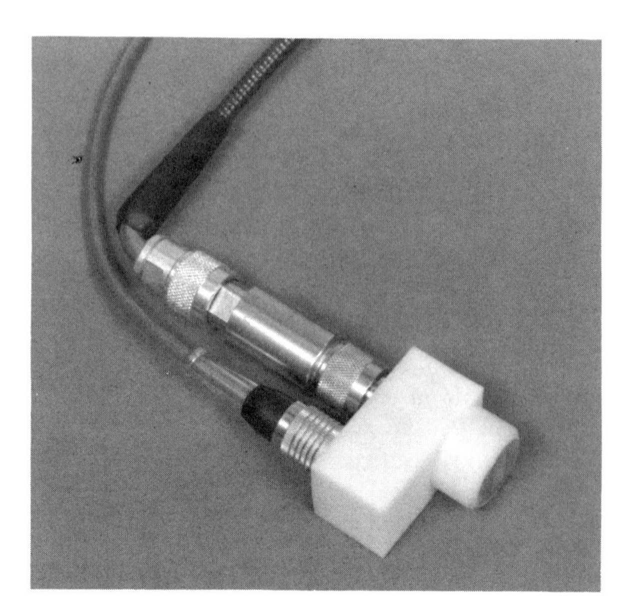

FIGURE 10.8 A 4.7-GHz contact antenna.

Equipment to date has employed waveguide rather than microstrip. A waveguide structure provides a fully shielded enclosure, taking advantage of the low loss waveguide transmission line.

Microstrip antennas[62–65] are small, lightweight and inexpensive and can be constructed on a flexible substrate material providing the ability to conform to the body surface. The use of microstrip, however, will reduce overall system performance (i.e., increased noise figure due to increased insertion loss when compared with waveguide and, hence, lower efficiency). For the most part, design data available for microstrip antennas pertain to mating to an air dielectric ($\varepsilon_r = 1$) rather than tissue. The effect on the design as a result of mating to lossy material having a high relative dielectric constant is dramatic. Bahl and Stuchly have discussed[63] the design of the microstrip covered with a lossy dielectric layer.

At the higher microwave frequencies, particularly at the millimeter-wave frequencies, waveguide antennas can be of convenient size. On the other hand, at the lower frequencies where greater depth can be achieved, the physical size of the antenna is significant and often unacceptable for clinical use, necessitating, as mentioned earlier, the need for dielectric loading.[66–69]

The reduction in aperture size is proportional to the $(\varepsilon_r)^{1/2}$, where ε_r is the relative dielectric constant of the material used. The geometrical dimensions of the aperture determine the amount of thermal energy received. Increasing the size of the aperture will, therefore, improve the signal:noise ratio. The size and shape of the aperture will determine the pattern directivity or beam.

A reduction in aperture size, however, can result in a decrease in effective detection depth. The beamwidth of the antenna[70] will increase with decreasing aperture width corresponding to reduced gain or, in this case, reduced depth of detection. Allowing the aperture width to approach zero creates, in essence, a point source at which the antenna becomes omnidirectional with minimal depth of penetration.

Figure 10.9 illustrates the calculated[71,72] effect of aperture size at a given frequency or depth of penetration when mated with ethanol or lower-water-content tissue. The larger aperture has an area of 10.5 cm^2, whereas the smaller unit has an area of 1.24 cm^2. When mated to a material having a high dielectric constant, such as water, the calculations show no appreciable difference in penetration. Correspondingly, in fats the wavelength at 4.7 GHz is 2.4 cm; in water, the wavelength is approximately 0.8 cm. These results indicate that for optimum performance the aperture must be larger than the wavelength in the mating tissue.

Figure 10.10 further illustrates the effect of aperture width on directivity and, in turn, spatial resolution.[73,74] To obtain these data, a small (0.41-cm)-diameter hole was located in a simulated layer of fat overlying a simulated layer of thick muscle. The fat layer thickness was 2.3 cm, with the top of the 0.41-cm diameter hole located 0.8 cm below the surface. The liquid flowing through the hole was 7 °C above the temperature of the fat-equivalent layer and the water bath. The fat-equivalent material is a laminate-based solid developed by Guy.[75] To limit heating of the material around the hole (thus effectively enlarging the diameter of the "hot line"), water from the underlying bath was recirculated through the hole immediately after each measurement at each antenna position, and the system was

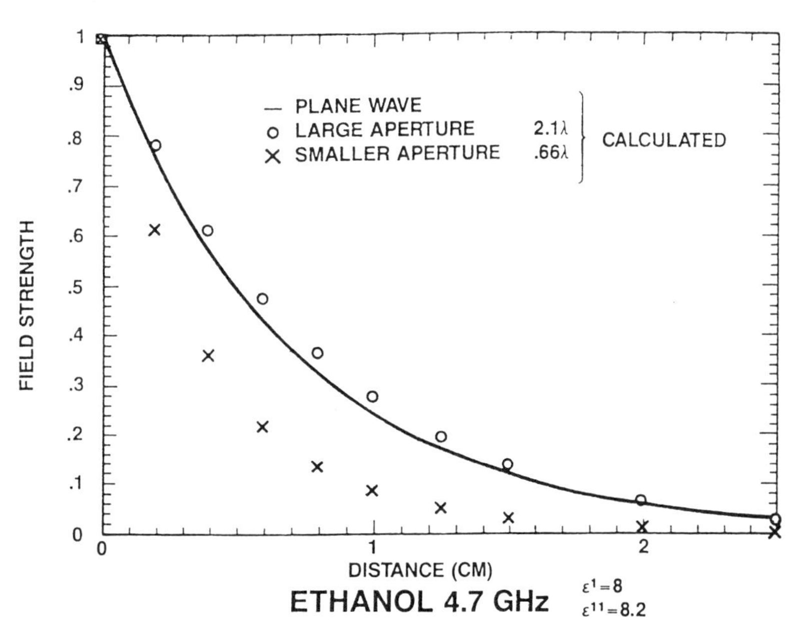

FIGURE 10.9 Calculated effect of aperture size on depth of penetration when mated with ethanol or lower-water-content tissue.

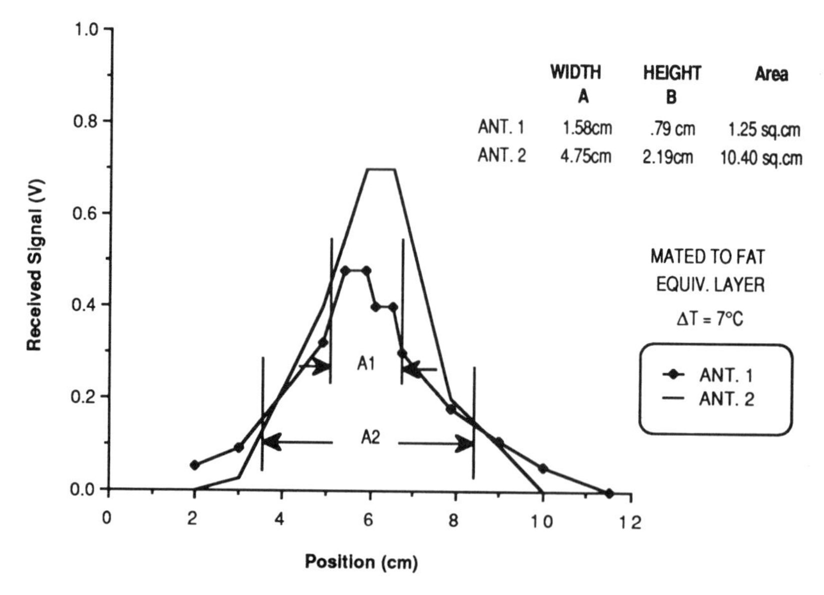

FIGURE 10.10 Effect of aperture width on directivity.

allowed to regain its initial temperature. The antenna was mechanically moved across the hot line. The aperture widths A1 and A2 as shown in Figure 10.10 are 1.58 m and 4.75 cm, respectively.

10.3 TEST RESULTS

Thermography at the higher frequencies, especially at the infrared frequencies due to poor penetration through biologic tissue, is limited to surface measurements, which, in turn, is altered significantly by factors such as menstruation. A healthy volunteer study was conducted[76] to determine the following:

1. Is the temperature variation due to menstruation a surface phenomenon, or does it affect thermal measurement at depth?
2. The consistency of microwave thermal patterns with increasing age.
3. Bilateral thermal symmetry (i.e., between right and left breast).

Volunteers were asked to undergo microwave examination beginning with the onset of flow during menstrual cycle, repeating examinations on a weekly basis for 4 consecutive weeks. It was found that the thermal patterns generated repeated to within ±0.2 °C over the measured period. There was, therefore, no apparent effect on core temperature resulting from menstrual cycle. Figure 10.11 illustrates the consistency of thermal patterns with increasing age. The radiometer operating at 4.7 GHz utilized for this study, as well as the following test results, is shown in Figure 10.12. The single antenna utilized is shown in Figure 10.8.

The apparent slight elevation (Fig. 10.11) in temperature in younger women is due primarily to the change in tissue characteristics rather than temperature. The higher fat content tissue associated with increasing age will result in slightly decreasing emissivity. It was also substantiated[77] that left–right symmetry was consistently within +0.2 °C and, therefore, is suitable for comparison.

The healthy volunteer study involved 16 subjects with an age distribution of 21–66 years. Figure 10.13 is representative of a microwave thermogram[78] using the 4.7-GHz radiometer described in the text. This case was a 62-year-old female having a palpable 3-cm mass on the left breast. The single dielectric-filled antenna is mechanically positioned as one would position a stethoscope. Utilizing the thermal symmetry that exists between the right and left breast, common points are compared (i.e., the upper inner quadrant of the right breast—position 1—is compared with the corresponding upper inner quadrant—position 2—of the left breast). Repeating this procedure through the various positions reveals a significant temperature differential between positions 5 and 6 (i.e., the upper outer quadrants). Similarly, there are temperature differences between positions 9 and 10 and, to a lesser degree, positions 7 and 8. This grouping of elevated temperatures on the left breast indicates a thermal anomaly that was found to peak at position 12. The temperature differential between position 12 of the left breast and the corresponding position 11 on the right breast was 2.5 °C. Obviously the number of positions is not sufficient; however, an increase in data points would lengthen the examination time to what is considered unacceptable.

In the clinical study conducted at the Nippon Medical School,[78] 183 volunteers with mammary gland disease were examined using the 4.7-GHz radiometer along with conventional techniques. In this study, an attempt was made to categorize the levels of temperature differentials measured and to use this information as a criterion for diagnosing the presence of malignancy. Using other methods available, including physical examination, x-ray mammography and biopsy, 142 cases were shown to be benign and 41 cases malignant. The performance of the microwave radiometric system was determined by assigning a positive malignancy for all ΔT values ≥ 0.5 °C. With this simplistic rating system, a correct diagnostic rate of 77.6% was found.

Thermography has been shown to be effective in the monitoring of the course of treatment.[79,80] In one particular instance involving a patient with Hodgkin's disease, with significant mass on both the left and right supraclavicular areas, there was a corresponding temperature differential, as shown in Figure 10.14. Following radiation therapy (Fig. 10.15), a microwave thermogram was taken substantiating a positive response to treatment. During the test procedure involving numerous antenna positions corresponding to the internal mammary and axillary gland areas, it was noted (antenna position 6) that a significant temperature differential of 0.9 °C existed. This was considered at that time, using

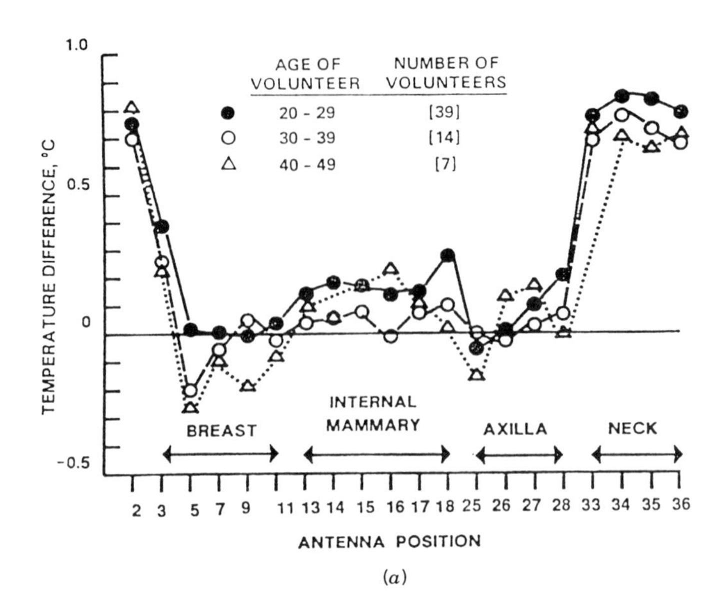

(*a*)

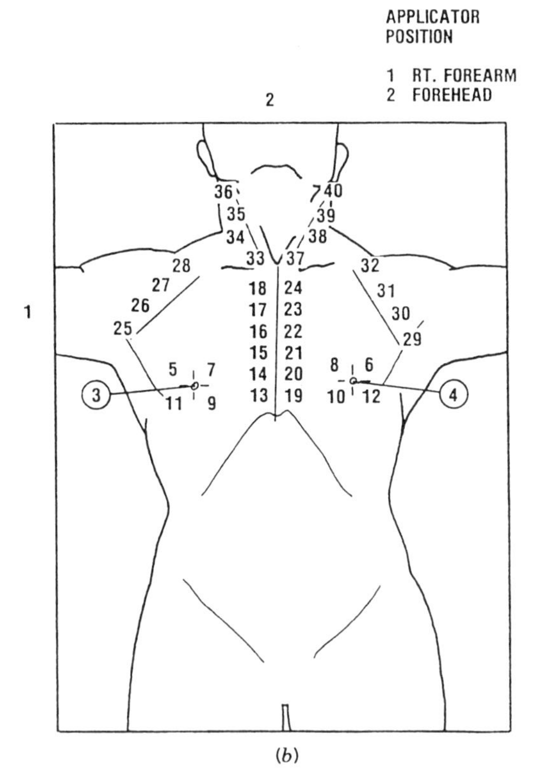

(*b*)

FIGURE 10.11 (*a*) Consistency of thermal patterns with increasing age; (*b*) Standard 40 position examination.

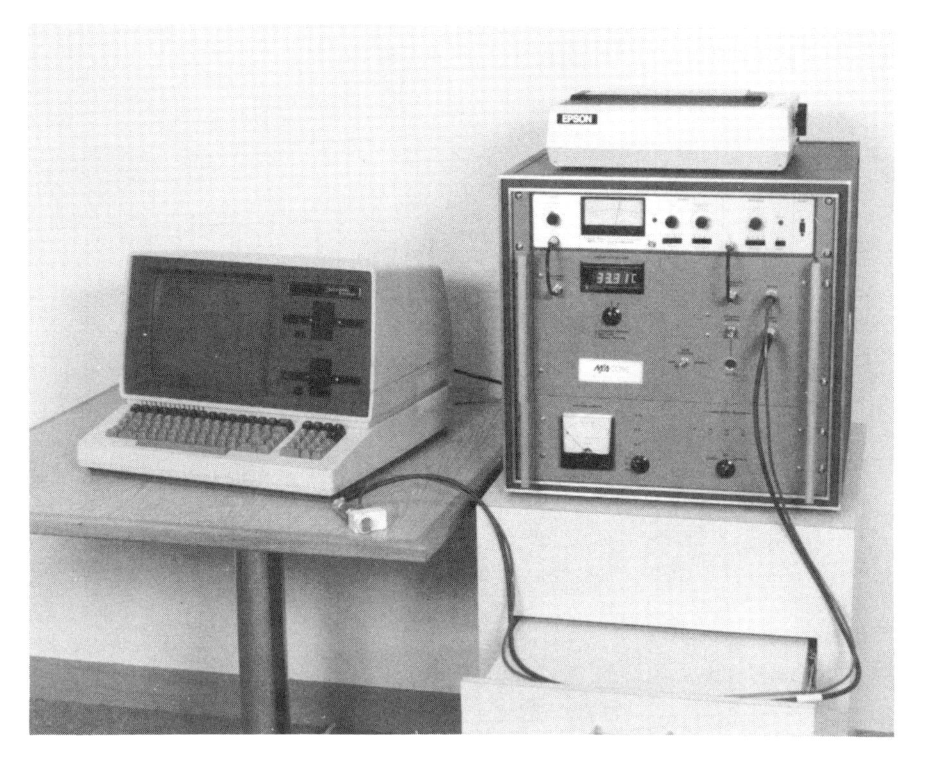

FIGURE 10.12 Radiometer (4.7 GHz).

diagnostic x-rays, to be a false positive; however, 2 years later this patient was found to have developed a palpable growth corresponding to that exact antenna position, suggesting the presence of thermal activity prior to the formation of significant mass.

Similarly, during a microwave examination of a patient (Fig. 10.16) having a known lesion on the right breast that was found to have a 1.2 °C temperature differential (the location of which agreed with that of a mammogram), a significant thermal anomaly was located on the left breast as well. The second primary site was not detected by clinical examination or mammography, but was later proved positive by biopsy. The interesting point here is that the significantly smaller tumor in the left breast exhibited a larger temperature differential, suggesting that thermal activity is primarily a function of metabolic heat rather than physical size. This further suggests the synergistic adjunctive role of thermography with respect to mammography and other diagnostic techniques.

Experience has shown using the single-antenna system that, while taking full advantage of thermal matching of the antenna to the mating body surface temperature, each placement of the antenna requires approximately 1–1.5 min

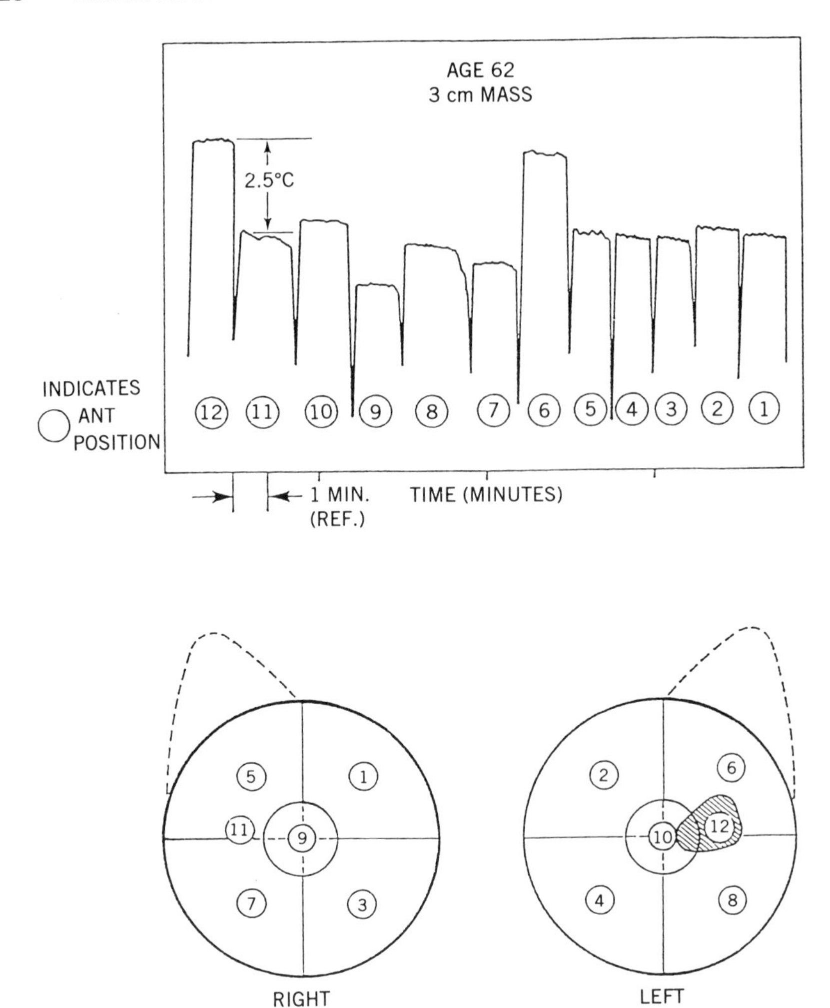

FIGURE 10.13 Microwave thermogram utilizing the 4.7-GHz radiometer.

per site to achieve stable and consistent data. Because of the number of data points required, resulting examination time is deemed excessive and basically unacceptable. The use of multiple antennas would allow simultaneous thermal stability of all antenna elements in the same time required for a single element. The use of multiple, site-optimized antennas will improve resolution and sensitivity and dramatically reduce examination time.

The multiple antenna system pictured in Figure 10.17 employs mechanical compression in a manner similar to that employed in mammography. Compression in this manner allows direct comparison of data with a mammogram. Measurement from two opposing surfaces further allows determination of relative depth. Obviously, signal strengths measured at two opposing antennas produced

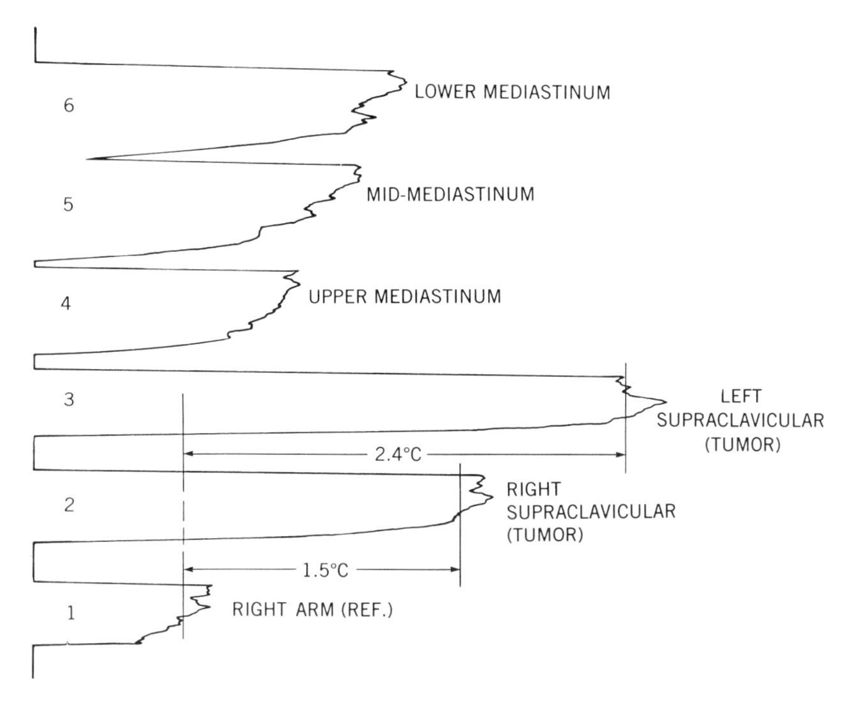

FIGURE 10.14 Microwave thermogram—patient with Hodgkin's Disease prior to radiation therapy.

by a hot spot equidistant from either surface will be equal in magnitude. Differentials in signal amplitude, coupled with knowledge of tissue characteristics, will allow depth determination. Physical compression, coupled with the ability to measure from opposing surfaces, will significantly reduce tissue thickness, allowing increased depth of detection and improved diagnostic capability.

The system pictured in Figure 10.17 is based on a simple commutation technique allowing the rapid selection and measurement of the individual antenna elements and does not involve correlation, commonly termed *phased-array* techniques, to be discussed briefly in this chapter. Each compression plate contains 12 antenna elements coupled to a single-pole six-throw mechanical switch allowing rapid selection of the individual thermally matched antennas. A mechanical switch is employed, taking advantage of the inherent low insertion loss (typically < 0.2 dB) with corresponding isolation between elements of > 60 dB. The low insertion loss will have a negligible effect on system noise figure. The upper and lower plate selection is controlled by a single-pole two-throw switch that, in turn, is coupled to the radiometer input.

In addition to a dramatic reduction in overall examination time, the rapid acquisition of data will improve performance through the elimination of drift, in both the equipment and the patient.

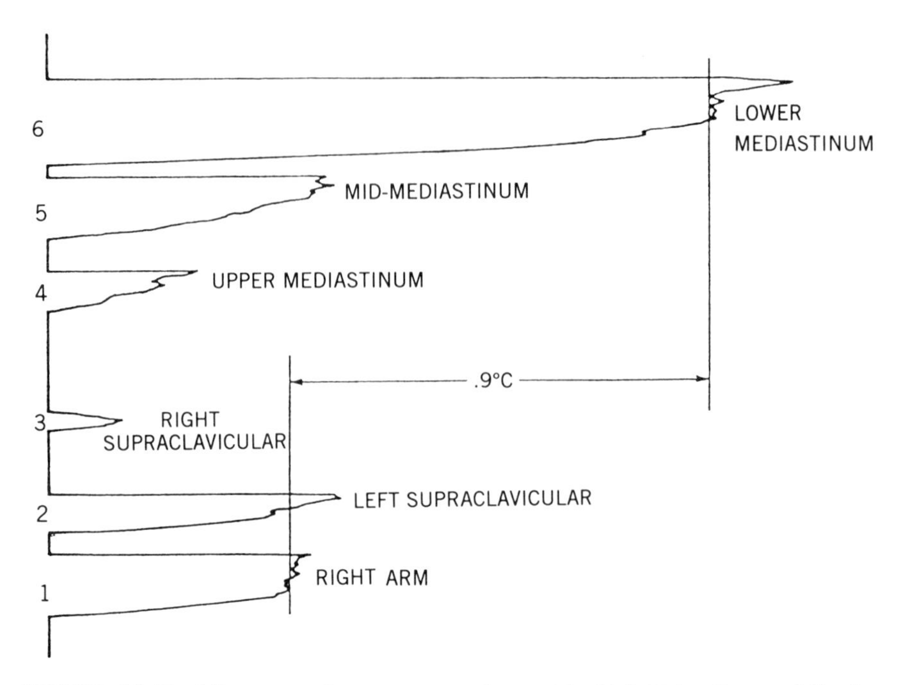

FIGURE 10.15 Microwave thermogram—patient with Hodgkin's disease following radiation therapy.

The high insertion loss of a switch matrix using solid-state technology would result in higher noise figure and significantly reduced sensitivity. For this reason, the mechanical switch was selected. The use of multiple radiometers creating, in essence, a passive array would eliminate the need for a switch matrix. The passive array would yield the lowest system noise figure; however, it introduces the need for amplitude and offset matching of the individual array elements. The increased complexity could result in a significant cost increase.

10.4 FUTURE DEVELOPMENT

Future development effort should include such areas as heat enhancement, multiple-frequency radiometry, and multiple-antenna correlation radiometry. With regard to heat enhancement, the application of heat, regardless of the technique or frequency used, would be considered active rather than passive radiometry. Also the elevation of temperature, regardless of the frequency or heating method used, will increase the intensity of emission at all frequencies proportionately. Thompson et al.[81] indicate that microwave heating is more effective than ultrasound, since with ultrasound normal tissue, because of the higher density of the tumor tissue, incurs greater elevations in temperature than tumor tissue.

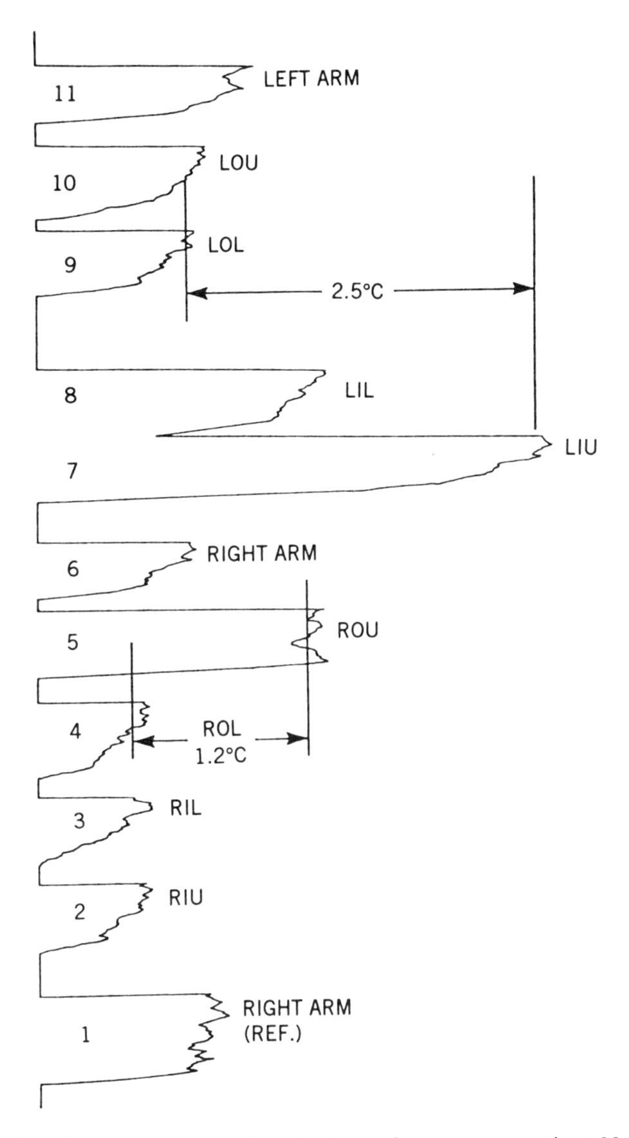

FIGURE 10.16 Microwave examination of primary breast cancer patient. Note: Figure 16 illustrates the temperature of the various quadrants of the breast, where LOU indicates left outer upper, RIL indicates right inner lower, and so on.

Experiments have been conducted[81–84] proving the feasibility of increasing tumor detectability (i.e., accentuating small temperature differentials through the application of electromagnetic energy), enhancing temperature differentials between the tumor and surrounding tissue. Thompson et al.[84] observed, using an infrared thermographic camera, that in situ irradiation of transplantable guinea

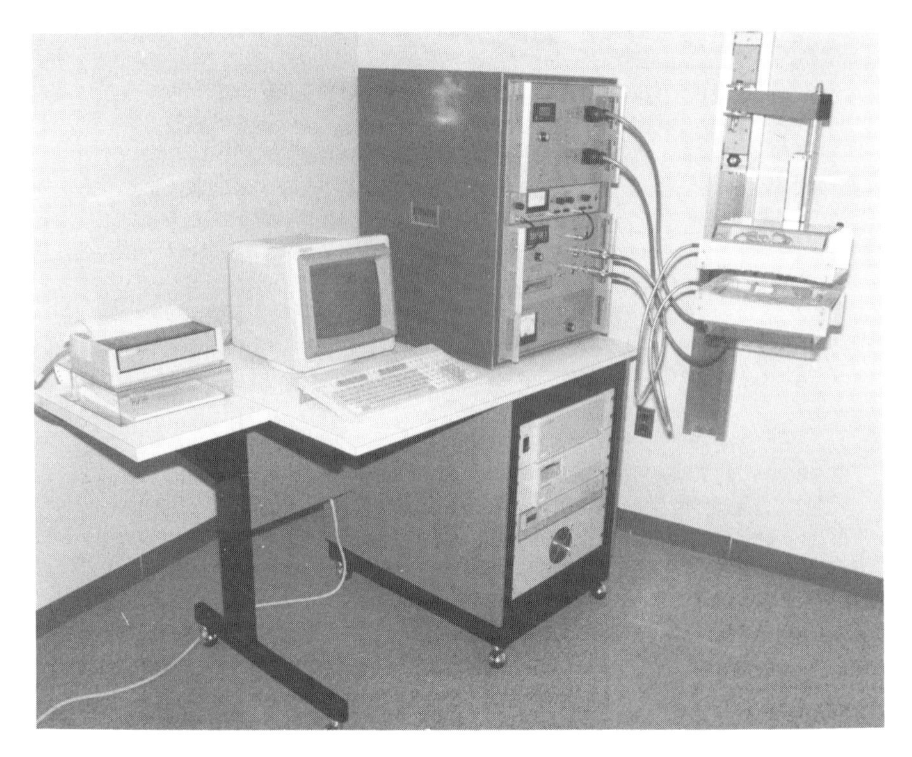

FIGURE 10.17 Microwave radiometer employing multiple antennas.

pig hepatoma using microwave heating at 2450 MHz induced a temperature rise of 5.5 °C compared to a rise of 2.5 °C in surrounding healthy tissue. This temperature differential of 3.0 °C is in contrast to an initial 0.5 °C difference prior to heat enhancement. The microwave energy was coupled from the transmitter to the subject through the use of a remote horn antenna rather than direct contact. This technique does not compensate for the large surface reflections resulting from the air–tissue interface requiring significantly higher transmitter power levels incident at the antenna. However, the experiment did prove that selective absorption of microwave energy produced a significant increase in surface temperature gradients. The selection absorption characteristics can be seen in Figure 10.5.

Further experiments[85] using the VX2 carcinoma growing subcutaneously in the ear of a New Zealand white rabbit show that the temperature differential between healthy tissue and tumor tissue can be increased substantially, taking advantage of the poor vascularity of the tumor with respect to the surrounding tissue. It can be seen in Figure 10.18 (antenna position 1 vs. antenna position 2) that a temperature differential of 1.2 °C existed prior to heating, while several minutes following microwave heating at 1600 MHz using a contact antenna, the temperature differential was 1.8 °C (antenna position 12 vs. antenna position 13). Detection was accomplished using a 4.7-GHz microwave radiometer.

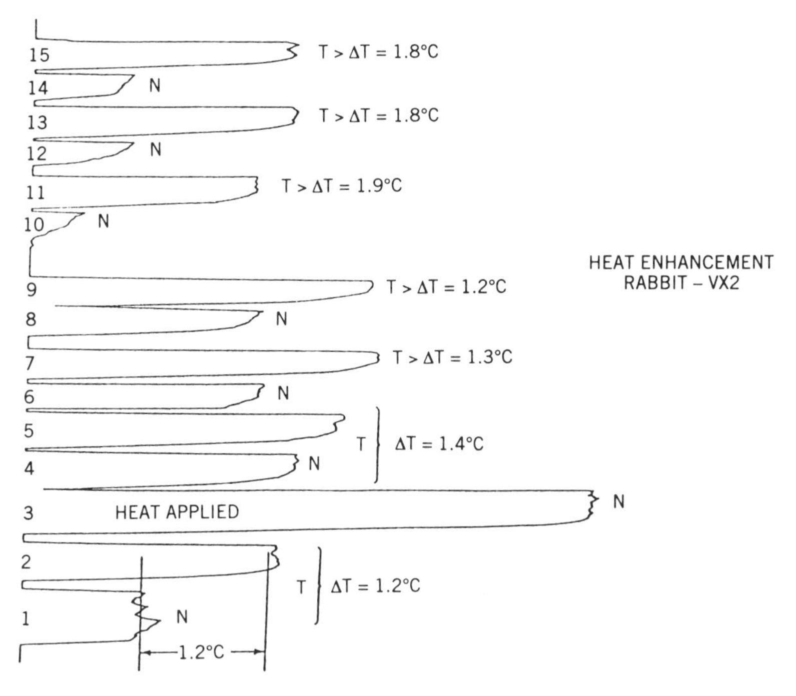

FIGURE 10.18 Radiometric detection employing heat enhancement.

The dual-mode antenna utilized is shown in Figure 10.19. The 4.7-GHz dielectric-filled antenna (1.83 × 0.92 cm) is metallized, forming the single ridge of the 1600 MHz dielectrically-filled single-ridge waveguide. The microwave power levels utilized are small, resulting in minimal heating of healthy tissue. Thermal data obtained, in addition to providing improved detectability, may prove to be useful in determining other parameters, such as vascularity. Conversely, improved detectability can also be enhanced by the cooling of the surrounding tissue with respect to the tumor tissue.

During the rabbit study discussed above,[86] it was observed (Fig. 10.20) that an initial temperature differential of 0.3 °C associated with a tumor less than 4 mm in diameter increased to greater than 1 °C during anesthesia. The anesthesia utilized was sodium pentobarbital. The increase in temperature differential was the result of a reduction in the temperature of the surrounding tissue, as shown in Figure 10.21. It is not suggested that anesthesia be utilized in the diagnostic procedure; however, it is suggested that detection can be enhanced through the reduction of normal adjacent tissue.

A second area of future development will involve multiple-frequency radiometry.[87–89] Barrett et al.[77] had suggested infrared combined with microwave since the complementary detection statistics are considerably improved, approaching that of mammography, possibly providing screening for subsequent

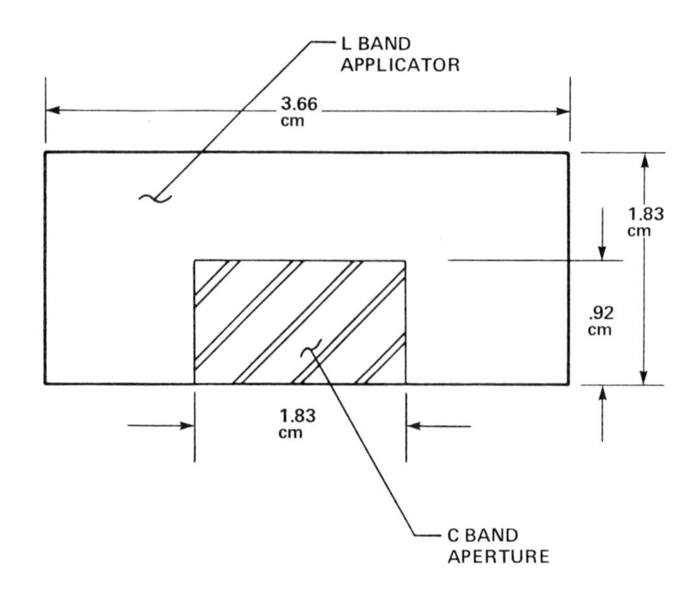

FIGURE 10.19 Dual-mode antenna.

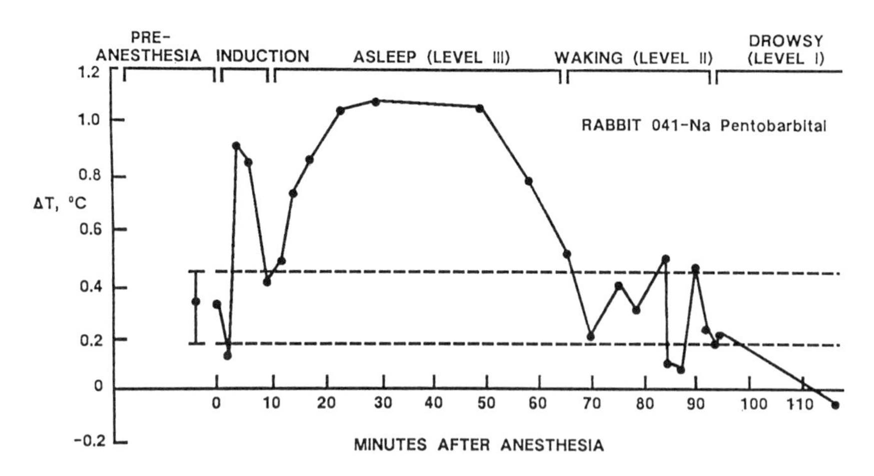

FIGURE 10.20 Effect of anesthesia on temperature differential.

mammography and thereby reducing the number of women exposed to x-rays. In addition, this approach to detection at different frequencies, coupled with the knowledge of tissue characteristics, in principle could provide depth information. The frequencies chosen must represent a compromise between penetration depth, radiation intensity, emissivity, and resolution. Lower frequencies exhibit greater penertration, but at the sacrifice of resolution and emissivity.

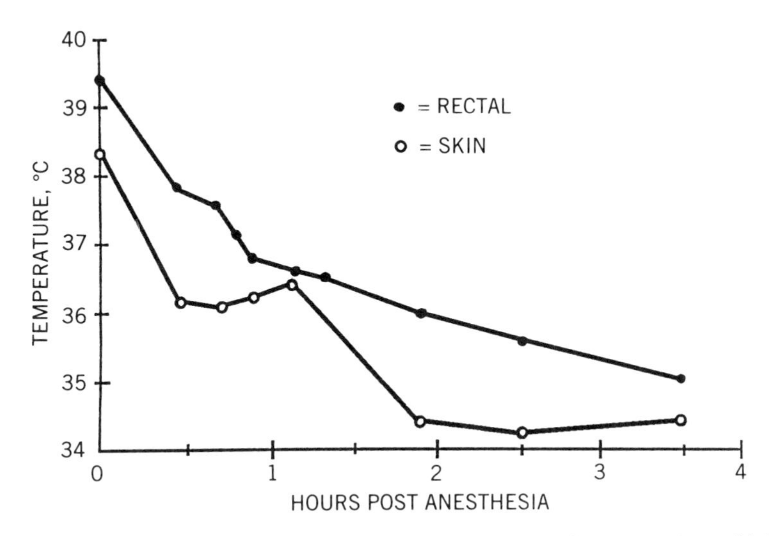

FIGURE 10.21 Rabbit temperature as a result of anesthesia (•, rectal; ○, skin).

A wide separation of microwave frequencies may preclude the use of a common antenna due to the inability to optimize antenna element performance over an appreciable bandwidth. This requirement for broadband performance will necessitate further development of printed configurations. The ridged waveguide configuration (Fig. 10.19) could be used to accommodate a dual-radiometer approach using two narrow-band waveguide apertures optimized at two widely separated frequencies integrated into a common transducer. For the same reason, this wide frequency separation may also preclude the use of common microwave componentry in the front end of the radiometer, adding additional complexity and cost. Prionas and Hahn,[89] with a detailed analysis of energy distributions versus depth and frequency, have established that multiple-frequency radiometry is a feasible technique of sensing one-dimensional temperature differentials noninvasively.

The multiple-antenna radiometer discussed earlier involves commutation in which individual antenna elements are time-shared with a common radiometer. Correlation radiometry, on the other hand (which also utilizes multiple antennas), is based on coherence theory widely used in phased array techniques and discussed extensively in the literature.[90–92] Mamouni et al.[93] have shown that in lossy homogeneous material, correlation radiometry can result in improved resolution of thermal gradients. In addition to the complications resulting from the introduction of nonhomogeneous and layered tissue, antenna element design must be reviewed.

A phased array, in the case of radiometric correlation, requires overlapping antenna patterns (Fig. 10.22) in which the overlapping pattern is in phase, or coherent, allowing additive beam forming. The need for overlapping antenna patterns will reduce the depth of penetration of the single element; however, beam forming associated with adjacent elements could offset this reduced depth of penetration. Figure 10.22 illustrates that directivity of the larger antenna is significantly better than that of the smaller antenna; however, the beamwidth of the smaller aperture is broader and would allow overlapping antenna patterns for closely spaced elements, which is important for correlation radiometry. If coherency in tissue can be achieved, it should be further noted that the electronic beam steering through electronic phase control or mechanical motion of the array with respect to the subject must be incorporated if complete coverage is to be achieved. The additional componentry to achieve correlation, primarily phase shifters and couplers, will result in additional circuit losses and, therefore, reduced sensitivity. If RF gain is added prior to the Dicke switching function to compensate for the increased loss, the amplifiers must track in both phase and amplitude. The improved resolution afforded by correlation radiometry has nevertheless spurned new and considerable interest.[88] The trend toward sophistication to accomplish improved thermal imaging involving multiple antenna commutation, phase coherency, multiple frequency, and enhancement techniques will increase system complexity and cost. This increased cost, however, will be offset to a great extent through current technology trends involving high-level microwave integration in both hybrid and monolithic circuitry, coupled with the development of multiple antenna arrays utilizing

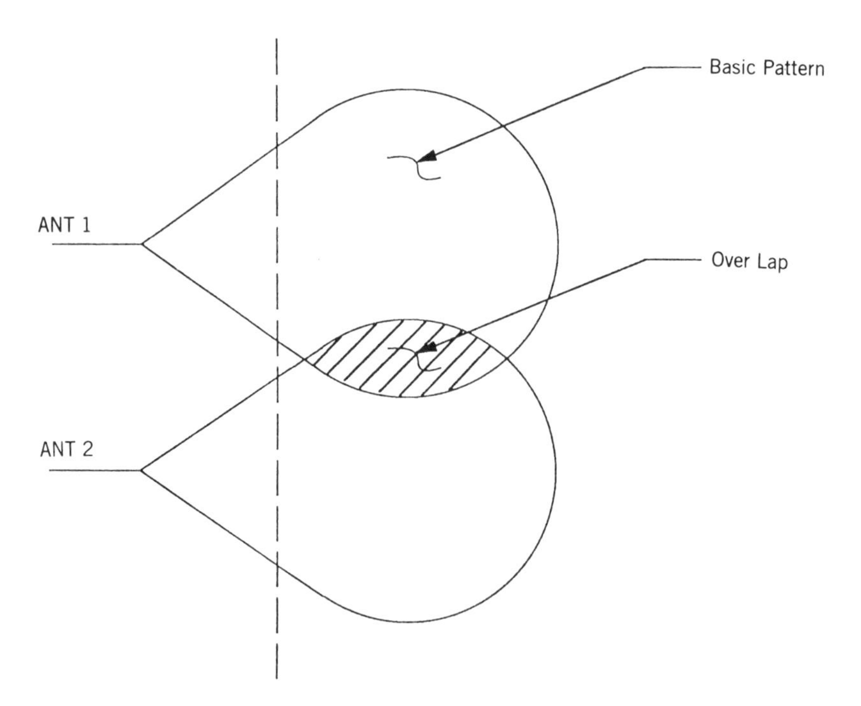

FIGURE 10.22 Overlapping antenna pattern.

conformal printed-circuit techniques on low-cost dielectric materials. Eventually microwave radiometry will be used to provide noninvasive thermometry to control noninvasive hyperthermia. This will occur, however, only after considerable knowledge has been gained through the use of invasive thermometry techniques. Thermography should not be considered as a competing technology, but rather an adjunctive procedure to mammography and clinical examination, particularly for early detection of breast cancer. The concept of a noninvasive diagnosis or screening procedure involving nonionizing radiation is attractive and, when used in combination with one or more modalities, will prove effective. The diagnostic potential of thermal knowledge will be of major importance in providing additional data on which primary treatment can be selected and the prognosis established.

REFERENCES

1. Boring CC, Squires TS, Tong T: Cancer statistics, 1992. *CA—A Cancer J Clinicians* 42:19, 1992.
2. Harris JR, Lippman ME, Umberto Veronesl U, et al: Breast cancer (first of three parts). *New Engl J Med* 327:319, 1992.
3. U.S. Department of Health, Education and Welfare: *Cancer Patient Survival*, Report No. 5, DHEW (NIH) 77-922; Washington, DC 1976, p. 163.
4. Berg JW: Clinical implications of risk factors for breast cancer. *Cancer* (Suppl), 53(3):589, 1984.
5. Strax P: Mass screening for cancer. *Cancer* 53:665, 1984.
6. Bedwani R, Vana J, Rosner D, et al: Management and survival of female patients with minimal breast cancer: As observed in the long-term and short-term surveys of the American College of Surgeons. *Cancer* 47:2769, 1981.
7. Nealon TF, Jr: *Management of the Patient With Cancer*. Saunders, Philadelphia, 1965.
8. Lundgren B: Observations on growth rate of breast carcinomas and its possible implications for lead time. *Cancer* 40:1722-1977.
9. Gullino PM: Natural history of breast cancer – progression from hyperplasia to neoplasia as predicted by angiogenesis. *Cancer* 39:2697, 1977.
10. Shapiro S, Venet W, Strax P, et al: Ten-to-fourteen-year effect on screening on breast cancer mortality. *J Natl Cancer Inst* 69:349, 1982.
11. Baker LH: Breast cancer detection demonstration project: Five year summary report. *CA—A Cancer J Clinicians* 32:194, 1982.
12. McLelland R: Responding to the challenge of breast cancer screening. *Diagn Imag*: 69, 1986.
13. Homer MJ: Breast imaging: Pitfalls, controversies and some practical thoughts. *Radiol Clin N Am* 23:459, 1985.
14. Gautherie M, Albert E: *Clinical Thermobiology*. Alan R Liss, New York, 1982.

15. Kraus JD: *Radio Astronomy*. McGraw-Hill, New York, 1966.

16. Harvey AF: *Microwave Engineering*, Academic Press, New York, 1963.

17. Lawson R: Implications of surface temperatures in the diagnosis of breast cancer. *Can Med Assoc J* 75:309, 1956.

18. Lloyd-Williams K: Infrared thermometry in the diagnosis of breast disease. *Lancet* 2:1378, 1961.

19. Lawson RN, Chughtai MS: Breast cancer and body temperature. *Can Med Assoc J* 88:68, 1963.

20. Edrich J, Jobe WE, Cacak RK, et al: Imaging thermograms at centimeter and millimeter wavelengths. *Ann NY Acad Sci* 335:456, 1980.

21. Gautherie M: Temperature and blood flow patterns in breast cancer during natural evolution and following radiotherapy. In Gautherie M, Albert E (eds): *Biomedical Thermology*. Alan R Liss, New York, 1982, pp. 21–64.

22. Gullino PM: Considerations on blood supply and fluid exchange in tumors. In Gautherie M, Albert E (eds): *Biomedical Thermology*. Alan R Liss, New York, 1982, pp. 1–22

23. Isard HJ, Sweitzer CJ, Edelstein GR: Breast thermography. A prognostic indicator for breast cancer survival. *Cancer* 62:484, 1982.

24. Myers PC, Barrett AH, Sadowsky NL: Microwave thermography of normal and cancerous breast tissue. *Ann NY Acad Sci* 335:443, 1980.

25. Shaeffer J, El-Mahdi AM, Carr KL: Thermographic detection of human cancers by microwave radiometry. In Gautherie M, Albert E (eds): *Biomedical Thermology*. Alan R Liss, New York, 1982, pp. 509–521.

26. Tenjin T, Oyama H, Maeda T, et al: Application of microwave thermography for the diagnosis of breast cancer. Nippon Medical School Report: First Meeting of Japan Thermography Association, June 1984.

27. Weidner JN, Semple JP, Welch WR, et al: Tumor angionesis and metastasis—correlation in invasive breast carcinoma. *New Engl Med* 324:1, 1991.

28. Folkman J, Watson K, Ingber D, et al: Induction of angiogenesis during the transition from hyperplasia to neoplasia. *Nature* 339:58, 1989.

29. U R, Noell K, Woodward K, Worde B, et al: Microwave-induced local hyperthermia in combination with radiotherapy of human malignant tumors. *Cancer* 45:638, 1980.

30. Folkman J: What is the evidence that tumors are angiogenesis dependent? *J Natl Cancer Inst* 82:4, 1990

31. Liotta L, Kleinerman J, Saidel G: Quantitative relationships of intravascular tumor cells, tumor vessels and pulmonary metastases following tumor implantation. *Cancer Res* 34:997, 1974.

32. Myers P, Barrett A: Microwave thermography of normal and cancerous breast tissue. *Ann NY Acad Sci* 335:443, 1979.

33. Edrich J: Centimeter- and millimeter-wave thermography—a survey on tumor detection. *J Microwave Power* 14:95, 1979.

34. Enander B, Larson G: Microwave radiometric measurements on the temperature inside the body. *Electron Lett* 10:317, 1974.

35. Mamouni A, Bliot F, Leroy Y, et al: A modified radiometer for temperature and microwave properties measurements of biological substances. *Conference Proceedings—7th European Microwave Conference*. Copenhagen, Denmark, 1977, pp. 703–707.

36. Carr KL, El-Mahdi AM, Shaeffer J: Dual-mode microwave system to enhance early detection of cancer. *IEEE Trans Microwave Theory Tech* MTT-29:256, 1981.

37. Schwan HP: Radiation biology, medical applications and radiation hazards. *Adv Biol Med Phys* 5:215, 1957.

38. Johnson CC, Guy AW: Nonionizing electromagnetic wave effects in biological material and systems. *Proc. IEEE* 60:692, 1972.

39. Barber PW, Gandhi OP, Hagmann MJ, et al: Electromagnetic absorption in a multi-layered model of man. *IEEE Trans Biomed Eng* BME-26:400, 1979.

40. Schwan HP: Electrical properties of tissues and cells. *Adv Biol Med Phys* 5:147, 1957.

41. Campbell AM, Land DV: Dielectric properties of female human breast tissue measured in vitro at 3.2 GHz. *Phys Med Biol* 37:193, 1992

42. Dicke RH: The measurement of thermal radiation at microwave frequencies. *Rev Sci Instrum* 17:268, 1946.

43. Ludeke KM, Kohler J: Microwave radiometric system for biomedical "true temperature" and emissivity measurements. *J Microwave Power* 18:277, 1983.

44. Carr KL, Bielawa RJ, Regan JF, et al: The effect of antenna match on microwave radiometric thermal patterns. *IEEE MTT DIG* F-5:189, 1983.

45. Shaeffer J, El-Mahdi AM, Bielawa RJ, et al: Thermal drift in microwave thermography. *IEEE MIT DIG* V-3:441, 1982.

46. Temme DH: Lithium ferrite for microwave latching ferrite devices. *IEEE Trans Magn* MAG-5:632, 1969.

47. EG&G Princeton Applied Research Corporation: *Lock-In Amplications Anthology*. EG&G Princeton Appl. Corp., Princeton, NJ.

48. Lawson RN: Thermography—a new look in the investigation of breast lesions. *Can Serv Med J* 13:517, 1957.

49. Dodd GD, Wallace JD, Freudich IM, et al: Thermography and cancer of the breast. *Cancer* 23:797, 1969.

50. Gershon-Cohen J: Medical thermography. *Sci Am* 216:94, 1967.

51. Gautherie M, Gros CM: Breast thermography and cancer risk prediction. *Cancer* 45:51, 1980.

52. Lloyd-Williams K: Temperature measurements in breast disease in thermography and its clinical application. *Ann NY Acad Sci* 121:278, 1964.

53. Gautherie M, Edrich J, Zimmer R, et al: Millimeter-wave thermography—application to breast cancer. *J Microwave Power* 14:123, 1979.

54. Cacak RK, Winans DE, Edrich J, et al: Millimeter wavelength thermographic scanner. *Med Phys* 8:462, 1981.

55. Myers P, Barrett A: Microwave thermography of normal and cancerous breast tissue. *Ann NY Acad Sci* 335:443, 1979.

56. Carr KL, El-Mahdi AM, Shaeffer J: Dual-mode microwave system to enhance early detection of cancer. Paper presented at IEEE International Microwave Symposium, Washington, DC, 1980.

57. Leroy Y: Microwave radiometry and thermography—present and prospect. Paper presented at Symposium International de Thermologie Biomedicale, Strasbourg, France, 1981.

58. Guy AW, Lehmann JF: On the determination of an optimum microwave diathermy frequency for a direct contact applicator. *IEEE Trans Biomed Eng* BME-13:67, 1966.

59. Myers PC, Sadowsky NL, Barrett AH: Microwave thermography: Principles, methods and clinical applications. *J Microwave Power* 14:105, 1979.

60. Carr KL, El-Mahdi AM, Shaeffer J: Dual-mode microwave system to enhance early detection of cancer. *IEEE Trans Microwave Theor Tech* MTT-29:256, 1981.

61. Harrington RF: *Time-Harmonic Electromagnetic Fields*. McGraw-Hill, New York, 1961, pp. 116–132.

62. Bahl IJ, Stuchly SS, Stuchly MA: New microstrip slot radiator for medical applications. *Electron Lett.* 16:731, 1980.

63. Bahl IJ, Stuchly SS: Analysis of a microstrip covered with a lossy dielectric. *IEEE Trans Microwave Theor Tech* MTT-28:104, 1980.

64. Iskander MF, Durney CH: An electromagnetic energy coupler for medical applications. *Proc IEEE* 67:1463, 1979.

65. Bahl IJ, Stuchly SS, Stuchly MA: A new microstrip radiator for medical applications. *IEEE Trans Microwave Theor Tech* MTT-28:1464, 1980.

66. Sterzer F, Paglione R, Nowogrodzki M, et al: Microwave apparatus for the treatment of cancer. *Microwave J* 23:39, 1980.

67. Myers PC, Barrett AH, Sadowsky NL: Microwave thermography of normal and cancerous breast tissue. *Ann NY Acad Sci* 335:443, 1979.

68. Guy AW, Lehmann JF, Stonebridge JB, et al: Development of a 915 MHz direct-contact applicator for therapeutic heating of tissues. *IEEE Trans Microwave Theory Tech* MTT-26:550, 1978.

69. Carr KL: Antenna: The critical element in successful medical technology. *Digest 1990 IEEE MTT-S*, Dallas, TX, May 1990, pp. 525–527.

70. Jasik H: *Antenna Engineering Handbook*. McGraw-Hill, New York, 1961.

71. Cheever EA: Capabilities of microwave radiometry for detecting subcutaneous targets. Ph.D. dissertation, University of Pennsylvania, 1989.

72. Cheever EA, Leonard JB, Foster KR: Depth of penetration of fields from rectangular apertures into lossy media. *IEEE Trans Microwave Theor Tech* MTT-35:865, 1987.

73. Behrman RH, Brodie MA, Sternick ES, et al: Spatial resolution measurements for passive microwave radiometry using a tissue-equivalent phantom. *Med Phys* 17:1064, 1990.

74. Carr KL: Microwave radiometry: Its importance to the detection of cancer. *IEEE Trans Microwave Theor Tech* 37:1862, 1989.

75. Johnson CC, Guy AW: Nonionizing electromagnetic wave effects in biological materials and systems. *Proc. IEEE* 60:692, 1972.

76. Shaeffer J, El-Mahdi AM, Carr KL: Microwave radiometry thermal profiles of breast and drainage lymph node areas. *Proc Ann Conf IEEE/Eng Med Biol Soc, 3rd,* Piscataway, NJ, 1986, pp. 102–104.

77. Myers PC, Sadowsky NL, Barrett AH: Microwave thermography: Principles, methods and clinical applications. *J Microwave Power* 14:105, 1979.

78. Tenjin T, Oyama H, Maeda T, et al: Application of microwave thermography for the diagnosis of breast cancer. Nippon Medical School Report, Japan Thermography Association, 1984.

79. Shaeffer J, El-Mahdi AM, Carr KL: Thermographic detection of human cancers by microwave radiometry. Paper presented at the International Symposium on Biomedical Thermology, Strasbourg, France, 1981.

80. Leroy Y: Microwave radiometry and thermography: Present and prospective. In Gautherie M, Albert E (eds): *Biomedical Thermology.* Alan R Liss, New York, pp. 485–499.

81. Thompson JE, Simpson TL, Caulfield JB: Thermographic tumor detection enhancement using microwave heating. *IEEE Trans Microwave Theor Tech* MTT-26:573, 1978.

82. Huhns MN, Faulk JC, Fellers RG, et al: *Computer Processing of Microwave-Enhanced Thermographic Images for Tumor Detection.* IEEE, New York, 1979, pp. 237–242.

83. Carr KL, El-Mahdi AM, Shaeffer J: Passive microwave thermography coupled with microwave heating to enhance early detection of cancer. *Microwave J* 25:125, 1982.

84. Thompson JE, Simpson TL, Caulfield JB: Thermographic tumor detection enhancement using microwave heating. *IEEE Trans MTT*, MTT-26, No. 8:573, August 1978.

85. Carr KL, El-Mahdi AM, Shaeffer J: Cancer detection and treatment by microwave techniques. *Medical Electron* 12:88, 1981.

86. Shaeffer J, Parker PE, El-Mahdi AM, et al: Detection of XV2 carcinoma in rabbits by passive microwave radiometry. Paper presented at IEEE Workshop, IEEE MTT-S, Boston, MA, 1983.

87. Hamamura Y, Mizushina S, Sugiura T: Noninvasive measurement of temperature-versus-depth profile in biological systems using a multiple-frequency-band microwave radiometer system. In Mizushina S (ed): *Aspects of Medical Technology*, Vol. I, *Noninvasive Temperature Measurement.* Gordon & Breach, New York, 1989, pp. 39–59.

88. Dubois L, Pribetich J, Fabre JJ, et al: Noninvasive microwave multifrequency radiometry used in microwave hyperthermia for bidimensional reconstruction of temperature patterns. *Internatl J Hyperthermia* 9:415, 1993.

89. Prionas SD, Hahn GM: Noninvasive thermometry using multiple-frequency-band radiometry: A feasibility study. *Bioelectromagnetics* 6:391, 1985.

90. Ko HC: Coherence theory of radio astronomical measurements. *IEEE Trans Antennas Propagation* AP-15:10, 1967.

91. Swenson G, Methur N: The interferometer in radio astronomy. *Proc IEEE* 56:2114, 1968.

92. Edrich J, Jobe WE: Correlated microwave thermography for breast cancer detection. *AAMI 20th Annual Meeting*. Boston, MA, May 1985, p. 26.

93. Mamouni A, Leroy Y, Van deVelde JC, et al: Introduction to correlation microwave thermography. *J Microwave Power* 18:285, 1983.

CHAPTER ELEVEN

Understanding the Danger of Very-Low-Frequency Electromagnetic Fields

RICHARD L. COREN, Ph.D., *Department of Electrical and Computer Engineering, Drexel University, Philadelphia, PA*

11.1 THE PROBLEM

It was learned, soon after their discovery early in this century, that the mysterious, invisible emissions, known as *x-rays*, could damage human tissue. Today we know that an x-ray photon has enough energy to ionize an atom or disturb a nucleus, causing changes to the organic molecules of which living matter is made. Electromagnetic radiation at other frequencies also affects living systems. Ultraviolet radiation causes chemical activity and is found to be biologically harmful. Infrared, which is "nonionizing," also has a direct heating effect, and at lower frequencies the accepted, limiting, whole-body exposure to intense microwave and radiowaves is set at 2 mW/cm^2, which is the body's ability to dissipate the energy delivered. At very low frequencies (VLF's), generally taken to be between 3 and 3000 Hz, the energy transported is so low that it has generally been believed there can be no biologic impact from exposure to such fields.

Then in 1979 a report was published by Wertheimer and Leper[1] showing an elevation of the cancer rate in children from homes with high magnetic fields at the 60-Hz power frequency. Despite cautionary statements from the scientific

New Frontiers in Medical Device Technology, Edited by Arye Rosen and Harel Rosen
ISBN 0-471-59189-0

community about accepting this result prima facie, the press was stimulated and the public aroused. This mood was exacerbated by the journalist Paul Brodeur who, in 1989, published alarmist magazine articles[2] and a book.[3] Today, nearly 15 years after the initial report, many additional studies have been reported, and analyses made. We have a better understanding of the problem of VLF electromagnetic fields (EMF) but no better evaluation of the conclusions of those studies. Although it is not yet possible to resolve the basic questions raised, this discussion will present some considerations that are relevant to developing our attitudes and actions with regard to this issue.

11.2 FIELD MAGNITUDES

To begin with, we must recognize that we are not considering radiation, as it is taken to mean in normal engineering and physics contexts. In electromagnetics class we generally consider two cases: "near" and "far" fields. The dividing parameter is $\beta r = 2\pi r/\lambda$, where r is the distance from the source and λ is the wavelength. At great distances $\beta r \gg 1$ we have the normal radiation fields, for example, with the ratio of electric to magnetic components, $E/H = 377\,\Omega$. However, at 60 Hz we find that $\beta r = 0.01$ when $r = 5$ miles. For such a small value the electric and magnetic fields are essentially decoupled; thus, we can treat them separately—like static fields. As a result the electric field strength produced by a power line depends on the voltage of the line and its distance, whereas the magnetic field depends on the current it carries and its distance; for instance, for a single line, $E \sim V/r$ and $H \sim I/r$.

For a pair of lines of opposite phase, separated by the relatively small distance s, we have electric and magnetic dipole configurations so that $H \approx Is/r^3$, leading to smaller field values. With three-phase lines the effective current is further reduced. It is generally found that these relations agree fairly well with measurements, although the presence of such dielectric objects as trees and buildings must be accounted for, along with steel frame buildings. The main point for us is that the power frequency electric field is primarily a function of the line potential (volts), and its distance, whereas the power frequency magnetic field depends on the line current (amperes), and its distance.

For comparison of the magnetic fields we note that at the latitude of New York and Philadelphia the Earth's field is approximately 600 milligauss, with a horizontal component of about 200 milligauss. While we will regard this as a steady field, it does, in fact, fluctuate in magnitude and direction in periods of seconds-through-millennia. In any particular environment there are several contributors to the ambient power frequency field, such as wiring in walls, nearby transmission and distribution lines, electrical machinery, or office or kitchen equipment. For reference as to the resulting 60-Hz field levels with which we are concerned in examining exposures, we have the following mean values, with their standard deviations.[4]

	E (V/m)	H (mG)
Commercial and retail facilities, offices	$\sim 5 \pm 4$	$\sim 2 \pm 2$
Home residences	$\sim 9 \pm 9$	$\sim 2 \pm 2$
Utility substations	$\sim 50 \pm 40$	$\sim 10 - 100$

Very near certain appliances, such as electric stove tops, radiant heaters, or power tools, these H values can be 100 times greater. And this does not describe components at other frequencies, resulting from electrical switching, rectification, faults, and so on. Although these may be relatively small and infrequent, some analysts suggest that they may be important.

Ever since 1786, when Galvani used an electric discharge to cause frogs' legs to twitch, it has been known that electric fields have biologic effects, with higher levels causing currents that burn and destroy tissue and with low levels stimulating nerve and muscle tissue. They are commonly used, for example, to facilitate tooth orientation changes and implantation, to enhance bone healing, and in treating tendinitis. Control and manipulation of electrically excitable tissue is widespread, ranging from cardiac pacemakers to muscle spasm controls. We are able to perceive a free-space electric field of about 9 kV/m and, perhaps because of our lack of perception, low-frequency magnetic fields have never been regarded as biologically significant. However, note that magnetic fields of about 100 G have been found to induce visual flashes under dark conditions.

11.3 IN VITRO AND IN VIVO STUDIES

In vitro studies of the effects of electric and magnetic fields on various isolated cells and tissues have been conducted. Because they are controllable, it has been hoped that laboratory studies would indicate mechanisms of field action, relevant parameters, and anticipated effects. Tests have been reported that deal with variations of frequency, waveshape, exposure duration, and field orientation.[5] Unfortunately the results have been inconclusive and, generally, irreproducible, and a common feature appears to be a lack of dose–response proportionality. Instead, many measurements indicate intensity windows of effect, for example, with smaller and higher values being less effective; this is also true of exposure time. Such behavior is not unexpected from nonlinear systems. Unfortunately, without understanding of the system dynamics of VLF–EMF interactions, whether nonlinear or not, interpreting these behaviors is extremely difficult. Although all types of effects have been reported, the most accepted in vitro observation is modification of the functioning of the cell membranes that control transport into and out of the living cell.

As a general rule one can question the applicability of in vitro studies to living systems. The dielectric and conducting properties of the whole, living body greatly confound the simple field geometries and levels used, and the living system is capable of repair of, adaptation toward, and compensation for many types of interferences. Furthermore, it is often uncertain how animal tissue resembles human tissue.

Some of these objections can also be raised with respect to in vivo studies, i.e., using whole animals.[6] Although animal studies are used to evaluate the effects of drugs and food additives, species response differences can be very great. It has been pointed out that if the Food and Drug Administration were to use dogs to test the effects of chocolate, its distribution would be disallowed, since it is poisonous to them. The closer the relation of the animal species to humans, the more significant are the animal study results; this makes the great apes primary subjects, although a few studies have also been carried out on human subjects. The results of in vivo measurements are also only poorly reproducible, although they indicate neural and neurendocrine influences.

From the in vitro and in vivo studies one can conclude that, although there is evidence linking biological effects to VLF–EMF, these are, at most, subtle, low-level, and nonlinear. In view of this, the Wertheimer–Leper study was a great surprise. They took the records of 344 children in Denver, who had died from cancer, and randomly selected a larger group of comparable cohorts without reported cancer histories. Comparison of the two groups, in terms of a measure of their exposure to power-line magnetic fields, indicated that there was an increased incidence of cancer in high-field environments. This increase was for all kinds of cancer, including leukemia, the most prevalent form of childhood cancer. While the scientific community was cautious, public concern was disproportionate. If we are to be balanced in our evaluation of this situation, we should understand the basis of these different attitudes.

11.4 RISK EVALUATION

The subject of risk evaluation distinguishes between mathematical and perceived risks. Potentially disastrous acts will generally be avoided even if the mathematical probability of the detrimental outcome is extremely small. You may contemplate whether you would play a single round of Russian roulette for $10 million. Or suppose there is some number, n, of revolvers on the table, only one of which is loaded with a single round, and you have to pick one. How great must be the value of n for you to take the gamble? In considering the subjective aspect of risk analysis it is acknowledged that three particular considerations skew the perceived willingness to accept gambling odds:

1. Exposure to involuntary risks are highly unacceptable relative to voluntary risks. We object to sitting in an asbestos-lined building, or to living in the flight path of an airport, all the while driving on a high-accident highway, or while smoking. The emotional feeling of control is important, even though we may intellectually acknowledge that a significant number of adversities occur despite our best efforts. In some cases an involuntary risk can be converted to a voluntary

one, thereby making it acceptable. For example, in some states, communities have accepted incentives (financial and otherwise) to allow the siting of toxic dumps, which are otherwise highly undesirable.

2. Invisible or poorly understood risks are disproportionally opposed. This applies to radiation of all sorts, and to pollutants of water and air. Scientific issues are generally mysterious to the public and are therefore regarded with special suspicion. This applies to VLF–EMF, to video display terminal (VDT) radiation, and to the use of modern technological tools such as police radar guns and microwave ovens. The ability to misinterpret the facts and to impart fantastic consequences to simple situations cannot be underestimated. To an American population in which 40% believe in the reality of ghosts and the supernatural, anything is possible. A recent newspaper article[7] purported to describe actual reports from an Italian village where new, nearby power lines caused the dead to leave their graves and wander through town.

3. Risks that affect children are particularly intolerable and are frequently the focal point of drives to resist the supposed threatening agents.

The VLF–EMF situation has all three elements. With such emotional, subjective content, the claims of experts that any danger is uncertain and, even if real, is very small, is not accepted. It is common in such cases that official and scientific disclaimers are taken by some to be the result of institutional and political collusion. Even obvious field reduction methods, such as splitting and alternating transmission-line phases, maybe opposed because it is assumed to be motivated by selfish interest, to avoid better, but more expensive solutions, such as moving electric distribution towers.

11.5 FINANCIAL CONSIDERATIONS

Regardless of the true significance of any VLF–EMF biologic effects, it has been pointed out that the issue already has considerable financial impact. There are several different causes for this and in considering proposed actions, and particularly in considering the urgency of such actions, we should be aware of how they arise.

A number of lawsuits involving claims of EMF-associated health loss have been filed against manufacturers, utilities and governmental agencies. In part because of the complexity of the issues and the emotional sympathies of juries, the defendants have chosen to settle most of these through out-of-court agreements. Of those that have been tried, most have eventually been decided in favor of the defendants. However, such actions are not without significant cost, and there is an awareness that the legal profession is gearing up to encourage and persue such claims.[8]

The rapidly increasing public clamor to avoid EMF exposure has had a number of other costly consequences.

- Delays in installation, or cancellation, of new lines, and refusal of right-of-way with the need for considerably longer rerouting, are costly because of the delay of efficient networking for the tranfer of power at the lowest cost, because of additional time and design expenses, and because of the necessity of acquiring more real estate and rights-of-way. Hundreds of millions of dollars may be involved.
- Local groups and regulatory agencies have undertaken to make measurements of magnetic fields at schools and other public meeting sites. Although there is no consensus of what constitutes a harmful dosage, these efforts have resulted in schoolrooms being sealed and distribution lines being moved or installed underground. Again, these actions involve nontrivial expenses.
- Some commercial businesses and construction firms have undertaken to construct shielded offices and buildings.
- A number of manufacturers of home and office equipment have developed "low-field" versions of their products, often to be sold at considerable extra cost. This is particularly true of electric blankets and VDTs.
- There is evidence that property along transmission-line routes has already been devalued. The acreage involved is tremendous, and the personal losses can be staggering.
- It has been suggested that medical insurance cost should reflect the client's home siting with respect to power-line locations.

Most of these costs are eventually passed to the consumer. Being acutely aware of these issues, and through prodding by regulatory agencies, power companies are seeking low-cost palliative responses. Figure 11.1 shows some tower-line alternatives that reduce stray fields. Some of these can be used to retrofit older lines. On new lines consideration is being given to routings that give a wide berth to populated areas, to paying for wider rights-of-way so that edge field levels are lower, and to placing lines underground, in shielded channels. While tower rephasing is relatively inexpensive to install, some of the other measures are certainly not. In addition, they carry large indirect costs in terms of enhanced corona discharge, more frequent arcing and breakdown, lower load capacities, more difficult maintenance, and other problems.

Such mounting costs have raised the issue of whether we would incur a considerable long-term savings by investing heavily, now, in studies to resolve the scientific questions. However, others have claimed that the "scientific method" can never really negate a hypothesis and that the irrationality of public fear in this case implies that no near-proof will be accepted. In addition, some claim that the available evidence is so weak that it is folly to cater to "ignorant" demands by investing in more studies. In spite of this, at present, EPRI, DOE, and NSF (Electric Power Research Institute, Department of Energy, and National Science Foundation) in the United States, as well as agencies in other countries, are undertaking extensive, long-range evaluations.

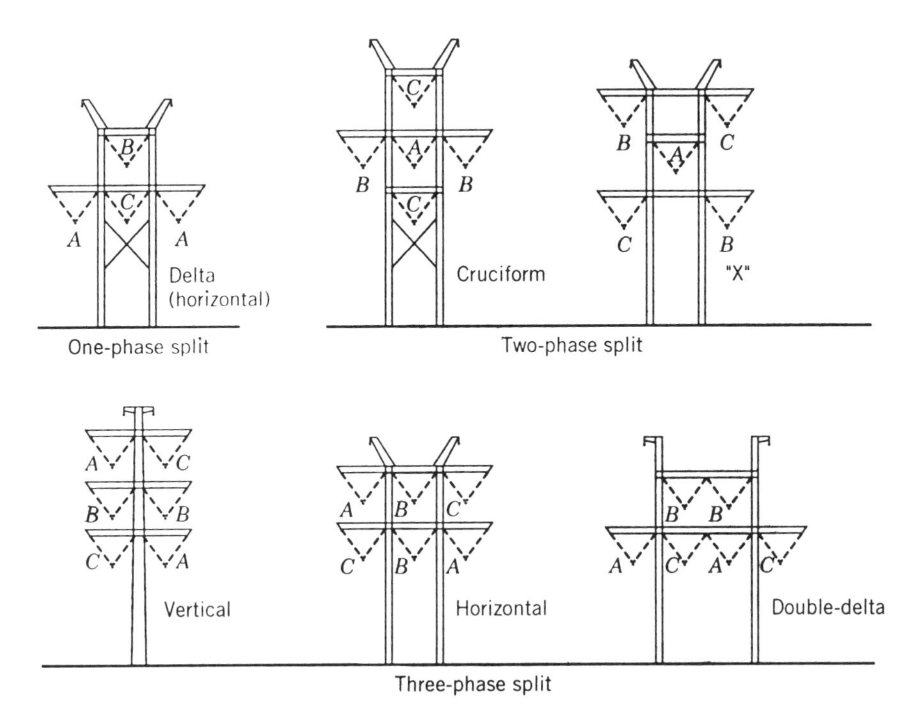

FIGURE 11.1 Split-phase single-circuit configurations.

11.6 EPIDEMIOLOGIC STUDIES

Up to this point we have described ambient power frequency electric and magnetic fields, related but indirect studies that have been made of their biologic effects, the public attitude toward and interpretation of their threat of harm, and the financial implications of that attitude. Ultimately, resolution of the problems connected with VLF–EMF revolve about the scientific evidence for the existence of a real, deleterious effect. In view of the conflicting claims for the necessity of such studies we should briefly examine the nature of some of the epidemiologic evidence that is presently available.

The Wertheimer–Leper study in 1979[1] showed a two to three fold increase of childhood cancer deaths, associated with the presence of high levels of ambient EMFs. Although that study has been greatly criticized[9] for technical reasons, it had a fairly large cohort of 344 cases of cancer deaths among children of Denver, CO, and corrected for a number of confounding influences (e.g., age, sex, local vehicle traffic density), some of which may parallel the field strength parameter. These investigators initiated an area of research that has grown considerably. Among their contributions is the *wiring code* used to estimate residential EMF exposure levels. The wiring code is based on an examination of the home location relative to transmission and distribution lines, the nature of those lines, the proximity to transformer stations, and the entrance wiring to the home. Wertheimer and Leper extrapolated wiring-code formulas and data to estimate the likely average current levels, and divided their sampled sites into low-current and high-current groups. Nearly all subsequent studies use, although some refine, this wiring code.

Several studies have also attempted to measure the in-home fields directly, both on a short time scale and over periods of a day to a few days. It is important to note that, even in cases where there is a consistent association of cancer with wiring-code levels there is seldom any association or correlation with the measured values. This is not entirely puzzling, because of changing line currents over the course of seasons and years of exposure. Further, some have interpreted this discrepancy to indicate that, since the exact parameter related to increased cancer incidence has not yet been identified—for instance, it might be the impulse frequency and amplitude rather than the average value—the use of this surrogate estimator, the wiring code, is somehow a more meaningful measure than short-term or steady-state field values.

Many studies have appeared since 1979 indicating or denying associations between residential EMF levels and cancer incidence. These are largely troubled by too small a sample size, poor controls, incomplete analysis, and other limitations. Recently a number of reports have appeared that seem to satisfy some criticism of earlier studies. As with the in vivo and in vitro studies, some found no VLF–EMF effect on cancer incidence, while some were positive, and at least one found a positive correlation at high current levels and a negative correlation (i.e., lower risk), at intermediate levels.

A study from Stockholm dealt with the association of congenital cancer incidence in the children of switchyard and other utility workers exposed to high EMFs. A two-fold increase in brain cancer was noted, with a reduction in leukemia, but the increase was not considered statistically significant. A reported study of cancer in 50,000 telephone workers in New York State[10] showed a lower rate of cancer in line workers, than among the general adult male population of New York State, but the rate was higher in cable splicers.

At this writing a study by the Swedish Karolinska Institute[11] is a recent, thorough, and widely accepted indicator of a positive correlation between cancer and ELF–EMF. It will pay for us to examine it in more detail.

This study included *all* persons having at least one year's residence within 300 m of any 220 kV or 440 kV line in Sweden, during 1960–1985. This population was over 500,000 people living in single-family houses and apartments. Of these, 123,420 were under the age of 16. Field exposure was estimated from detailed computer calculation of each wiring configuration and residential location, and using the detailed records of line loads, maintained by utility companies. Extensive spot measurements of field values were also made but, as in previous studies, showed little correspondence to either calculated values or outcomes. The availability and use of extensive residence and hospital records allowed separation of such factors as socio-economic position, environment (urban vs. rural), age, and gender. Exposure levels were divided into three groups, and the relative risk (RR) was calculated for the higher exposures relative to the lowest level. (At each exposure level one takes the number of cancer incidents divided by the number of controls at that level; RR is the ratio of that figure to the same figure for the lowest exposure level).

The unique finding of this study, relative to other studies, is the existence of a dose–response relation for single-family homes. It is believed that the variability of apartment building structures and the difficulty to pinpoint exact residence locations in such a building confounded the data taken there. Data are presented from several analytical vantage points. In one presentation, for single-residence children under the age of 16, there was a relative risk (RR) of 2.7 with respect to leukemia. A less touted result is that RR = 1.7 for myeloid leukemia in adults.

These results have achieved a good degree of acceptance because of the magnitude of the population considered, the completeness of the field exposure determination, the thoroughness of its data and analysis, and the satisfying finding of a dose–risk relation. However, before we fully accept these results as having settled the issue of VLF–EMF exposure effects, we should be aware that even this study reveals some of the questionable relations of other similar work. For example, not only was no risk found for brain tumors or other cancers in children, but the risk of brain tumor seemed to decline with increased exposure level, and other cancers exhibited erratic variations. The numbers involved in these individual categories were significantly smaller, and this leads us to consider one of the main weaknesses of most studies in this area—their quantitative evaluations—rather than the trends they seem to indicate.

11.7 QUANTITATIVE CONSIDERATIONS

One presentation of Karkolinska Institute data is shown on Table 11.1, for single residence children under the age of 16. We note that of the tables given in that report, this one is probably the most definitive. Figure 11.2 shows a graphical representation of the data for the RR. While it is not the careful numerical treatment performed by the reporters from the Karolinska Institute, let us engage in a crude analysis to indicate a weakness one may encounter in analyzing such reports.

The last line in Table 11.1 divides the number of occurrences by the RR, thereby reducing the number at each dosage to an equivalent number at the lowest exposure level. The sum of these numbers is 26.5. Since extrapolating the curve on Figure 11.2 to the zero-exposure level, gives an RR of about 2/3, this implies an expected, zero field incidence of cancer to be 18 cases out of the relevant population of 123,420. The normal, uninfluenced probability of cancer is, therefore, $p = 18/123420 = 3.1 \times 10^{-4}$. This is very small, indeed; a child on city streets, these days, may have a greater chance of dying by random violence. For such unlikely random events in large populations it is known that

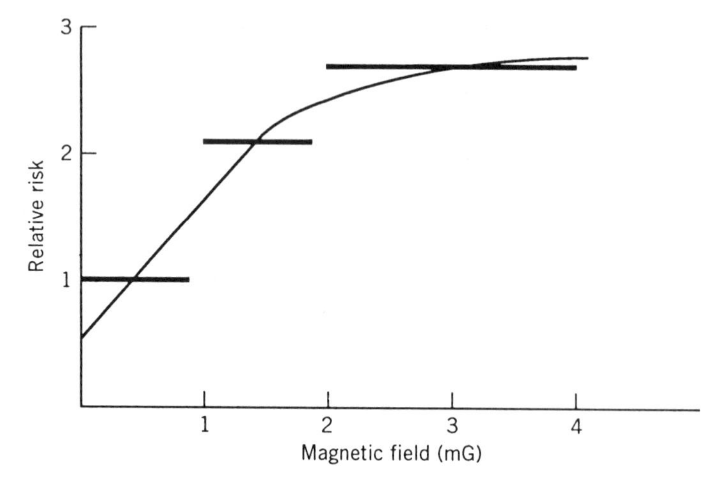

FIGURE 11.2 Correlation between VLF–EMF and cancer incidence reported by Swedish Karolinska Institute for years 1960–1985.

TABLE 11.1 Analysis of Karolinska Report[11]

	Magnetic Field (mG)			
	0–0.9	1–1.9	> 2	Σn
Leukemia cases (n)	22	4	7	33
Number of controls	475	33	46	
Risk ratio (RR)	1	2.1	2.7	
n/(RR)	22	1.9	2.6	26.5

the frequency of occurence of an outcome, n, follows the Poisson distribution law.

$$P(n) = \frac{m^{-n}e^{-m}}{n!}$$

where m is the mean, or expectation, value; here $m = 18$. Figure 11.3 shows a plot of this $P(n)$. The probability of having exactly 18 events is $P(18) = .094$ and, for the actual number found, $P(33) = 4.7 \times 10^{-4}$. We therefore have $P(33)/P(18) = .005$; clearly this anticipated relative probability is so small that the occurrence of the 33 cases cannot be considered to be a random event, thus supporting the conclusion that the great number of occurrences is due to some deterministic influence, supposedly the VLF–EMF.

For comparative purposes let us adopt a different epidemiological model. We speculate that the lowest level of exposure (0–0.9 G) represents a threshold below which no deleterious effect occurs. Such threshold behavior is common

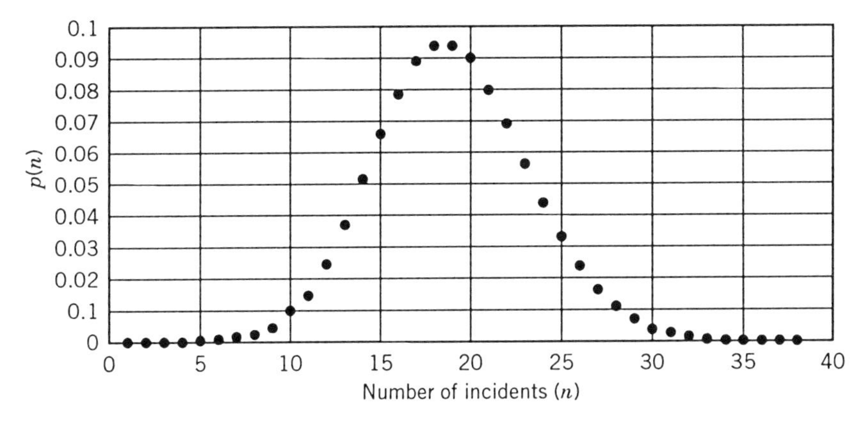

FIGURE 11.3 Poisson distribution: $P(n)$, $m = pN = 18$.

in toxicologic phenomena, where small exposures to an agent are harmless, or may even be advantageous, despite toxicity at higher levels. Such is the case, for example, with vitamins and with solar exposure. In this event, the equivalent 26 cases can be taken to be the normal, uninfluenced occurrence of cancer in the population considered; this replaces m in the Poisson distribution. We now find that $P(33)/P(26) = .37$, implying that random incidence would be expected to generate 33 cases in about a third of equivalent studies. This is a figure that would receive little support as indicating a definitive magnetic field effect. Before leaving this illustration we must note, in all fairness, that there is no reason to expect such a threshold effect for the influence of magnetic fields. On the other hand, as discussed below, at present we have no reason to expect *any* effect at all by the magnetic fields considered here; one conjecture may be as good or as bad as the other.

Most other studies that have been reported involve much smaller population numbers so that, with the very small probabilities, the expected number of incidents, $m = pN < 1$. Generally these studies present relative numbers rather than absolutes, so the significance of their smallness is not revealed. However, one result of their small size is that, even with no EMF influence, there is a significant chance of finding nearly any small number of cases that just happen to fall in the exposed group. This is indicated in the Karolinska data for apartment buildings, where the investigators found an RR of 1.1–1.3. In their conclusions they commented:

> One can speculate about the reasons why the excess risk is concentrated to one family homes. It is noteworthy, however, that it would only take the shifting of one case from the lowest to the highest exposure category, among those in apartment homes, in order to produce a relative risk of 1.8.

Although this statement is probably intended to strengthen the near correlation of exposure to cancer, it indicates the sensitivity of the conclusions to the smallness of the numbers involved. This, and sensitivity to features of the model used, as indicated by our example, are potential weaknesses of all studies of highly improbable events.

11.8 CAUSAL MECHANISMS

We should not leave without an appreciation of one major consideration leading to some of the scientific scepticism over the phenomenon of VLF–EMF effects in living systems: the lack of mechanism. It was pointed out at the beginning of this review that the EMFs of concern here are nonionizing, that is, they do not have sufficient energy to disrupt chemical bonds and, therefore, cannot initiate or interfere with chemical processes in the body. However, modern theories of cancer development contend that the initiation and promotion processes are distinct; so magnetic fields could, conceivably, act in the second role.

The supposition is that, for any number of genetic or causal reasons, potential cancer sites develop. How and whether these actually become carcinomas or sarcomas depends on a different set of influences, such as body chemistry (intrinsic) and foreign (extrinsic) influences, including VLF–EMF. The greater influence in children supposedly is due to the relative immaturity of their normal body protective immune system and repair mechanisms and because the higher metabolic process rates exhibited in children renders these influences more critical.

There is little doubt that sufficiently great VLF fields can affect living cells in vitro and living systems in vivo. However, with few exceptions, fields of the magnitudes of concern here are too small for substantiation of or extrapolation to any VLF–living system interaction. Furthermore, the dielectric nature of the body causes it to polarize in an electric field, and its depolarizing field reduces the internal field from its external value. Also, current flow from this changing polarization charge, due to the body conductivity, shields the body internally from an alternating external magnetic field. One would normally expect these effects to be small and, in addition, the nonuniform internal and external body structures should randomize the otherwise uniform effects of the external fields.

Furthermore, the normal electrical activity of nerve and muscle tissue is known to generate intimate, local fields that are greater than those that are suspect, and thermal fluctuations (Nyquist noise) of ion concentrations also cause fields that are greater. To counter these arguments it has been pointed out that, while they apply on the level of an individual cell, tissues are larger, and the presence of a field that is moderately uniform in direction and relatively slowly varying, might cause an average effect beyond the fluctuational fields just mentioned. Of course, living systems are not stationary in an external field, so these should also average to smaller values.

One rather subtle mechanism has been mentioned frequently, although there is only weak and indirect evidence linking it to effects of the type seen in the epidemiologic studies. This is the pineal–melatonin effect. The *pineal* is a small endocrine gland at the base of the brain, whose principal hormone is *melatonin*. In the body, melatonin acts on several other endocrine glands, and in vitro and in vivo experiments have shown that it supresses certain cancers.

Maximum melatonin production occurs in the early morning hours, and it is known that pineal function is controlled by the visual signals of the day/night–light/dark cycle that set the body's circadian rhythm. It has been found that, after several weeks of exposure to combined fields (26 V/m and 1 G), monkeys and rats experience a time shift of their natural, diurnal rhythms, that is, of their daily hormone production and sleep and awake periods. Similarly, experiments with human volunteers indicate that VLF–EMFs can reduce the nighttime melatonin peak. While this VLF–EMF effect may result from direct action on the pineal, one can also conjecture that the mechanism is by subliminal stimulation of the optic system, which, at least at very high levels, is sensitive to VLF–EMF. It must be recalled here that, as mentioned above, the statistics of these associations are certainly subject to critical errors, so that any connection of this sort is still highly speculative. In any event it has been suggested that a reduced or phase-shifted pineal cycle is a factor in the enhancement of cancer cases. Since it is primarily the night cycle production of melatonin that is depressed, this supposition places electric blankets and water-bed heaters as particularly strong potential VLF–EMF influences.

11.9 RECOURSE

In view of the uncertainties of the existence of ambient VLF–EMF effects on human populations, what, if any, actions are appropriate for the individual, for industry, and for governing agencies? Epidemiologists, in cases such as these, give a standard cautionary warning that reasonable care should be taken to avoid exposure. This implies that there is no crisis at hand but one may choose to be wary. The fact is that any danger appears to be less than that for most other pollutants, and is certainly less than for other dangers to which we are regularly exposed.

Few people appear to be concerned with fields from clocks, radios, washing machines, electric ranges, and TVs, to mention a few household appliances, or from Xerox (Photocopy) or Fax (Facsimile) machines or computers, to mention a few office appliances. However, the exposure levels from electric blankets and waterbeds have caused consternation, and the high visibility of – and publicity received about – power lines has produced near hysteria in some cases. As a cautionary measure some people are moving their children's rooms away from the side of the house nearest any neighborhood power lines or home power entry points; other families are moving away entirely or, reportedly, having their homes physically relocated. School officials have closed classrooms with "high" ambient fields, without any understanding of what "high" means or implies. A few, who claim to be exceptionally sensitive to power-line fields, have had portions of their homes enclosed in magnetic shielding; here it might be noted that incomplete shielding can result in focused hot spots, exacerbating the situation rather than remedying it.

Utilities and manufacturers have begun to take measures to reduce the fields they generate. Phased transmission lines have been mentioned, and bifilar wound electric blankets have appeared on the market. Electronics manufacturers have long included circuit guards, such as transformer shields and closely coupled lines, against internal, electronic interference and, to the extent that these are successful, they reduce external fields as well. EPRI, NSF, and DOE in the United States, as well as agencies in other countries, are undertaking significant studies, hopefully to delineate the dangers, if any—this inspite of their all having already issued evaluations concluding that there is little evidence for concern.[12]

11.10 CONCLUSIONS

Because the ongoing efforts, described just above, are costly to individuals and governments, and any real effects are continuing to harm individuals in our society, some have urged accelerating the epidemiologic and laboratory investigations. Others point to the lack of convincing reason for concern and feel that this is throwing good money after bad, besides which it is likely that an "official" negative finding will not be believed by the public. More likely, they may claim, is that the news media and society will tire of this issue and that it will eventually fade from prominence.

The belief in real, adverse health effects is expressed by David A. Savitz:[13]

> When the question is posed, Is there theoretical or empirical evidence that exposure to electric and magnetic fields at commonly encountered levels poses a threat to health?, the answer must be a firm yes. Epidemiologic evidence linking power lines near residences and elevated magnetic fields to childhood cancer continues to accrue; employment in selected electrical occupations seems to confer an increased risk of leukemia, brain cancer, and perhaps breast cancer; and there have been numerous laboratory studies indicative of influences on circadian rhythms, calcium efflux from nerve cells, and a hypothesized pathway linking such exposures to cancer.

The skeptic's view is expressed by the British National Radiation Protection Board.[14]

> Review of the 29 studies . . . leads to the conclusion that a tendency for the selective publication of results that suggest an increased risk is perhaps the most likely explanation of the very small excess incidence from the disease that the totality of the data implies.

REFERENCES

1. Wertheimer N, Leper E: Electrial wiring configurations and childhood cancer. *Am J Epidemiol* 109:273, 1979.
2. Brodeur P: Annals of radiation: The hazards of electromagnetic fields. *The New Yorker*: June 12, 19, 26; July 9, 1989.
3. Brodeur P: *Currents of Death: Power Lines, Computer Terminals, and the Attempt to Cover Up Their Threat to Your Health*. Simon & Schuster, 1989.
4. Bracker TD: Exposure assessment for power frequency electric and magnetic fields. *Am Ind Hyg Assoc* 54:178, 1993.
5. Cleary SF: A review of in vitro studies: Low-frequency electric and magnetic fields. *Am Ind Hyg Assoc* 54:165, 1993.
6. Anderson LE: Biological effects of extremely low-frequency electromagnetic fields: In vivo studies. *Am Ind Hyg Assoc* 54:186, 1993.
7. *The Sun*: Jan. 12, 1993.
8. *The Wall Street Journal*: Feb. 5, 1993.
9. Savitz DA: Overview of epidemiologic research on electric and magnetic fields and cancer. *Am Ind Hyg Assoc* 54:197, 1993.
10. Matanoski GM, et al: Quoted in Wartenberg D, Greenberg M (eds): Epidemiology, the Press, and the EMF Controversy. *Public Understand Sci* 1:383, 1992.
11. Karolinska Institutet for Miljomedicin: Magnetic fields and cancer in people residing near Swedish high voltage power lines, Stockholm, 6, 92.
12. (a) U.S. Congress Office of Technology Assessment, Biological Effects of Power Frequency Electric and Magnetic Fields (May 1989); (b) U.S. Committee on Interagency Radiation Research, Oak Ridge Associated Universities Panel, Health Effects of Low-Frequency Electric and Magnetic Fields (June 1989); (c) British National Radiological Protection Board, Electromagnetic Fields and the Risk of Cancer (1992).
13. Savitz DA: Health effects of low-frequency electric and magnetic fields. *Environ Sci Technol* 27:52, 1993.
14. British National Radiological Protection Board: *Electromagnetic Fields and the Risk of Cancer*. 1992.

Index